CRC HANDBOOK OF

MEDICINAL MINTS

(AROMATHEMATICS)

Phytochemicals *and* Biological Activities

CRC HANDBOOK OF MEDICINAL MINTS (AROMATHEMATICS)

Phytochemicals *and* Biological Activities

STEPHEN M. BECKSTROM-STERNBERG
JAMES A. DUKE

CRC Press
Taylor & Francis Group
Boca Raton London New York

CRC Press is an imprint of the
Taylor & Francis Group, an **informa** business

First published 1996 by CRC Press
Taylor & Francis Group
6000 Broken Sound Parkway NW, Suite 300
Boca Raton, FL 33487-2742

Reissued 2018 by CRC Press

Library of Congress Cataloging-in-Publication Data

Duke, James A.
Handbook of aromathematics (mints) / by James A. Duke, Stephen M. Beckstrom-Stemberg
p. cm.
Includes bibliographical references.
ISBN 0-8493-2665-6 (alk. paper)
1. Mints (Plants)—Therapeutic use—Handbooks, manuals, etc. 2. Mints (Plants)—Composition—Handbooks, manuals, etc. 3. Aromatherapy—Handbooks, manuals, etc. 4. Aromatic compounds—Handbooks, manuals, etc. I. Beckstrom-Sternberg, Stephen M. II. Title.
RS165.M38D85 1996
615'.32387—dc20
DNLM/DLC
for Library of Congress

96-8974

A Library of Congress record exists under LC control number: 96008974

ISBN 13: 978-1-315-89358-7 (hbk)
ISBN 13: 978-1-351-07268-7 (ebk)

Visit the Taylor & Francis Web site at http://www.taylorandfrancis.com and the
CRC Press Web site at http://www.crcpress.com

Dedication

This book is dedicated to Steve's three sons (Fred MacMurray not withstanding), Marc, Jamie, and David, and Jim's son and daughter, John and Celia, all budding botanists.

The Authors

Stephen M. Beckstrom-Sternberg, Ph.D., is President, StarMountain Enterprises, Inc., Vice-President, Duke's Herbal Vineyard, Inc., and Faculty Research Associate, Department of Plant Biology, University of Maryland, College Park.

Dr. Beckstrom-Sternberg graduated in 1979 from University of California, Riverside, with a B.S. degree in Plant Science. He obtained his M.S. degree in Botany in 1981 from University of California, Riverside, and received a Ph.D. in Botany in 1988 from Claremont Graduate School, Claremont, California. He served as a Postdoctoral Scientist in Molecular Systematics from 1988 through 1991 at Rancho Santa Ana Botanic Garden. During 1991 he taught courses in Botany at California State University at Los Angeles. Towards the close of 1991 he and his family moved to Beltsville, Maryland, to do postdoctoral work with the coauthor at the National Germplasm Resources Laboratory, Agricultural Research Service, U.S. Department of Agriculture, making taxonomic, ecological, geographic, phytochemical, and ethnobotanical databases available in relational database format, and later on the World Wide Web (WWW). In 1994 Dr. Beckstrom-Sternberg began working as a computational biologist with the Genome Informatics Group at the National Agricultural Library for University of Maryland, College Park, interconnecting the various databases of the Agricultural Genome Information Server, as well as with databases around the world via WWW. In 1996, he formed StarMountain Enterprises, Inc., and began serving as VP of Duke's Herbal Vineyard, Inc. He is a contributing columnist on the topic of the WWW for Probe, a newsletter for the plant genome community.

Dr. Beckstrom-Sternberg is a member of the American Scientific Affiliation, the American Association for the Advancement of Science, and the Society of Economic Botany.

Among his other publications, Dr. Beckstrom-Sternberg has published over 14 databases on the WWW, including PhytochemDB, EthnobotDB, MPNADB, FoodPlantDB, and Ecosys, as well as four book chapters, several papers and posters. He has spoken at several scientific meetings and chaired a session at the 1995 Molecular Biology Conference in San Diego. His current interests include making a number of other databases available via WWW, as well as developing WWW data mining tools for plant biologists.

James A. Duke, Ph.D. (Botany), retired in 1995 after more than 35 years as Economic Botanist, mostly with the USDA (Beltsville MD). Curently he is Senior Scientific Adviser to Nature's Herbs (Spanish Forks, UT), Medical Adviser to Herbalife (Los Angeles, CA), on the Scientific Advisory Board of Shaman Pharmaceuticals, (San Francisco,CA), one of five directors, Botanical Products, Inc., Hakilua Hawaii, growing tropical medicinal species, and Director of Duke's Herbal Vineyard, Fulton MD, growing several temperate aromatic and medicinal species.

Dr. Duke graduated in 1952 from the University of North Carolina (UNC) at Chapel Hill, also taking his M.A. in botany there (1955). Following military service, mostly at Ft. Detrick MD, Duke returned to UNC to take his Ph.D. under Dr. A. E. Radford and C. R. Bell. Then he moved directly to the Missouri Botanical Garden as a post-doctoral student of R. E. Woodson and Edgar Anderson. Following four years work there, as assistant curator at the Garden and associate professor at Washington University, he moved to the USDA at Beltville MD, in 1963 where he served until 1995, except for a seven-year absence to join Battelle Columbus Laboratories for a resident ethnobotanical study in Panama and Colombia. It was here that he was "converted" from taxonomy to ethnobotany.

Dr. Duke is a life member of several societies, including the Association of Plant Taxonomists, Society for Conservation Biology, Society of Economic Botany, and a member of the American Chemical Society. He serves as officer or adviser to the Amazonian Center for Environmental Education and Research, the American Botanical Council, Herb Research Foundation, and Microbotanica Inc.

As author of over 15 books in the field of economic botany, including the CRC Handbook of Medicinal Herbs, the CRC Handbook (and database) of Phytochemical Constituents of GRAS Herbs and other Economic Plants, and the CRC Handbook of Biological Activities, Duke is in retirement preparing an encyclopedic Green Pharmacy, parts of which are under contract to be published by Rodale Press (estimated 1997). These and other databases are being improved and expanded as one of the main activities of Herbal Vineyard, Inc.

Introduction

Since the publication of Duke's first phytochemical works (Duke 1992a, 1992b), we have almost doubled the phytochemicals, biological activities and species in our database. With this new contribution, we include much new data in the aromatic mint family, Lamiaceae. Data on the mint family are presented here in much the same format and with many of the same caveats and interpretations as before.

Aggregated vs. Non-Aggregated Entries

This volume differs from earlier volumes in that we maintain the integrity of many entries, publishing the phytochemical quantitative data for a given analysis, rather than strictly aggregating the data. Over the years it has become clear to us that the aggregated data, which report the highs and lows for a given phytochemical in parts per million (ppm), are most useful for showing the range of variation of these individual phytochemicals, which can be quite striking. For some species of thyme, this variation can be as much as 13,900-fold.

One common mistaken interpretation of the aggregated data is that a species might appear to be high in two closely related compounds, for example thymol and carvacrol. However, the non-aggregated analyses show that, rather than positive correlations between levels of closely related compounds, there is often compensation - when one compound is raised in quantity, another is lowered.

Synergy

We have learned in the last five years, that most species have many phytochemicals with many biological activities and that many of the phytochemicals we use for medicines, especially the antibiotics (anthelmithics, bactericides, fungicides, viricides, vermifuges, etc.), serve as natural pesticides for the species containing them. As we would expect, evolution seems to favor a synergy among such pesticides, which often carries over into their medical potential.

By examining the non-aggregated entries, one can, through various computational techniques, ask which of the mint analyses cited in this volume have the greatest reported variety or total concentration of phytochemicals with a specific activity (for instance which mints have the greatest number of antispasmodic compounds and which have the greatest total reported concentration of one or all of these spasmolytic compounds). A number of tools have been developed by the authors to allow these types of queries to be made for the entire phytochemical database (including the mints) on the Internet (Beckstrom-Sternberg and Duke) at the following URL: http://www.ars-grin.gov/~ngrlsb/.

Properties of Aromatic Compounds

Because of the unique properties of aromatic compounds, which are important for both herbal medicine and aromatherapy, there's more to aromatherapy than meets the nose. It has been clearly demonstrated that many of the aromatic compounds are biologically active whether inhaled, ingested, or applied topically. Cineole for example, via its CNS activities, can improve a rodent's ability to work its way through a maze, whether ingested or inhaled. Likewise, cineole, as well as other aromatic compounds, can speed up and increase the transdermal absorption of other compounds, sometimes by as much as 100-fold. The implications of this are inspiring as well as sobering. The increased absorbtion allows smaller amounts of an active compound to be used, and puts it directly into the bloodstream rather than passing through the gut, where it could be altered or inactivated. On the down side, too much of a good thing could be fatal, pointing to the need for standardization of topically applied compounds and their carriers, especially in light of the huge variation in the concentrations of plant chemical constituents.

Ecotypes

One interesting speculation to us is that rosemary, the herb of remembrance, may in fact be preventively active, perhaps even transdermally, against Alzheimer's disease. Rosmarinic acid, namesake of rosemary, has three different activities that might be useful in Alzheimer's disease: anticomplement activity, antioxidant activity, and choline sparing activity. There are over a dozen antioxidants in rosemary and more than five anticholinesterase compounds. By analyzing the non-aggregated, individual assay data we can see which ecotype or variety is best endowed with these biologically active compounds. This ecotypic variation could form the basis of an industry dedicated to cultivating specialized ecotypes of the same species for different medicinal applications.

Food and Drug Administration (FDA)

With the new labeling laws signed by President Clinton on October 15, 1994, it seems legal and possible for herbalists to say that rosemary contains these compounds, but not to say that rosemary will prevent or decelerate Alzheimer's. We believe that rosemary, the herb of remembrance, can be proved useful in Alzheimer's, and that it may be more useful than some FDA-approved drugs for Alzheimer's. However, we doubt that anyone will invest the required 500 million to prove that rosemary is safe and efficacious for Alzheimer's. So for economic reasons, we may not be getting the best medicine for Alzheimer's. It would be difficult to secure a patent for rosemary for Alzheimer's. Strangely though, one could analyze different varieties of rosemary to find which one was richest in the anticholinesterase compounds and patent that variety as unique phytochemically, to be clonally reproduced, under ecological conditions that increase the quantities of the anticholinesterase compounds.

Aromathematics

Of course, as new data accumulates, other edible and non-edible mints may prove even richer in antialzheimeran phytochemicals. Mints other than lavender and melissa may prove richer in sedative compounds. Mints other than peppermint and spearmint may prove richer in carminative compounds and antipruritic phytochemicals. Mints other than perilla may prove richer in breast-cancer preventive phytochemicals like carvone, limonene, and perillyl-alcohol. That's what's so exciting about this new field we call aromathematics (so as not to offend long-term advocates of aromatherapy), which is here defined as the study of aromatic compounds and their biological activities; with each new detailed analysis published, a new candidate may emerge for several diseases.

Standardized Extracts vs. the Silver Bullet

We feel that in many cases, standardized extracts of these potent mints, may be safer and just as efficacious as many of the more expensive synthetic options. Consider the following example. If your physician has diagnosed you correctly (with Lyme disease, the physician is wrong nearly half the time), and if you do not have any co-morbid factors (most of us do), and if you are not deficient in some mineral, vitamin, or vital phytochemical that has not yet proven vital (and most of us are deficient in at least one), then the physician's silver bullet may help. But if all three conditions are not satisfied, then the safe herbs may have more to offer. The homeostatic human body is good at sequestering from an herb tea those phytochemicals that it needs and rejecting things it doesn't need. Thus the menu of thousands of phytochemicals in an herb tea may give the human body opportunity to select those that it needs, rejecting those that might be harmful.

The Asthma Challenge

"Three months of 5-lipoxygenase inhibition produced a significant improvement in asthma control" (JAMA#275(12):931.1996). This tells us that the American Medical Association believes that 5-lipoxygenase inhibition can help with asthma. So we pose to you, the reader, the question, which of the mints mentioned herein has the greatest variety and/or quantity of lipoxygenase inhibitors reported? Would that be worth taking for asthma in a desperate situation: no physician, no medicine? The above cited paper in the Journal of the American Medical Association states that inhibition of 5-lipoxygenase is associated with improved asthma control as evidenced by reduced asthma exacerbations, objective measures of lung function, quality of life, and medication use. We suspect that antileukotrienic activity in many of our mints could likewise be channeled into antiasthmatic utility.

Further, we suspect that, on average, through synergy, lower doses of total mixed LO-inhibitors will be more effective than an equal dose of a single LO-inhibitor, and possibly less likely to have serious side effects, especially from mints that are GRAF (Generally Recognized As Food).

With fewer and fewer Americans able to afford the synthetic silver bullets offered us by the Pharmaceutical Industry, we anticipate that North Americans will be clamoring for a program like the TRAMIL program in the Caribbean countries. Experts in the fields of botany, chemistry, medicine, and pharmacy review the folk medicines available and select the safer and/or more effective among these herbs. We believe that data like those presented in this book will help select the best and safest medicinal mints for those of us who prefer herbal alternatives for economic, medicinal or mystical reasons.

Caveats, Details and Abbreviations

In order to make this book as useful as possible, it has been divided into two distinct parts. Part 1 lists the chemicals found in each plant, with quantities listed in parts per million (ppm), unless otherwise stated. Part 2 is a list of some reported biological activities of those chemicals. The astute reader may notice that a number of rather common chemicals are missing from these pages; this is done by design. Chemicals which the authors deemed ubiquitous, or common to all or most plants, are omitted when there is no quantitative data, since they communicate little of real value when quantitative data are lacking. The quantitative data for many of the minerals are often more a mirror of the soil's mineral content rather than an indication that the plant is specially endowed, since few plants are true accumulators.

The list of non-quantitative ubiquitous chemicals excluded from this work is found below:

Non-Quantitative Ubiquitous Chemicals Excluded from the Analysis

ACETALDEHYDE
ACETONE
ALANINE
ALPHA-TOCOPHEROL
ALUMINUM
AMMONIA(NH3)
ANTHOCYANINS
ARACHIDIC-ACID
ARACHIDONIC-ACID
ARGININE
ARSENIC
ASCORBIC-ACID
ASH
ASPARTIC-ACID
BARIUM
BEHENIC-ACID
BETA-CAROTENE
BETA-SITOSTEROL
BETA-SITOSTEROL-GLYCOSIDE
BETA-TOCOPHEROL
BORON
BROMINE
BUTYRIC-ACID
CADMIUM
CALCIUM
CAPROIC-ACID
CARBOHYDRATES
CELLULOSE
CHLORINE
CHLOROPHYLL
CHOLESTEROL
CHOLINE
CHROMIUM
CITRIC-ACID
COBALT
COPPER
CYSTEINE
CYSTINE
DOCOSAHEXAENOIC-ACID
DOCOSAPENTAENOIC-ACID
EICOSAPENTAENOIC-ACID
EO
ERUCIC-ACID
ETHANOL

FAT
FIBER
FLUORINE
FOLACIN
FRUCTOSE
GADOLEIC-ACID
GALACTOSE
GLUCOSE
GLUTAMIC-ACID
GLUTAMINE
GLYCINE
GOLD
HISTIDINE
INDOLE-ACETIC-ACID
IODINE
IRON
ISOLEUCINE
KILOCALORIES
LAURIC-ACID
LEAD
LECITHIN
LEUCINE
LIGNOCERIC-ACID
LINOLEIC-ACID
LINOLENIC-ACID
LITHIUM
LYSINE
MAGNESIUM
MANGANESE
MERCURY
METHIONINE
MOLYBDENUM
MUFA
MYRISTIC-ACID
NIACIN
NICKEL
NITROGEN
OCTADECATETRAENOIC-ACID
OLEIC-ACID
PALMITIC-ACID
PALMITOLEIC-ACID
PANTOTHENIC-ACID
PECTIN
PEROXIDASE
PHENYLALANINE
PHOSPHORUS
POTASSIUM
PROLINE
PROTEIN
PUFA
RESIN
RHAMNOSE
RIBOFLAVIN
RIBONUCLEASE
RUBIDIUM
SELENIUM
SERINE
SFA
SHIKIMIC-ACID
SILICON
SITOSTEROL
SODIUM
STARCH
STEARIC-ACID
STIGMASTEROL
STRONTIUM
SUCCINIC-ACID
SUCROSE
SULFUR
SYRINGIC-ACID
TANNIN
TARTARIC-ACID
THIAMIN
THREONINE
TIN
TITANIUM
TOCOPHEROL
TOCOTRIENOL
TRYPTOPHAN
TYROSINE
URONIC-ACID
VALINE
VANADIUM
VIT-B12
VIT-B6
VIT-D
VIT-E
WATER
XANTHOPHYLLS
XYLOSE
ZINC
ZIRCONIUM

Table of Contents

Acknowledgements

The authors express their appreciation to Mary Jo Bogenschutz-Godwin, without whose tireless efforts in data entry, error correction, and tracking of elusive leads, this book would not have been possible. We also wish to thank the National Germplasm Resources Laboratory of the Agricultural Research Service (US Department of Agriculture) for providing us with the excellent computing facilities required to build and maintain our database.

Important Phytochemicals in Mints

Acinos alpinus var. *meridionalis* (NYMAN) P.W.BALL "Te de Sierra Nevada"
Unaggregated A

ALLO-AROMADENDRENE 0.5 Shoot FFJ8:127
ALPHA-BISABOLOL 6 Shoot FFJ8:127
ALPHA-CADINOL 1.5 Shoot FFJ8:127
ALPHA-COPAENE 0.5 Shoot FFJ8:127
ALPHA-CUBEBENE 0.5 Shoot FFJ8:127
ALPHA-ELEMENE 1.5 Shoot FFJ8:127
ALPHA-HUMULENE 16 Shoot FFJ8:127
ALPHA-PHELLANDRENE 1.5 Shoot FFJ8:127
ALPHA-PINENE 12 Shoot FFJ8:127
ALPHA-TERPINENE 0.5 Shoot FFJ8:127
ALPHA-TERPINEOL 1 Shoot FFJ8:127
ALPHA-THUJENE 1 Shoot FFJ8:127
ALPHA-THUJONE 2.5 Shoot FFJ8:127
AROMADENDRENE 1 Shoot FFJ8:127
BETA-BOURBONENE 5 Shoot FFJ8:127
BETA-CUBEBENE 0.5 Shoot FFJ8:127
BETA-ELEMENE 10 Shoot FFJ8:127
BETA-GURJUNENE 3 Shoot FFJ8:127
BETA-MYRCENE 29 Shoot FFJ8:127
BETA-PINENE 7.5 Shoot FFJ8:127
BICYCLOGERMACRENE 60 Shoot FFJ8:127
BORNEOL 1 Shoot FFJ8:127
CADINENOL 4.5 Shoot FFJ8:127
CAMPHENE 2 Shoot FFJ8:127
CAMPHOR 1 Shoot FFJ8:127
CARYOPHYLLENE-OXIDE 3 Shoot FFJ8:127
CIS-BETA-OCIMENE 10 Shoot FFJ8:127
DELTA CADINENE 4.5 Shoot FFJ8:127
EO 500 Shoot FFJ8:127
GAMMA-MUUROLENE 0.5 Shoot FFJ8:127
GAMMA-TERPINENE 2.5 Shoot FFJ8:127
GERMACRENE-D 215 Shoot FFJ8:127

LINALOL 1 Shoot FFJ8:127
SABINENE 1.5 Shoot FFJ8:127
SPATHULENOL 2 Shoot FFJ8:127
T-CADINOL 1.5 Shoot FFJ8:127
T-MUUROLENE 4 Shoot FFJ8:127
T-MUUROLOL 4 Shoot FFJ8:127
TERPINEN-4-OL 3 Shoot FFJ8:127
TERPINOLENE 0.5 Shoot FFJ8:127
TRANS-BETA-CARYOPHYLLENE 75 Shoot FFJ8:127
TRANS-BETA-OCIMENE 3 Shoot FFJ8:127
TRANS-CALAMENENE 2 Shoot FFJ8:127
VIRIDIFLOROL 1 Shoot FFJ8:127

Acinos alpinus var. *meridionalis* (NYMAN) P.W.BALL "Te de Sierra Nevada" Unaggregated B

ALLO-AROMADENDRENE 4.5 Shoot FFJ8:127
ALPHA-BISABOLOL 2 Shoot FFJ8:127
ALPHA-CADINOL 2.5 Shoot FFJ8:127
ALPHA-COPAENE 4.5 Shoot FFJ8:127
ALPHA-CUBEBENE 1.5 Shoot FFJ8:127
ALPHA-HUMULENE 3 Shoot FFJ8:127
ALPHA-PHELLANDRENE 18 Shoot FFJ8:127
ALPHA-PINENE 0.5 Shoot FFJ8:127
ALPHA-TERPINENE 0.5 Shoot FFJ8:127
ALPHA-TERPINEOL 0.5 Shoot FFJ8:127
ALPHA-THUJONE 1 Shoot FFJ8:127
AROMADENDRENE 1.5 Shoot FFJ8:127
BETA-BOURBONENE 20 Shoot FFJ8:127
BETA-CUBEBENE 0.5 Shoot FFJ8:127
BETA-GURJUNENE 4 Shoot FFJ8:127
BETA-MYRCENE 1 Shoot FFJ8:127
BETA-PINENE 0.5 Shoot FFJ8:127
BICYCLOGERMACRENE 30 Shoot FFJ8:127
BORNEOL 2 Shoot FFJ8:127
CADINENOL 1 Shoot FFJ8:127
CAMPHENE 0.5 Shoot FFJ8:127
CAMPHOR 1 Shoot FFJ8:127
CARYOPHYLLENE-OXIDE 6 Shoot FFJ8:127
CIS-BETA-OCIMENE 4 Shoot FFJ8:127
DELTA-CADINENE 8.5 Shoot FFJ8:127
EO 500 Shoot FFJ8:127
GAMMA-MUUROLENE 1 Shoot FFJ8:127
GAMMA-TERPINENE 0.5 Shoot FFJ8:127
GERMACRENE-D 285 Shoot FFJ8:127
LINALOL 1.5 Shoot FFJ8:127
SABINENE 0.5 Shoot FFJ8:127
SPATHULENOL 6 Shoot FFJ8:127
T-CADINOL 2.5 Shoot FFJ8:127
T-MUUROLOL 12 Shoot FFJ8:127
TERPINENE-4-OL 2.5 Shoot FFJ8:127
TERPINOLENE 0.5 Shoot FFJ8:127
TRANS-BETA-CARYOPHYLLENE 31 Shoot FFJ8:127
TRANS-BETA-OCIMENE 3 Shoot FFJ8:127
TRANS-CALAMENENE 2 Shoot FFJ8:127
VIRIDIFLOROL 14 Shoot FFJ8:127

Acinos suaveolens (SIBT. & SMITH) G. DON F.

1,8-CINEOLE 5 Shoot JEO3:191
1-8-CINEOL 3,700 Shoot NAP
1-HEXEN-3-OL 7 Shoot JEO3:191
2-ETHYL-CYCLOBUTANOL 200 Shoot NAP
3-6-DIMETHYL-4-5-6-7-OXYGEN
HETEROCYCLE-BENZO-B-FURAN 1,300 Shoot NAP
3-METHYL-CYCLOHEXANONE 1,900 Shoot NAP
3-METHYL-CYCLOPENTANONE 300 Shoot NAP
ACETIC ACID 100 Shoot NAP
ALPHA-PINENE 45 Shoot JEO3:191
ALPHA-PINENE 9,500 Shoot NAP
ALPHA-TERPINEOL 8 Shoot JEO3:191
APIGENIN 23.3 Shoot NAP
BENZALDEHYDE 200 Shoot NAP
BETA-MYRCENE 2,400 Shoot NAP
BETA-PINENE 50 Shoot JEO3:191
BETA-PINENE 11,300 Shoot NAP
BORNEOL 25 Shoot JEO3:191
CAMPHENE 10 Shoot JEO3:191
CAMPHENE 900 Shoot NAP
CARVACROL 15 Shoot JEO3:191
CARVONE 5 Shoot JEO3:191
CHRYSOERIOL 6.6 Shoot NAP
COSMOSIIN 50 Shoot NAP
CYNAROSIDE 123.3 Shoot NAP
DELTA-CADINENE 20 Shoot JEO3:191
DIDYMIN 20 Shoot NAP
EO 6,400 Shoot JEO3:191
GAMMA-TERPINENE 3 Shoot JEO3:191
GERANIAL 50 Shoot JEO3:191
HESPERIDIN 1,993.3 Shoot NAP
HEX-1-EN-3-OL 700 Shoot NAP
ISOMENTHONE 50.86% Essential Oil NAP
ISOMENTHONE 1,090-3,255 Shoot JEO3:191
ISOPIPERITONE 2,800 Shoot NAP
ISOPULEGONE-I 15 Shoot JEO3:191
ISOPULEGONE-I 3,800 Shoot NAP
ISOPULEGONE-II 35 Shoot JEO3:191
ISOPULEGONE-II 10,100 Shoot NAP
ISOQUERCITRIN 430 Shoot NAP
KAEMPFEROL 16.6 Shoot NAP
LIMONENE 180 Shoot JEO3:191
LIMONENE 3,700 Shoot NAP

LINALOOL 1,800 Shoot NAP
LUTEOLIN 66.6 Shoot NAP
LUTEOLIN-7-O-RUTINOSIDE 20 Shoot NAP
MENTHOL 3 Shoot JEO3:191
MENTHONE 150 Shoot JEO3:191
MENTHONE 13,600 Shoot NAP
METHYL-BENZENE 300 Shoot NAP
MYRCENE 15 Shoot JEO3:191
N-UNDECANE 600 Shoot NAP
NAPHTHALENE 3,500 Shoot NAP
NARIRUTIN 26.6 Shoot NAP
NONANAL 10 Shoot JEO3:191
OCT-1-EN-3-OL 4,800 Shoot NAP
OCTAN-3-OL 2,800 Shoot NAP
OCTAN-3-ONE 300 Shoot NAP
P-CYMENE 4 Shoot JEO3:191
PARA-8-HYDROXY-MENTH-(4-5)-EN-3-ONE 2,500 Shoot NAP
PARA-CYMENE 1,800 Shoot NAP
PERILLALDEHYDE 710 Shoot JEO3:191
PIPERITENONE 45 Shoot JEO3:191
PIPERITONE 20 Shoot JEO3:191
PIPERITONE 1,300 Shoot NAP
PULEGONE 332,200 Essential Oil NAP
PULEGONE 2,125-6,200 Shoot JEO3:191
PULEGONE 690,400 Shoot NAP
QUERCETIN 430 Shoot NAP
SABINENE 20 Shoot JEO3:191
TERPINEN-4-OL 8 Shoot JEO3:191
TERPINEN-4-OL 1,400 Shoot NAP
TETRAHYDRO-BETA-BOURBONENE 1,100 Shoot NAP
THYMOL 5 Shoot JEO3:191

Aeolanthus myriantha BAKER "Ninde"

BETA-PINENE Flower WO2
CITRONELLAL Flower WO2
EO 20,200 Flower WO2
GERANIOL Flower WO2
LIMONENE Flower WO2
LINALOL Flower WO2
NEROL Flower WO2

Agastache foeniculum (PURSH) KUNTZE
"Giant Hyssop"

(Z)-BETA-OCIMENE 3-120 Plant BML JEO4:295
3-OCTANOL 3-35 Plant BML JEO4:295
3-OCTANONE 0-30 Plant JEO4:295
4-CADENEN-OL 0-5 Plant JEO4:295
ACACETIN Inflorescence NAP
ACACETIN Leaf NAP
ALPHA-PINENE 1-35 Plant BML JEO4:295
ANISALDEHYDE 0-1 Plant JEO4:295
BENZALDEHYDE 1-135 Plant JEO4:295
BETA-PINENE 2-25 Plant BML JEO4:295
BETA-SABININE 1-125 Plant BML JEO4:295
CAMPHENE 1-35 Plant BML JEO4:295
CARYOPHYLLENE 8-100 Plant BML JEO4:295
CIS-OCIMENE 3-35 Plant BML JEO4:295
COSMOSIIN Inflorescence NAP
COSMOSIIN Leaf NAP
CYNAROSIDE Inflorescence NAP
CYNAROSIDE Leaf NAP
D-LIMONENE 5-10% Plant NAP
DIMETHYL-HEXANAL 1-60 Plant JEO4:295
DIOSMETIN-7-O-BETA-D-GLUCOSIDE Inflorescence NAP
DIOSMETIN-7-O-BETA-D-GLUCOSIDE Leaf NAP
EO 1,000-12,500 Plant BML JEO4:295
ESTRAGOLE 820,000-930,000 Essential Oil NAP
ESTRAGOLE Plant JBH
ETHYLDIMETHYLHEPTENE 1-75 Plant JEO4:295
GERMACRENE-D 25-2,525 Plant BML JEO4:295
LIMONENE 2-25 Plant BML JEO4:295
LINALOL 0-15 Plant JEO4:295
METHYL-CHAVICOL 555-12,160 Plant BML JEO4:295
MYRCENE 0-30 Plant JEO4:295
PARA-METHOXY CINNAMALDEHYDE 5-10% Plant NAP
TILIANIN Inflorescence NAP
TILIANIN Leaf NAP
TRANS-ANETHOLE 0-20 Plant JEO4:295
TRANS-OCIMENE 15-1,150 Plant BML JEO4:295

Agastache nepetoides (L.) KUNTZE

3-OCTANOL 4 Plant BML
ACACETIN Inflorescence NAP
ACACETIN Leaf NAP
ALPHA-PINENE 20 Plant BML
BETA-PINENE 2 Plant BML
CAMPHENE 1 Plant BML
CARYOPHYLLENE 184,000 Essential Oil NAP
CARYOPHYLLENE 90-100 Plant BML
CIS-OCIMENE 4 Plant BML
COPAENE 6 Plant BML
COSMOSIIN Inflorescence NAP
COSMOSIIN Leaf NAP
CYNAROSIDE Inflorescence NAP
CYNAROSIDE Leaf NAP
DELTA-CADINENE 11 Plant BML
EO 500 Plant BML
GERMACRENE-D 240-260 Plant BML
LIMONENE 2 Plant BML
LINALOL 9 Plant BML
MYRCENE 2 Plant BML
TILIANIN Inflorescence NAP
TILIANIN Leaf NAP
TRANS-OCIMENE 36,000 Essential Oil NAP
TRANS-OCIMENE 20 Plant BML

Agastache pallidiflora (HELLER) RYDBERG

DIOSMETIN-7-O-BETA-D-GLUCOSIDE Plant NAP
EO 1,550-3,160 Flower BML
EO 830 Leaf BML
KAEMPFEROL-3-O-ALPHA-L-GALACTOSIDE Plant NAP
KAEMPFEROL-3-O-ALPHA-L-RHAMNOSIDE Plant NAP
PULEGONE Leaf BML
QUERCITRIN Plant NAP
TILIANIN Plant NAP

Agastache rugosa (Fischer & C. Meyer) Kuntze

(-)-CARYOPHYLLENE Tissue Culture NAP
3-ALLYL-5-6-DIMETHOXY PHENOL 0.05% Shoot NAP
3-O-ACETYL-OLEANOLIC ALDEHYDE Plant NAP
3-O-ACETYL-OLEANOLIC ALDEHYDE 4.4 Root NAP
4-METHOXY-CINNAMALDEHYDE 0.14% Shoot NAP
4-METHOXY-CINNAMIC ALCOHOL 0.08% Shoot NAP
ABIETA-4(18)-8-11-13-TETRAEN-7-ONE 26.6 Root NAP
ACACETIN Inflorescence NAP
ACACETIN Leaf NAP
ACACETIN 6.8 Root NAP
ACACETIN 285 Shoot NAP
AGASTACHIN 19 Shoot NAP
AGASTACHOSIDE 9.2 Root NAP
AGASTACHOSIDE Shoot NAP
AGASTANOL 35.5 Root NAP
ALPHA CADINOL 0.04% Shoot NAP
ALPHA HUMULENE 0.02% Shoot NAP
ALPHA LIMONENE Tissue Culture NAP
ALPHA PINENE Essential Oil NAP
ALPHA PINENE Tissue Culture NAP
ALPHA YLANGENE Tissue Culture NAP
ANETHOLE 20,000 Essential Oil NAP
ANETHOLE Tissue Culture NAP
ANISALDEHYDE 30,000 Essential Oil NAP
ANISALDEHYDE 0.26% Shoot NAP
ANISALDEHYDE Tissue Culture NAP
BETA BOURBONENE 0.36% Shoot NAP
BETA CARYOPHYLLENE 0.06% Shoot NAP
BETA ELEMENE Essential Oil NAP
BETA ELEMENE Tissue Culture NAP
BETA FARNESENE Tissue Culture NAP
BETA HUMULENE Tissue Culture NAP
BETA OPLOPENTONE 0.09% Shoot NAP
BETA PINENE Essential Oil NAP
BETA PINENE Tissue Culture NAP
BETA SITOSTEROL Root NAP
CALAMENE Tissue Culture NAP
CARYOPHYLLA-4(14)-8(15)-DIEN-5-ALPHA-OL 0.35% Shoot NAP
CARYOPHYLLENE Essential Oil NAP
CARYOPHYLLENE OXIDE 0.89% Shoot NAP
CARYOPHYLLENOL I 0.19% Shoot NAP
CARYOPHYLLENOL II 0.21% Shoot NAP
CHAVIBETOL 0.13% Shoot NAP

CHAVICOL 0.14% Shoot NAP
CIS CARVEOL 0.05% Shoot NAP
CIS JASMONE 0.08% Shoot NAP
COSMOSIIN Inflorescence NAP
COSMOSIIN Leaf NAP
CYNAROSIDE Inflorescence NAP
CYNAROSIDE Leaf NAP
D LIMONENE 50,000 Essential Oil NAP
D LIMONENE 90,000 Essential Oil NAP
DAUCOSTEROL 5.2 Root NAP
DEHYDRO-AGASTOL 6.4 Root NAP
DEHYDRO-AGASTOL 13.3 Root NAP
DELTA CADINENE Essential Oil NAP
DIHYDRO-ACTINIDIOLIDE 15% Shoot NAP
DIOSMETIN-7-O-BETA-D-GLUCOSIDE Inflorescence NAP
DIOSMETIN-7-O-BETA-D-GLUCOSIDE Leaf NAP
ERYTHRODIOL-3-O-ACETATE Root NAP
ESSENTIAL OIL Plant NAP
ESTRAGOLE Essential Oil tr NAP
ESTRAGOLE Essential Oil NAP
ESTRAGOLE 895,000 Essential Oil NAP
ESTRAGOLE 900,000 Essential Oil NAP
ESTRAGOLE Plant JBH
ESTRAGOLE 91% Shoot NAP
ESTRAGOLE Tissue Culture NAP
EUGENOL 0.01% Shoot NAP
EUGENOL METHYL ETHER Essential Oil NAP
EUGENOL METHYL ETHER 0.01% Shoot NAP
GAMMA CADINENE Tissue Culture NAP
GERMACRENE D 0.17% Shoot NAP
HEX-1-EN-3-OL ACETATE 0.56% Shoot NAP
HUMULENE Essential Oil NAP
ISO-AGASTACHOSIDE 210 Shoot NAP
LIMONENE Essential Oil NAP
LINALOOL Essential Oil NAP
LINALOOL 0.02% Shoot NAP
LINALOOL Tissue Culture NAP
MASLINIC ACID 12.4 Root NAP
OCT-1-EN-3-OL Essential Oil NAP
OCT-1-EN-3-OL ACETATE 0.81% Shoot NAP
OCT-1-EN-3-ONE Tissue Culture NAP
OCTAN-3-OL Essential Oil NAP
OCTAN-3-OL Tissue Culture NAP
OCTAN-3-ONE Tissue Culture NAP
OLEANOLIC ACID 10 Root NAP

OPLOPANONE 0.05% Shoot NAP
P-ANISALDEHYDE Plant JBH
P-METHOXYCINNAMALDEHYDE Shoot JBH
PACHYPODOL Leaf JBH
PARA CYMENE Essential Oil NAP
PARA CYMENE Tissue Culture NAP
PARA-METHOXY-CINNAMALDEHYDE 10,000 Essential Oil NAP
PARA-METHOXY-CINNAMALDEHYDE 1-2.5% Plant NAP
PARA-METHOXY-CINNAMALDEHYDE Tissue Culture NAP
ROSMARINIC ACID 9,000 Leaf NAP
SAFROLE Essential Oil NAP
SPATHULENOL 0.66% Shoot NAP
TILIANIN Inflorescence NAP
TILIANIN Leaf NAP
TILIANIN 84 Root NAP
TILIANINE 46 Shoot NAP
TRANS CARVEOL 0.04% Shoot NAP
TRANS JASMONE 0.09% Shoot NAP

Agastache urticifolia (BENTH.) KUNTZE
"Nettle-Leaf Giant Hyssop"
Unaggregated A

1-OCTEN-3-OL 26 Plant BML
3-OCTANOL 91 Plant BML
ALPHA-PINENE 32 Plant BML
BETA-PINENE 46 Plant BML
CIS-OCIMENE 72 Plant BML
EO 6,500 Plant BML
GERMACRENE-D 26 Plant BML
ISOMENTHONE 962 Plant BML
LIMONENE 2,464 Plant BML
MENTHONE 546 Plant BML
MYRCENE 266 Plant BML
PULEGONE 1,138 Plant BML
TRANS-OCIMENE 370 Plant BML

Ajuga bracteosa WALL.
"Nilkanthi"

BETA SITOSTEROL Shoot NAP
CEROTIC-ACID Plant WO2
CERYL-ALCOHOL Plant WO2
DAUCOSTEROL Shoot NAP
GAMMA-SITOSTEROL Plant WO2
HEXACOSAN-1-OL Shoot NAP
LIGNOCERIC ACID Shoot NAP
TRIACONTAN-1-OL-DOCOSANOATE Shoot NAP

Ajuga chamaepitys (L.) Schreber
"Bugle"

14-15-DIHYDRO-AJUGAPITIN 15.2 Plant NAP
14-HYDRO-15-HYDROXY-AJUGAPITIN 267.5 Plant NAP
15-ETHOXY-14-HYDRO-AJUGAPITIN 15 Plant NAP
ACETYL HARPAGIDE Fruit NAP
ACETYL HARPAGIDE Shoot NAP
AJUGALACTONE 42.5 Plant NAP
AJUGALACTONE Shoot JBH
AJUGAPITIN 0-36.2 Plant NAP
BETA ECDYSONE 37.5 Plant NAP
CHAMAEPITIN 200 Plant NAP
CYASTERONE Plant JBH
CYASTERONE 160 Plant NAP
MAKISTERONE A 40 Plant NAP
MAKISTERONE-B Plant JBH

Ajuga iva (L.) Schreber
"Ivy Bugle"

14-15-DIHYDRO AJUGAPITIN 35.1 Plant NAP
2-ACETYL IVAIN I 8.8 Plant NAP
22-OXO-CYASTERONE Plant NAP
24(28)-DEHYDRO-MAKISTERONE A Plant NAP
8-O-ACETYL HARPAGIDE Shoot NAP
AJUGA STERONE C Plant NAP
APIGENIN-7-O-NEOHESPERIDOSIDE Shoot NAP
BETA ECDYSONE 2,200 Leaf NAP
BETA ECDYSONE 50 Plant NAP
BETA ECDYSONE 120 Plant NAP
BETA ECDYSONE 12,000 Plant NAP
BETA ECDYSONE Shoot NAP
BETA ECDYSONE 520-800 Stem NAP
CYASTERONE 440 Leaf NAP
CYASTERONE Plant NAP
CYASTERONE 0.7 Plant NAP
CYASTERONE 1,330 Root NAP
CYASTERONE Shoot NAP
CYASTERONE 530 Stem NAP
FERULIC ACID 1,000 Plant NAP
FERULIC-ACID Shoot JBH
HARPAGIDE Shoot NAP
IVAIN I 310 Plant NAP
IVAIN II 10 Plant NAP
IVAIN III 40 Plant NAP
IVAIN IV 80 Plant NAP
MAKISTERONE A 440 Leaf NAP
MAKISTERONE A 1,070 Plant NAP
MAKISTERONE A 280 Root NAP
MAKISTERONE A Shoot NAP
MAKISTERONE A 640 Stem NAP
NARINGIN Shoot NAP

Ajuga remota
"Bugle"

AJUGALACTONE Plant NAP
AJUGARIN B Leaf NAP
AJUGARIN I 500 Leaf NAP
AJUGARIN II 20 Leaf NAP
AJUGARIN III 10 Leaf NAP
AJUGARIN IV Leaf NAP
AJUGARIN V Leaf NAP
AJUGARIN-I Leaf JBH
AJUGARIN-II Leaf JBH
AJUGARIN-III Leaf JBH
AJUGASTERONE C Leaf NAP
AJUGASTERONE C Plant NAP
BETA ECDYSONE Leaf NAP
BETA ECDYSONE Plant NAP
BETA ECDYSONE Root NAP
CYASTERONE Leaf NAP
CYASTERONE Root NAP
PONASTERONE A Plant NAP

Ajuga reptans L.
"Bugle"

24-NORCYASTERONE Plant JSG
29-NORSENGOSTERONE Plant JSG
AJUGALACTONE Plant JBH JSG
AJUGAREPTANSONE Plant JSG
AJUGASTERONE Plant JSG
AJUGOL Plant JSG
AJUGOSIDE Plant JSG
ASPERULIN Plant HHB
AUCUBIN Plant YYB
BETA-ECDYSONE Plant JSG
CYANIDIN Plant CRC
CYASTERONE Plant JBH
DELPHINIDIN Plant CRC
HARPAGIDE Plant JSG
HARPAGIDE-ACETATE Plant JSG
POLYPODINE-B Plant JBH JSG
REPTIOSIDE Plant JSG
TANNIN 150,000 Plant CRC HHB

Anisomeles indica (L.) KUNTZE
"Kalabhangra"

ISOOVATODIOLIDE Leaf WO2
MUCILAGE 16,000 Seed WO2
OVATODIOLIDE Leaf WO2
PENTOSANS 130,000 Seed WO2
PROTEIN 169,000 Seed WO2
WATER 82,000 Seed WO2

Anisomeles malabarica R.BR.
"Malabar Catmint"

ANISOMELIC-ACID Plant WO2
BETULINIC-ACID 100 Plant NAP
BETULINIC-ACID 100 Plant PHYT13:2002
CITRAL Plant WO2
EO 700 Flower WO2
EO 250 Leaf WO2
EO 50 Root WO2
EO 50 Stem WO2
GERAINIC-ACID Plant WO2
OVATODIOLIDE Plant WO2

Ballota nigra L.
"Black Horehound"

1-OCTEN-3-OL 4 Plant BML
7-ACETOXYMARRUBIIN Plant 411/
ALPHA-HUMULENE 8 Plant BML
ALPHA-PINENE 3 Plant BML
BALLONIGRIN Plant 411/
BALLOTENOL Plant 411/
BALLOTINONE Plant 411/
BETA-PINENE 1 Plant BML
CARYOPHYLLENE 35 Plant BML
COPAENE 19 Plant BML
DELTA-CADINENE 10 Plant BML
EO 300 Plant BML
GERMACRENE-D 127 Plant BML
LINALOL 3 Plant BML
MARRUBIIN Plant 411/
SABINENE 4 Plant BML

Calamintha nepeta ssp. *glandulosa* (REQ.) P.W.BALL
"Turkish Calamint"
Unaggregated A

1,8-CINEOLE 80 Leaf JEO3:7
1-OCTEN-3-OL <1 Leaf JEO3:7
3-OCTANOL <1 Leaf JEO3:7
ALPHA-COPAENE 25 Leaf JEO3:7
ALPHA-PINENE 90 Leaf JEO3:7
BETA-BOURBONENE 8 Leaf JEO3:7
BETA-PINENE 95 Leaf JEO3:7
CAMPHENE Leaf JEO3:7
CAMPHOR <1 Leaf JEO3:7
CIS-SABINENE-HYDRATE <1 Leaf JEO3:7
CUMINALDEHYDE Leaf JEO3:7
EO 8,000 Leaf JEO3:7
ISOMENTHONE 465 Leaf JEO3:7
LIMONENE 80 Leaf JEO3:7
LINALOL <1 Leaf JEO3:7
MENTHOL <1 Leaf JEO3:7
MENTHONE <1 Leaf JEO3:7
MYRCENE 55 Leaf JEO3:7
P-CYMENE <1 Leaf JEO3:7
PIPERITENONE 3,230 Leaf JEO3:7
PIPERITONE 150 Leaf JEO3:7
PULEGONE 3,360 Leaf JEO3:7
SABINENE 70 Leaf JEO3:7
TRANS-SABINENE-HYDRATE <1 Leaf JEO3:7

Calamintha nepeta ssp. *glandulosa* (REQ.) P.W.BALL "Turkish Calamint"
Unaggregated B

(Z)-2-HEXEN-1-OL 0.4 Shoot JEO4:189
(Z)-BETA-OCIMENE 9 Shoot JEO4:189
(Z,Z)-ISOPROPENYL-2-METHYL-3-CYCLOHEXYL-ACETATE 17 Shoot JEO4:189
1,8-CINEOLE 18 Shoot JEO4:189
1-OCTEN-3-OL 0.9 Shoot JEO4:189
2-HEXANOL 0.9 Shoot JEO4:189
2-METHYL-6-(2-PROPENYL)-PHENOL 15 Plant JEO4:189
6-METHYL-3-HEPTANOL 75 Plant JEO4:189
ACETONE <0.1 Plant JEO4:189
ALPHA-PINENE 13 Shoot JEO4:189
ALPHA-TERPINEOL 16 Shoot JEO4:189
BETA-BOURBONENE 5 Shoot JEO4:189
BETA-PINENE 21 Shoot JEO4:189
BORNEOL 16 Shoot JEO4:189
BORNYL-ACETATE 1 Shoot JEO4:189
CAMPHENE 1 Shoot JEO4:189
CARVACROL 3 Shoot JEO4:189
CIS-GERANYL-ACETONE 6 Shoot JEO4:189
CIS-PIPERITONE-OXIDE 55 Shoot JEO4:189
CIS-SABINENE-HYDRATE 4 Shoot JEO4:189
DELTA-CADINENE 3 Shoot JEO4:189
EO 4,500 Shoot JEO4:189
ETHYL-2-METHYL-BUTYRATE 0.9 Plant JEO4:189
ETHYL-3-METHYL-BUTYRATE <0.1 Plant JEO4:189
EUGENOL 8 Shoot JEO4:189
GAMMA-TERPINENE 31 Shoot JEO4:189
GERMACRENE-D 11 Shoot JEO4:189
HEXENAL 0.9 Shoot JEO4:189
ISOPIPERITENONE 40 Shoot JEO4:189
LIMONENE 585 Shoot JEO4:189
LINALOL 34 Shoot JEO4:189
METHYL-2-METHYL-BUTYRATE 0.4 Plant JEO4:189
MYRCENE 32 Shoot JEO4:189
P-CYMEN-8-OL 1 Shoot JEO4:189
P-CYMENE 9 Shoot JEO4:189
PATCHOULANE 7 Shoot JEO4:189
PIPERITENONE 40 Shoot JEO4:189
PIPERITENONE-OXIDE 1,970 Shoot JEO4:189
SABINENE 13 Shoot JEO4:189
TERPINEN-4-OL 105 Shoot JEO4:189

TERPINENE 14 Shoot JEO4:189
TERPINOLENE 0.9 Shoot JEO4:189
THYMOL 14 Shoot JEO4:189
TRANS-PIPERITONE-OXIDE 1,135 Shoot JEO4:189
TRANS-SABINENE-HYDRATE 17 Shoot JEO4:189

Calamintha nepeta (L.) SAVI "Calamint"

1,8-CINEOL 25 Plant JE39:167
ALPHA-PINENE 40 Plant JE39:167
ALPHA-TERPINEOL 30 Plant JE39:167
BETA-CARYOPHYLLENE 135 Plant JE39:167
BETA-PINENE 35 Plant JE39:167
D-LIMONENE 385 Plant JE39:167
EO 6,000 Plant JE39:167
ISOMENTHONE 10 Plant JE39:167
MENTHOL 290 Plant JE39:167
MENTHONE 590 Plant JE39:167
P-CYMENE 65 Plant JE39:167
PIPERITENONE 120 Plant JE39:167
PIPERITENONE-OXIDE 150 Plant JE39:167
PIPERITONE-OXIDE 135 Plant JE39:167
PONCIRIN Plant JBH
PULEGONE 2,760 Plant JE39:167

Cleonia lusitanica (L.) L.
"Spanish Heal-All"

1,8-CINEOLE 2-10 Leaf JEO3:441
ALLOOCIMENE-ISOMER <1-6 Leaf JEO3:441
ALPHA-CADINOL <1-1 Leaf JEO3:441
ALPHA-COPAENE <1 Leaf JEO3:441
ALPHA-HUMULENE <1 Leaf JEO3:441
ALPHA-PHELLANDRENE 1-2 Leaf JEO3:441
ALPHA-PINENE 90-170 Leaf JEO3:441
ALPHA-TERPINEOL <1-5 Leaf JEO3:441
ALPHA-THUJONE <1-1 Leaf JEO3:441
BETA-CARYOPHYLLENE 2-5 Leaf JEO3:441
BETA-PINENE 15-35 Leaf JEO3:441
BORNEOL 1-5 Leaf JEO3:441
BORNYL-ACETATE <1 Leaf JEO3:441
CAMPHENE 1-5 Leaf JEO3:441
CAMPHOR 5-7 Leaf JEO3:441
CARYOPHYLLENE-EPOXIDE 4-6 Leaf JEO3:441
CIS-VERBENOL 3-10 Leaf JEO3:441
DIHYDROCARVEOL 1-8 Leaf JEO3:441
EO 500 Leaf JEO3:441
FARNESOL-ACETATE-ISOMER <1-1 Leaf JEO3:441
FARNESOL-ISOMER <1-1 Leaf JEO3:441
FENCHONE <1-1 Leaf JEO3:441
FENCHYL-ALCOHOL 4-6 Leaf JEO3:441
GAMMA-TERPINENE 1-3 Leaf JEO3:441
GERMACRENE-D <1-2 Leaf JEO3:441
HUMULENE-EPOXIDE <1 Leaf JEO3:441
ISOBORNEOL 1-5 Leaf JEO3:441
ISODIHYDROCARVEOL 2-3 Leaf JEO3:441
LIMONENE 165-325 Leaf JEO3:441
LINALOL <1-2 Leaf JEO3:441
MYRCENE 8 Leaf JEO3:441
NEODIHYDROCARVEOL 2-7 Leaf JEO3:441
NEOISODIHYDROCARVEOL 1-5 Leaf JEO3:441
P-CYMENE 2-3 Leaf JEO3:441
SABINENE 10-12 Leaf JEO3:441
TERPINEN-4-OL 2-3 Leaf JEO3:441
TERPINOLENE <1-1 Leaf JEO3:441
VERBENONE 2 Leaf JEO3:441

Coleus barbatus BENTH.
"Forskohl's Coleus"

(+)-FERRUGINOL Leaf RAA
1,9-DIDEOXY-7-DEACETYL-FORSKOHLIN Plant PC29:821
1,9-DIDEOXYFORSKOHLIN Plant PC29:821
20-DEOXOCARNOSOL Stem PC23:1677
3BETA-HYDROXY-3-DEOXYBARBATUSIN Leaf EB41:474
6BETA,7BETA,9ALPHA-TRIHYDROXY-8,13-EPOXY-LABD-14-EN-11-ONE Plant PC29:821
6BETA-HYDROXY-8,13-EPOXY-LABD-14-EN-11-ONE Plant
PC29:821 6BETA-HYDROXY-CARNOSOL Leaf RAA
7BETA-ACETOXY-6BETA,9ALPHA-DIHYDROXY-8,13-EPOXY-LABD-14-EN-11-ONE Plant PC29:821
8,13-EPOXY-LABD-14-EN-11-ONE Plant PC29:821
9-DEOXYFORSKOHLIN Plant PC29:821
BARBATUSIN Plant ABS
BARBATUSOL Leaf RAA
COLEOL Plant PC29:821
COLEONOL Leaf JBH
COLEONOLS Tuber EB41:474
COLEONS Plant JSG
COLEOSOL Plant JSG
CYCLOBUTATUSIN Leaf RAA
DEACETYL-FORSKOHLIN Plant PC29:821
FORSKOLIN 4,630 Tuber JBH EB41:474

Coleus blumei BENTH.
"El Nene"

AROMADENDRIN Plant CRC
CYANIDIN-3,5-DI-O-BETA-D-GLUCOSYL-P-COUMARATE Plant CRC
CYANIDIN-3,5-DIGLUCOSIDE Plant JSG
DIHYDROKAEMPFEROL Plant JSG
PHYTOSTEROL Plant CRC

Coleus forskohlii BRIQ
"Forskohl's Coleus"

(+)-FERRUGINOL Leaf FNF
1,9-DIDEOXY-7-DEACETYL-FORSKOHLIN Plant FNF
1,9-DIDEOXYFORSKOHLIN Plant FNF
20-DEOXOCARNOSOL Stem FNF
3BETA-HYDROXY-3-DEOXYBARBATUSIN Leaf FNF
6BETA,7BETA,9ALPHA-TRIHYDROXY-8,13-EPOXY-LABD-14-EN-11-ONE Plant FNF
6BETA-HYDROXY-8,13-EPOXY-LABD-14-EN-11-ONE Plant FNF
6BETA-HYDROXY-CARNOSOL Leaf FNF
7BETA-ACETOXY-6BETA,9ALPHA-DIHYDROXY-8,13-EPOXY-LABD-14-EN-11-ONE Plant FNF
8,13-EPOXY-LABD-14-EN-11-ONE Plant FNF
9-DEOXYFORSKOHLIN Plant FNF
BARBATUSIN Plant FNF
BARBATUSOL Leaf FNF
COLEOL Plant FNF
COLEONOLS Tuber FNF
COLEONS Plant FNF
COLEOSOL Plant FNF
CYCLOBUTATUSIN Leaf FNF
DEACETYL-FORSKOHLIN Plant FNF
FORSKOLIN 4,630 Tuber FNF

Collinsonia canadensis L.
"Stone Root"

ALPHA-NAGITANINE 8 Plant FNF
BETA-ELEMENE 16 Plant FNF
CARYOPHYLLENE Tuber FNF
DELTA-CADINENE 13 Plant FNF
ELEMICIN 18 Plant FNF
EO 500 Plant FNF
GERMACRENE-D 230 Plant FNF

Coridothymus capitatus (L.) REICHB. f.
"Spanish Oregano"
Unaggregated A

4-TERPINEOL 155 Shoot ZLU197:20
ALPHA-CARYOPHYLLENE Shoot ZLU197:20
ALPHA-HUMULENE <0.1 Shoot ZLU197:20
ALPHA-MUUROLENE Shoot ZLU197:20
ALPHA-PHELLANDRENE 22 Shoot ZLU197:20
ALPHA-PINENE 44 Shoot ZLU197:20
ALPHA-TERPINENE 130 Shoot ZLU197:20
ALPHA-TERPINEOL 44 Shoot ZLU197:20
ALPHA-THUJENE 22 Shoot ZLU197:20
AROMADENDRENE Shoot ZLU197:20
BETA-BISABOLENE 44 Shoot ZLU197:20
BETA-CARYOPHYLLENE 200 Shoot ZLU197:20
BETA-PINENE Shoot ZLU197:20
BORNEOL 130 Shoot ZLU197:20
BORNYL-ACETATE <0.1 Shoot ZLU197:20
CAMPHENE 22 Shoot ZLU197:20
CARVACROL 18,105 Shoot ZLU197:20
CARVACROL-METHYL-ETHER 22 Shoot ZLU197:20
CARVONE Shoot ZLU197:20
CIS-OCIMENE Shoot ZLU197:20
CIS-PINANE 175 Shoot ZLU197:20
CIS-THUJANOL 110 Shoot ZLU197:20
DELTA-3-CARENE <0.1 Shoot ZLU197:20
EO 22,000 Shoot ZLU197:20
GAMMA-CADINENE <0.1 Shoot ZLU197:20
GAMMA-GURJUNENE 44 Shoot ZLU197:20
GAMMA-TERPINENE 595 Shoot ZLU197:20
LIMONENE 110 Shoot ZLU197:20
MYRCENE 285 Shoot ZLU197:20
P-CYMENE 1,275 Shoot ZLU197:20
TERPINOLENE 22 Shoot ZLU197:20
THYMOL 65 Shoot ZLU197:20
TRANS-OCIMENE Shoot ZLU197:20
TRANS-THUJANOL 240 Shoot ZLU197:20

Cunila origanoides (L.) BRITTON
"Mountain Dittany"
Unaggregated A

1,8-CINEOLE 28 Shoot BML
1-OCTEN-3-OL 924 Shoot BML
3-OCTANOL 56 Shoot BML
ALPHA-PINENE 812 Shoot BML
ALPHA-TERPINENE 644 Shoot BML
BETA-PINENE 28 Shoot BML
CARVACROL 504 Shoot BML
CARYOPHYLLENE 140 Shoot BML
EO 28,000 Shoot BML
GAMMA-TERPINENE 7,560 Shoot BML
LIMONENE 140 Shoot BML
METHYL-CARVACROL 1,092 Shoot BML
MYRCENE 672 Shoot BML
P-CYMENE 3,388 Shoot BML
SABINENE 56 Shoot BML
THYMOL 10,612 Shoot BML

Dracocephalum parviflora **(NUTT.) BRITT.** **"Small-Flowered Moldavica"**

BICYCLOGERMACRENE 15 Plant FNF
CARYOPHYLLENE 50 Plant FNF
EO 1,000 Plant FNF
GERMACRENE-D 310 Plant FNF
MYRCENE 160 Plant FNF

Dracocephalum thymiflora RYDB. "Thyme-Flowered Moldavica"

1,8-CINEOLE 134 Plant FNF
ALPHA-PINENE 2 Plant FNF
ALPHA-TERPINEOL 48 Plant FNF
BETA-ELEMENE 148 Plant FNF
BETA-PINENE 2 Plant FNF
CARYOPHYLLENE 250 Plant FNF
CARYOPHYLLENE-OXIDE 46 Plant FNF
CIS-OCIMENE 84 Plant FNF
EO 2,000 Plant FNF
GERMACRENE-D 140 Plant FNF
ISOMENTHONE 216 Plant FNF
LIMONENE 10 Plant FNF
LINALOL 32 Plant FNF
MENTHONE 28 Plant FNF
MYRCENE 20 Plant FNF
PULEGONE 280 Plant FNF
SABINENE 8 Plant FNF
TRANS-OCIMENE 96 Plant FNF

Elsholtzia blanda BENTH.
"Bantaluki, Bantulsi"

(E)-BETA-OCIMENE 38 Shoot JEO4:121
(E)-NEROLIDOL 2 Shoot JEO4:121
(Z)-BETA-OCIMENE 18 Shoot JEO4:121
(Z,E)-ALPHA-FARNESENE 2 Shoot JEO4:121
1,8-CINEOLE 12 Shoot JEO4:121
ACETOPHENONE 15 Shoot JEO4:121
ALPHA-BERGAMOTENE 22 Shoot JEO4:121
ALPHA-HUMULENE 4 Shoot JEO4:121
ALPHA-PINENE 0.4 Shoot JEO4:121
BENZALDEHYDE 13 Shoot JEO4:121
BETA-CARYOPHYLLENE 23 Shoot JEO4:121
BETA-PINENE 0.5 Shoot JEO4:121
BORNEOL 55 Shoot JEO4:121
BORNYL-ACETATE 4 Shoot JEO4:121
CAMPHENE 0.4 Shoot JEO4:121
CETRONELLOL 12 Shoot JEO4:121
CITRONELLYL-ACETATE 9 Shoot JEO4:121
DELTA-CADINENE 0.4 Shoot JEO4:121
ELEMOL 0.4 Shoot JEO4:121
EO 1,300 Shoot JEO4:121
EUGENOL 0.9 Shoot JEO4:121
GERANIAL 5 Shoot JEO4:121
GERANIOL 49 Shoot JEO4:121
GERANYL-ACETATE 930 Shoot JEO4:121
GERMACRENE-D 0.1 Shoot JEO4:121
LIMONENE 55 Shoot JEO4:121
LINALOL 65 Shoot JEO4:121
LINALYL-ACETATE 14 Shoot JEO4:121
MYRCENE 4 Shoot JEO4:121
NERAL 3 Shoot JEO4:121
NERYL-ACETATE 2 Shoot JEO4:121
P-CYMENE 0.5 Shoot JEO4:121

Elsholtzia cristata WILLD. "Furan Chemotype" Unaggregated A

(E,E)-ALPHA-FARNESENE 8 Shoot PM1987:268
1,8-CINEOLE 5 Shoot PM1987:268
1-OCTEN-3-OL 26 Shoot PM1987:268
2-ISOVALERYL-3-METHYLFURAN 6 Shoot PM1987:268
3-METHYL-2(3-METHYLBUT-2-ENOYL)-FURAN 3,550 Shoot PM1987:268 3-METHYL-2(3-METHYLBUT-3-ENOYL)-FURAN 36 Shoot PM1987:268
ACETOPHENONE 8 Shoot PM1987:268
ALPHA-BERGAMOTENE 8 Shoot PM1987:268
ALPHA-PINENE <0.1 Shoot PM1987:268
BENZENE Shoot PM1987:268
BETA-PINENE <0.1 Shoot PM1987:268
BOURBONENE 11 Shoot PM1987:268
CAMPHENE <0.1 Shoot PM1987:268
CAMPHOR 4 Shoot PM1987:268
CARYOPHYLLENE 38 Shoot PM1987:268
EO 4,000 Shoot PM1987:268
HUMULENE 55 Shoot PM1987:268
METHYL-ACETATE Shoot PM1987:268
MYRCENE <0.1 Shoot PM1987:268
SABINENE <0.1 Shoot PM1987:268
TOLUENE Shoot PM1987:268

Elsholtzia eriostachya var. *pusilla*
"'Dwarf' Bush Mint"
Unaggregated A

6-METHYL-5-HEPTEN-2-ONE 9.5 Shoot JEO4:547
ACETOPHENONE <0.1 Shoot JEO4:547
ALPHA-HUMULENE 3 Shoot JEO4:547
BETA-CARYOPHYLLENE 9 Shoot JEO4:547
CARYOPHYLLENE-OXIDE 0.8 Shoot JEO4:547
EO 1,000 Shoot JEO4:547
GERANIAL 525 Shoot JEO4:547
GERANIOL 11 Shoot JEO4:547
GERMACRENE-D 1.3 Shoot JEO4:547
LIMONENE <0.1 Shoot JEO4:547
LINALOL 6.2 Shoot JEO4:547
MYRCENE <0.1 Shoot JEO4:547
NERAL 390 Shoot JEO4:547
NEROL 11 Shoot JEO4:547
SABINENE 0.6 Shoot JEO4:547

Elsholtzia nipponica OHWI
"Hsiang Ju"

BETA-DEHYDROELSCHOLTZIAKETONE Plant CCO
CITRAL Plant HEG
ELSCHOLTZIAKETONE Plant CCO
GERANIOL Plant CCO
LIMONENE Plant HEG
NERYL-ACETATE Plant CCO
P-CYMENE Plant CCO
PHENOL Plant JBH

Elsholtzia pilosa GARKE
"Hairy Mint Shrub"
Unaggregated A

(E)-OCIMENE 16 Shoot ZN43:370
1,5-P-MENTHADIEN-7-OL 8 Shoot ZN43:370
1,8-CINEOLE 1,000 Shoot ZN43:370
4-PHENYLBUT-3-EN-2-OL 4 Shoot ZN43:370
ALPHA-PINENE 7.6 Shoot ZN43:370
ALPHA-TERPINENE 10 Shoot ZN43:370
ALPHA-TERPINOLENE 1.2 Shoot ZN43:370
ALPHA-TERPINYL-ACETATE 44 Shoot ZN43:370
ALPHA-THUJENE 1.4 Shoot ZN43:370
AROMADENDRENE 4 Shoot ZN43:370
BETA-PINENE 34 Shoot ZN43:370
CAMPHENE <0.1 Shoot ZN43:370
CARYOPHYLLENE 12 Shoot ZN43:370
CIS-VERBENOL 4 Shoot ZN43:370
EO 2,000 Shoot ZN43:370
GAMMA-CADINENE 36 Shoot ZN43:370
GAMMA-TERPINENE 105 Shoot ZN43:370
HUMULENE 6 Shoot ZN43:370
LINALOL 46 Shoot ZN43:370
MYRCENE 10 Shoot ZN43:370
MYRCENOL 21 Shoot ZN43:370
NEROL 1 Shoot ZN43:370
NERYL-ACETATE 6 Shoot ZN43:370
P-CYMENE 0.6 Shoot ZN43:370
PINOCARVONE 42 Shoot ZN43:370
SABINENE 15 Shoot ZN43:370
TERPINEN-4-OL 6 Shoot ZN43:370
THYMOL 85 Shoot ZN43:370
TRANS-PINOCARVEOL 10 Shoot ZN43:370
TRANS-SABINENE-HYDRATE 12 Shoot ZN43:370
VERBENENE <0.1 Shoot ZN43:370

Elsholtzia polystachya BENTH.
"Bush Mint"
Unaggregated A

(E)-OCIMENE 365 Leaf PM58:376
(Z)-OCIMENE 12 Leaf PM58:376
1,8-CINEOLE 745 Leaf PM58:376
1-OCTEN-3-OL 1.4 Leaf PM58:376
3-HEXYL-ACETATE 0.4 Leaf PM58:376
ALLOAROMADENDRENE Leaf PM58:376
ALPHA-CUBEBENE 0.4 Leaf PM58:376
ALPHA-GURJUNENE Leaf PM58:376
ALPHA-MUUROLENE Leaf PM58:376
ALPHA-PHELLANDRENE 2.4 Leaf PM58:376
ALPHA-PINENE 10 Leaf PM58:376
ALPHA-TERPINENE 8 Leaf PM58:376
ALPHA-TERPINEOL 24 Leaf PM58:376
AROMADENDRENE 4 Leaf PM58:376
BENZALDEHYDE 0.2 Leaf PM58:376
BENZYL-ISOVALERATE 5 Leaf PM58:376
BETA-CURCUMENE 24 Leaf PM58:376
BETA-EUDESMOL 1.2 Leaf PM58:376
BETA-PINENE 65 Leaf PM58:376
BETA-SELINENE Leaf PM58:376
CADINOL Leaf PM58:376
CAMPHENE 1.8 Leaf PM58:376
CARYOPHYLLENE 165 Leaf PM58:376
CIS-PINOCARVEOL 4.4 Leaf PM58:376
CIS-PINOCARVYL-ACETATE 42 Leaf PM58:376
CIS-SABINENE-HYDRATE 0.2 Leaf PM58:376
CIS-VERBENOL 7 Leaf PM58:376
COPAENE 0.4 Leaf PM58:376
DELTA-CADINENE 8.8 Leaf PM58:376
EO 2,000 Leaf PM58:376
GAMMA-CADINENE 0.4 Leaf PM58:376
GAMMA-CURCUMENE 20 Leaf PM58:376
GAMMA-MUUROLENE Leaf PM58:376
GAMMA-TERPINENE 46 Leaf PM58:376
GERANIAL Leaf PM58:376
GERANIOL Leaf PM58:376
GERANYL-ACETATE Leaf PM58:376
GERMACRENE 2.4 Leaf PM58:376
HEXAHYDROFARNESYLACETONE 1 Leaf PM58:376
HUMULENE 23 Leaf PM58:376
ISOAMYL-ISOVALERATE 0.2 Leaf PM58:376

JASMONE 1.8 Leaf PM58:376
JUNIPER-CAMPHOR Leaf PM58:376
LAVANDULOL 0.2 Leaf PM58:376
LIMONENE 6.6 Leaf PM58:376
LINALOL 17 Leaf PM58:376
LINALYL-ACETATE 0.2 Leaf PM58:376
METHYL-OCTANOATE 0.2 Leaf PM58:376
MYRCENE 24 Leaf PM58:376
NERAL Leaf PM58:376
OCIMENOL 14 Leaf PM58:376
OCTAN-3-ONE 1 Leaf PM58:376
OCTENYL-ACETATE 4.4 Leaf PM58:376
P-CYMENE 4.2 Leaf PM58:376
PERILLENE 8 Leaf PM58:376
PINOCARVONE 0.2 Leaf PM58:376
SABINENE 24 Leaf PM58:376
TERPINEN-4-OL 12 Leaf PM58:376
TERPINOLENE 5.8 Leaf PM58:376
THUJENE 4 Leaf PM58:376
THYMOL Leaf PM58:376
TRANS-PINOCARVEOL 1.6 Leaf PM58:376
TRANS-PINOCARVYL-ACETATE 11 Leaf PM58:376
TRANS-SABINENE-HYDRATE 4.4 Leaf PM58:376
VERBENENE 8 Leaf PM58:376
VERBENONE 0.2 Leaf PM58:376
YLANGENE Leaf PM58:376

Elsholtzia polystachya BENTH.
"Bush Mint"
Unaggregated B

(E)-OCIMENE 13 Leaf PM58:376
(Z)-OCIMENE 0.2 Leaf PM58:376
1,8-CINEOLE 260 Leaf PM58:376
1-OCTEN-3-OL 4 Leaf PM58:376
3-HEXYL-ACETATE Leaf PM58:376
ALLOAROMADENDRENE 6 Leaf PM58:376
ALPHA-CUBEBENE 0.4 Leaf PM58:376
ALPHA-GURJUNENE 3.6 Leaf PM58:376
ALPHA-MUUROLENE 6 Leaf PM58:376
ALPHA-PHELLANDRENE Leaf PM58:376
ALPHA-PINENE 1.2 Leaf PM58:376
ALPHA-TERPINENE 0.2 Leaf PM58:376
ALPHA-TERPINEOL 10 Leaf PM58:376
AROMADENDRENE Leaf PM58:376
BENZALDEHYDE Leaf PM58:376
BENZYL-ISOVALERATE Leaf PM58:376
BETA-CURCUMENE Leaf PM58:376
BETA-EUDESMOL 1.8 Leaf PM58:376
BETA-PINENE 6.4 Leaf PM58:376
BETA-SELINENE 22 Leaf PM58:376
CADINOL 15 Leaf PM58:376
CAMPHENE Leaf PM58:376
CARYOPHYLLENE 38 Leaf PM58:376
CIS-PINOCARVEOL Leaf PM58:376
CIS-PINOCARVYL-ACETATE Leaf PM58:376
CIS-SABINENE-HYDRATE Leaf PM58:376
CIS-VERBENOL Leaf PM58:376
COPAENE 0.2 Leaf PM58:376
DELTA-CADINENE 16 Leaf PM58:376
EO 2,000 Leaf PM58:376
GAMMA-CADINENE 10 Leaf PM58:376
GAMMA-CURCUMENE Leaf PM58:376
GAMMA-MUUROLENE 3 Leaf PM58:376
GAMMA-TERPINENE 4.6 Leaf PM58:376
GERANIAL 15 Leaf PM58:376
GERANIOL 0.2 Leaf PM58:376
GERANYL-ACETATE 23 Leaf PM58:376
GERMACRENE Leaf PM58:376
HEXAHYDROFARNESYLACETONE Leaf PM58:376
HUMULENE 7.4 Leaf PM58:376
ISOAMYL-ISOVALERATE Leaf PM58:376

JASMONE Leaf PM58:376
JUNIPER-CAMPHOR 4 Leaf PM58:376
LAVANDULOL Leaf PM58:376
LIMONENE Leaf PM58:376
LINALOL 0.2 Leaf PM58:376
LINALYL-ACETATE Leaf PM58:376
METHYL-OCTANOATE Leaf PM58:376
MYRCENE 1.6 Leaf PM58:376
NERAL 4 Leaf PM58:376
OCIMENOL 3.2 Leaf PM58:376
OCTAN-3-ONE 4.4 Leaf PM58:376
OCTENYL-ACETATE 0.6 Leaf PM58:376
P-CYMENE 3.6 Leaf PM58:376
PERILLENE 1,225 Leaf PM58:376
PINOCARVONE Leaf PM58:376
SABINENE 0.4 Leaf PM58:376
TERPINEN-4-OL 2.6 Leaf PM58:376
TERPINOLENE Leaf PM58:376
THUJENE Leaf PM58:376
THYMOL 6.6 Leaf PM58:376
TRANS-PINOCARVEOL 0.2 Leaf PM58:376
TRANS-PINOCARVYL-ACETATE Leaf PM58:376
TRANS-SABINENE-HYDRATE 0.4 Leaf PM58:376
VERBENENE Leaf PM58:376
VERBENONE Leaf PM58:376
YLANGENE 1 Leaf PM58:376

Galeopsis segetum NECK.
"Downy Hemp Nettle"

8-HYDROXYFLAVONES Shoot BIS
8-O-ACETYLHARPAGIDE Shoot BIS
ANTIRRHINOSIDE Shoot BIS
ANTIRRHINOSIDE-5-O-GLUCOSIDE Shoot BIS
FLAVONOIDS Shoot BIS
HARPAGIDE Shoot BIS
HYPOLAETIN-4'-METHYL ETHER-7-(2"-ALLOSYL)-GLUCOSIDE Shoot BIS
IRIDOIDS Shoot BIS
SILICA 6,000-10,000 Shoot BIS
TANNINS 50,000 Shoot BIS

Galeopsis tetrahit L.
"Hemp Nettle"

3-OCTANOL 4 Shoot BML
ALPHA-HUMULENE 6 Shoot BML
CARYOPHYLLENE 40 Shoot BML
CARYOPHYLLENE-OXIDE 128 Shoot BML
COPAENE 5 Shoot BML
EO 400 Shoot BML
GALIROSIDE Plant JSG
GERMACRENE-D 22 Shoot BML
GLUROSIDE Plant JSG
LINALOL 12 Shoot BML

Glechoma hederacea L.
"Alehoof"

1,8-CINEOLE 6-37 Plant BML
1-OCTEN-3-OL 1-7 Plant BML
1-OCTEN-3-YL-ACETATE 1-2 Plant BML
3-OCTANOL 1-2 Plant BML
3-OCTANONE 3-16 Plant BML
ALPHA-CADINOL 1-6 Plant BML
ALPHA-PINENE Plant OMM
ALPHA-TERPINEOL 1-7 Plant BML
APIGENIN Plant POT
ASIATIC-ACID Plant CCO
ASPARAGIC-ACID Plant CRC
BARIUM 45-89 Plant CPB38:2205
BETA-BOURBONENE 1-6 Plant BML
BETA-ELEMENE 9-53 Plant BML
BETA-PINENE 3-17 Plant BML
BORNEOL 1-5 Plant BML
BORNYL-ACETATE 1-4 Plant BML
BROMINE 18-19 Plant CPB38:2205
CAFFEIC-ACID Plant JLS58:156
CALCIUM 21,000-23,000 Plant CPB38:2205
CAMPHENE 1-3 Plant BML
CHLORINE 8,000-11,000 Plant CPB38:2205
CHROMIUM 3-4 Plant CPB38:2205
CIS-OCIMENE 9-55 Plant BML
COPPER 11 Plant CPB38:2205
ELEMOL 1-3 Plant BML
EO 100-600 Plant BML POT
EPSILON-BULGARENE 2-11 Plant BML
EPSILON-MUUROLENE 1-6 Plant BML
FERULIC-ACID Plant JLS58:156
GAMMA-ELEMENE 1-7 Plant BML
GERMACRENE-B 14-83 Plant BML
GERMACRENE-D 20-116 Plant BML
GLECHOMAFURAN Plant POT
GLECHOMANOLIDE Plant JBH
GLECHOMINE Plant POT
HYPEROSIDE Plant POT
IRON 290-500 Plant CPB38:2205
ISOFURANOGERMACRENE 1 Plant BML
ISOMENTHONE Plant CCO
ISOPINOCAMPHONE Plant CCO
ISOQUERCITIN Plant POT

L-MENTHONE Plant OMM
L-PINOCAMPHONE Plant OMM
L-PULEGONE Plant CCO
LIMONENE 1-6 Plant BML
LINALOL 1-4 Plant BML
LUTEOLIN Plant POT
MANGANESE 82-100 Plant CPB38:2205
MARRUBIIN Plant CRC
MENTHOL Plant CCO
MYRCENE 3-20 Plant BML
MYRTENAL 1-1 Plant BML
OLEANOLIC-ACID 360 Plant ABS
P-COUMARIC-ACID Plant JLS58:156
P-CYMENE Plant CCO
PHOSPHORUS 2,100-3,500 Plant CPB38:2205
POTASSIUM 32,000-46,000 Plant CPB38:2205
POTASSIUM-NITRATE Plant OMM
ROSMARINIC-ACID 25,000 Leaf FT62:166
RUBIDIUM 10-11 Plant CPB38:2205
RUTIN Plant CRC
SABINENE 1-9 Plant BML
SAPONIN Plant POT
STRONTIUM 51-110 Plant CPB38:2205
SULFUR 3,100-3,400 Plant CPB38:2205
TERPINEN-4-OL 1-8 Plant BML
TITANIUM 24-50 Plant CPB38:2205
URSOLIC-ACID 166 Plant CCO
ZINC 46-53 Plant CPB38:2205

Hedeoma drummondii BENTH.
"Drummond's Pennyroyal"

1,8-CINEOLE 1-2,342 Plant BML
ALPHA-PINENE 7-6,780 Plant BML
BETA-PHELLANDRENE 9-31 Plant FNF
BETA-PINENE 0-540 Plant BML
BORNEOL 1-3,800 Plant BML
CAMPHENE 0-900 Plant BML
CAMPHOR 1-786 Plant BML
CITRONELLAL 228-2,180 Plant BML
EO 10,000 Plant FNF
GERANIAL 3,065-5,150 Plant BML
ISOMENTHONE 13-3,809 Plant BML
ISOPULEGONE 0-320 Plant BML
LIMONENE 1-410 Plant BML
MENTHOL 10 Plant BML
MENTHONE 1-126 Plant BML
MYRCENE 6-232 Plant BML
NERAL 36-2,690 Plant BML
PULEGONE 5,500-7,240 Plant BML
SABINENE Plant tr BML
SABINENE-HYDRATE 20-127 Plant BML
TERPINEN-4-OL 100 Plant BML
TRANS-OCIMENE 1-606 Plant BML

Hedeoma hispida PURSH.
"Hispid Pennyroyal"
Unaggregated A

1-OCTEN-3-YL-ACETATE 36 Plant BML
ALPHA-PINENE 18 Plant BML
BETA-PINENE 22 Plant BML
BORNEOL 214 Plant BML
BORNYL-ACETATE 94 Plant BML
CAMPHENE 66 Plant BML
CARYOPHYLLENE 474 Plant BML
CIS-OCIMENE 4 Plant BML
EO 2,000 Plant BML
GERMACRENE-D 354 Plant BML
LIMONENE 36 Plant BML
MYRCENE 164 Plant BML
P-CYMENE 10 Plant BML
SABINENE 4 Plant BML
TERPINOLENE 4 Plant BML
TRANS-OCIMENE 48 Plant BML

Hedeoma pulegioides (L.) PERS "American Pennyroyal"

(-)-MENTHONE 36-420 Plant FNF
1,8-CINEOLE 6-30 Plant BML
1-OCTEN-3-OL 48-240 Plant BML
3-OCTANOL 18-90 Plant BML
3-OCTYL-ACETATE 18-90 Plant BML
ACETIC-ACID Plant HHB
ALPHA-HUMULENE 30-150 Plant BML
ALPHA-PINENE 24-210 Plant BML
BETA-PINENE 12-180 Plant BML
CARYOPHYLLENE 30-150 Plant BML
DECYLIC-ACID Plant HHB
DIOSMIN 10,000 Plant HHB
DIPENTENE Plant HHB
EO 6,000-30,000 Plant HHB
FORMIC-ACID Plant HHB
GERMACRENE-D 36-180 Plant BML
ISOHEPTYLIC-ACID Plant HHB
ISOMENTHONE 48-9,300 Plant BML
LIMONENE 36-570 Plant BML
LINALYL-ACETATE 36-180 Plant BML
MENTHOFURAN 1 Plant BML
MENTHOL 6-30 Plant BML
METHYL-SALICYLATE Plant HHB
METHYLCYCLOHEXANONE Plant FNF
MYRCENE 30-150 Plant BML
OCTYLIC-ACID Plant FNF
PIPERITENONE 1-1,410 Plant BML
PIPERITONE 12-270 Plant BML
PULEGONE 3,678-27,600 Plant JBH
SABINENE 18-90 Plant BML
SALICYLIC-ACID Plant FNF
TRANS-BETA-FARNESENE Plant BML

Hedeoma reverchonii GRAY "Reverchon's Pennyroyal" Unaggregated A

1,8-CINEOLE 447 Plant BML
ALPHA-PINENE 525 Plant BML
BETA-PINENE 651 Plant BML
BORNEOL 2,528 Plant BML
CAMPHENE 460 Plant BML
CAMPHOR 31 Plant BML
EO 10,000 Plant FNF
GAMMA-TERPINENE 509 Plant BML
ISOPULEGONE 422 Plant BML
LIMONENE 368 Plant BML
MYRCENE 461 Plant BML
SABINENE 651 Plant BML
TERPINEN-4-OL 422 Plant BML
TERPINOLENE 222 Plant BML

Hyptis suaveolens POIT.
"Wild Hops"

1,1,3-TRIMETHYLDECAHYDROCYCLOPROPAZULENE 140 Plant QJC28:74
1,3,3-TRIMETHYLBICYCLO{2.2.1}-HEPTAN-2-OL 80 Plant QJC28:74
1,4-DIMETHYL,1,2,3,3A,4,5,6,7-OCTAHYDROAZULENE 195 Plant QJC28:74
1,8-CINEOLE 130-4,555 Plant QJC28:74
2,5-DIMETHYL-3-METHYLENE-1,5-HEPTADIENE 125 Plant QJC28:74
2,6-DIMETHYL-6-(4-METHYL)BICYCLO{3.1.1}-HEPT-2-ENE 110 Plant QJC28:74
3,7-DIMETHYL-1,6-OCTADIEN-3-OL 180 Plant QJC28:74
3-CYCLOHEXEN-1-CARBOXALDEHYDE 840 Plant QJC28:74
4,11,11-TRIMETHYL-8-METHYLENE-BICYCLO{7.2.0}-UNDEC-4-ENE 3,320 Plant QJC28:74
4-METHYL-1-(1-METHYLETHYL)-3-CYCLOHEXEN-1-OL 260 Plant QJC28:74
5ALPHA-ANDROST-2,11-DIONE 390 Plant QJC28:74
5ALPHA-ANDROST-9(11)-EN-12-ONE 270 Plant QJC28:74
5BETA,8BETA,H-9BETA,H-10ALPHA-LAB-14-ENE 435 Plant QJC28:74 ALPHA-CADINOL 125 Plant QJC28:74
ALPHA-CARYOPHYLLENE 480 Plant QJC28:74
ALPHA-CARYOPHYLLENE-ALCOHOL 85 Plant QJC28:74
ALPHA-CYMENE 250 Plant QJC28:74
ALPHA-PHELLANDRENE 285 Plant QJC28:74
ALPHA-PINENE 215 Plant QJC28:74
ALPHA-TERPINENE 130 Plant QJC28:74
ALPHA-TERPINEOL 110 Plant QJC28:74
BETULINIC-ACID Root NAP
ELEMENE 260 Plant QJC28:74
EO 15,000 Plant QJC28:74
FAT 143,000-244,000 Seed CRC
GAMMA-TERPINENE 175 Plant QJC28:74
LIMONENE 390 Plant QJC28:74
LINALOL 80 Plant QJC28:74
MENTHOL Plant JFM
PROTEIN 188,000-219,000 Seed CRC
THUJANE 325 Plant QJC28:74

Hyptis suaveolens POIT.
"Wild Hops"
Unaggregated A

(E)-BETA-OCIMENE 1.4 Shoot JEO5:321
(Z)-BETA-OCIMENE 1.4 Shoot JEO5:321
1,8-CINEOLE 475 Shoot JEO5:321
ALPHA-BERGAMOTENE 8 Shoot JEO5:321
ALPHA-COPAENE 0.6 Shoot JEO5:321
ALPHA-FENCHOL 2 Shoot JEO5:321
ALPHA-GUAIENE 10 Shoot JEO5:321
ALPHA-HUMULENE 0.8 Shoot JEO5:321
ALPHA-PHELLANDRENE 15 Shoot JEO5:321
ALPHA-PINENE 19 Shoot JEO5:321
ALPHA-TERPINENE 10 Shoot JEO5:321
ALPHA-TERPINEOL 3 Shoot JEO5:321
ALPHA-THUJENE 2.5 Shoot JEO5:321
BETA-BOURBONENE 4 Shoot JEO5:321
BETA-CARYOPHYLLENE 150 Shoot JEO5:321
BETA-PINENE 75 Shoot JEO5:321
BORNEOL 0.6 Shoot JEO5:321
CAMPHENE 0.6 Shoot JEO5:321
CAMPHOR 3 Shoot JEO5:321
CARYOPHYLLENE-OXIDE 8 Shoot JEO5:321
DELTA-3-CARENE 5 Shoot JEO5:321
EO 1,500 Shoot JEO5:321
FENCHONE 3.5 Shoot JEO5:321
GAMMA-TERPINENE 19 Shoot JEO5:321
LINALOL 190 Shoot JEO5:321
MYRCENE 11 Shoot JEO5:321
P-CYMENE 1 Shoot JEO5:321
SABINENE 225 Shoot JEO5:321
TERPINEN-4-OL 55 Shoot JEO5:321
TERPINOLENE 2.5 Shoot JEO5:321
THYMOL 8 Shoot JEO5:321

Hyptis suaveolens POIT.
"Wild Hops"
Unaggregated B

(E)-BETA-OCIMENE Shoot JEO5:321
(Z)-BETA-OCIMENE 3 Shoot JEO5:321
1,8-CINEOLE 885 Shoot JEO5:321
ALPHA-BERGAMOTENE 28 Shoot JEO5:321
ALPHA-COPAENE 1.5 Shoot JEO5:321
ALPHA-FENCHOL 17 Shoot JEO5:321
ALPHA-GUAIENE 6 Shoot JEO5:321
ALPHA-HUMULENE 6 Shoot JEO5:321
ALPHA-PHELLANDRENE 2 Shoot JEO5:321
ALPHA-PINENE 48 Shoot JEO5:321
ALPHA-TERPINENE Shoot JEO5:321
ALPHA-TERPINEOL 9 Shoot JEO5:321
ALPHA-THUJENE 21 Shoot JEO5:321
BETA-BOURBONENE 15 Shoot JEO5:321
BETA-CARYOPHYLLENE 7 Shoot JEO5:321
BETA-PINENE 80 Shoot JEO5:321
BORNEOL 2 Shoot JEO5:321
CAMPHENE 4 Shoot JEO5:321
CAMPHOR 12 Shoot JEO5:321
CARYOPHYLLENE-OXIDE 265 Shoot JEO5:321
DELTA-3-CARENE Shoot JEO5:321
EO 2,500 Shoot JEO5:321
FENCHONE 85 Shoot JEO5:321
GAMMA-TERPINENE 10 Shoot JEO5:321
LINALOL 34 Shoot JEO5:321
MYRCENE 1.5 Shoot JEO5:321
P-CYMENE 90 Shoot JEO5:321
SABINENE 240 Shoot JEO5:321
TERPINEN-4-OL 100 Shoot JEO5:321
TERPINOLENE 6 Shoot JEO5:321
THYMOL 4 Shoot JEO5:321

Hyssopus officinalis L.
"Hyssop"

(1S,2R,5R)-3-PINANONE Plant JSG
1,8-CINEOLE 488-610 Shoot IP35:51
1,8-EPOXY-2-P-MENTHENE 1 Shoot IP35:51
1-(1,4-DIMETHYL-3-CYCLOHEXEN-1-YL)ETHANONE <0.1-260 Shoot JAF42:776
10-EPI-ALPHA-CADINOL 1 Plant BML
2-HEXANONE 5-6 Shoot IP35:51
3-CARENE 1 Shoot IP35:51
3-OCTANOL Plant PM55:226
3-OCTANONE 1 Shoot IP35:51
ALLO-AROMADENDRENE <0.1-140 Shoot JAF42:776
ALPHA-CARYOPHYLLENE 100-580 Shoot JAF42:776
ALPHA-GURJUNENE 30-240 Shoot JAF42:776
ALPHA-HUMULENE 1-12 Flower JAD FFJ6:69
ALPHA-HUMULENE 1-140 Leaf AYL JAD FFJ6:69
ALPHA-PHELLANDRENE 12-120 Flower FFJ6:69
ALPHA-PHELLANDRENE 5-320 Leaf BML JAD FFJ6:69
ALPHA-PINENE <0.1-280 Leaf JAD FFJ6:69 JAF42:776
ALPHA-TERPINENE 1-411 Leaf BML JAD FFJ6:69
ALPHA-TERPINEOL 0.3-3 Flower FFJ6:72
ALPHA-TERPINEOL 0.1-360 Leaf BML JAD FFJ6:72
ALPHA-THUJENE <0.1-120 Shoot JAF42:776
ALPHA-THUJONE <0.1 Shoot IP35:51 JAF42:776
BENZALDEHYDE Shoot tr IP35:51
BENZYL-ALCOHOL 3-32 Flower FFJ6:72
BENZYL-ALCOHOL 0.1-30 Leaf FFJ6:72
BETA-CARYOPHYLLENE 40-415 Flower JAD FFJ6:69
BETA-CARYOPHYLLENE 3-660 Leaf BML JAD FFJ6:69 JAF42:776
BETA-IONONE 0.6-6 Flower FFJ6:72
BETA-IONONE 0.3-4 Leaf FFJ6:72
BETA-PHELLANDRENE 34-348 Flower FFJ6:69
BETA-PHELLANDRENE 1-800 Leaf JAD FFJ6:69
BETA-PINENE 170-4,580 Leaf BML JAD FFJ6:69
BETA-SELINENOL <0.1-80 Shoot JAF42:776
BETA-THUJONE <0.1-40 Shoot IP35:51 JAF42:776
BICYCLOGERMACRENE 8-1,520 Shoot IP35:52 JAF42:776
BORNEOL 3-4 Shoot IP35:52
BORNYL-ACETATE Plant 411/
CAFFEIC-ACID Plant JLS58:156
CAMPHENE 6-80 Leaf BML JAD FFJ6:69
CAMPHOR 125-1,240 Flower JAD FFJ6:69
CAMPHOR 10-3,280 Leaf JAD FFJ6:69

CARVACROL 1 Plant BML IP35:52
CIS-BETA-OCIMENE 0.6-6 Flower JAD FFJ6:69
CIS-BETA-OCIMENE <0.1-720 Leaf BML JAD FFJ6:69 JAF42:776
CIS-CALAMENENE Plant FNF
CIS-NEROLIDIOL 2-20 Leaf FFJ6:72
CIS-PINIC-ACID Plant FNF
CIS-PINOIC-ACID Plant FNF
CRYPTONE 7-9 Shoot IP35:52
CUMINALDEHYDE 4-7 Shoot IP35:52
D-2-HYDROXYISOPINOCAMPHONE 6-140 Plant AYL IP35:52
D-TERPINEOL 12-15 Shoot IP35:52
DELTA-CADINENE 14-420 Plant BML
DIOSMIN 30,000-60,000 Plant AYL
ELEMOL 21-215 Flower FFJ6:72
ELEMOL 0.4-608 Leaf FFJ6:72
EO 300-20,000 Plant AYL BML HHB FFJ6:69
ESTRAGOLE 2-18 Flower FFJ6:69
ESTRAGOLE 1-80 Leaf FFJ6:69
EUGENOL 62-624 Flower FFJ6:72
EUGENOL 2-443 Leaf FFJ6:72 IP35:52
EUGENOL-METHYL-ETHER 10-100 Shoot JAF42:776
FERULIC-ACID Plant JLS58:156
FURFURAL 1-2 Shoot IP35:51
GAMMA-TERPINENE 1-60 Leaf JAD FFJ6:69
GERANIOL 0.5-6 Flower FFJ6:72
GERANIOL 0.1-2 Leaf FFJ6:72
GERMACRENE-D 10-200 Flower JAD FFJ6:69 IP35:52
GERMACRENE-D 4-3,100 Leaf BML JAD FFJ6:69 JAF42:776
GUM Plant FNF
HEDYCARYOL 10-105 Flower JAD FFJ6:69
HEDYCARYOL 8-1,700 Leaf JAD FFJ6:69
HESPERIDIN 50,000-60,000 Plant AYL JBH
HEXAN-1-OL 1-10 Flower FFJ6:72
HEXAN-1-OL 1-15 Leaf FFJ6:72
HYSSOPIN Plant AYL
IODINE 0.014 Plant FNF
ISOPINOCAMPHOME 5-55 Flower JAD FFJ6:69
ISOPINOCAMPHONE 3-6,520 Leaf BML JAD FFJ6:69
LEDOL 20-120 Shoot JAF42:776
LIMONENE 5-54 Flower FFJ6:69
LIMONENE 13-240 Leaf BML JAD FFJ6:69
LINALOL 0.1-1 Flower FFJ6:69
LINALOL 0.2-160 Leaf FFJ6:72 JAF42:776
MARRUBIIN Plant 411/
METHYL-CHAVICOL 1-260 Shoot BML JAD IP35:52

METHYL-EUGENOL 7-100 Plant FNF
METHYL-MYRTENATE 5-54 Flower JAD FFJ6:69
METHYL-MYRTENATE 4-480 Leaf BML JAD FFJ6:69
MYRCENE 27-400 Shoot BML IP35:51
MYRTENAL 24-30 Shoot IP35:52
MYRTENIC-ACID-METHYL-ESTER 8-11 Shoot IP35:52
MYRTENIC-CIS-PINIC-ACID Plant FNF
MYRTENIC-CIS-PINOIC-ACID Plant FNF
MYRTENOL 16-160 Flower JAD FFJ6:69
MYRTENOL 0.5-520 Leaf JAD FFJ6:69
MYRTENYL-METHYL-ETHER 12-120 Flower JAD FFJ6:69
MYRTENYL-METHYL-ETHER 1-380 Leaf BML JAD FFJ6:69
NEROL 0.8-8 Flower FFJ6:72
NEROL 0.8-9 Leaf FFJ6:72
NONANOIC-ACID Plant FNF
NOPINONE 6-8 Shoot IP35:51
O-VANILLIN 0.1-21 Leaf FFJ6:72
O-VANILLIN Plant PM55:226
OCT-1-EN-3-OL 0.1-1 Flower FFJ6:72
OCT-1-EN-3-OL 0.5-48 Leaf FFJ6:72
OCTAN-3-OL 1 Leaf FFJ6:72
OLEANOLIC-ACID Plant 411/
P-CYMENE 6-180 Plant BML
P-CYMENOL-8 1 Shoot IP35:52
P-CYMOL 3-4 Shoot IP35:51
P-VANILLIN 1-38 Leaf FFJ6:72
PHENETHYL-ALCOHOL Plant PM55:226
PHENETHYLETHANOL 2-21 Flower FFJ6:72
PHENETHYLETHANOL 0.2-40 Leaf FFJ6:72
PINOCAMPHEOL Plant FNF
PINOCAMPHONE 200-2,060 Flower JAD FFJ6:69
PINOCAMPHONE 24-12,040 Leaf JAD FFJ6:69 IP35:51
PINOCARVEOL 32-40 Shoot IP35:52
PINOCARVONE 120-5,240 Shoot IP35:51 JAF42:776
ROSMARINIC-ACID 5,000 Plant FT62:166
SABINENE 10-100 Flower JAD FFJ6:69
SABINENE 28-380 Leaf BML JAD FFJ6:69
SALICYLIC-ACID-METHYL-ESTER 1-2 Shoot IP35:52
SPATHULENOL Flower tr JAD FFJ6:69
SPATHULENOL 1-440 Leaf BML JAD FFJ6:69
T-CADINOL <0.1-60 Leaf FFJ6:72
T-CADINOL Shoot JAF42:776
TANNIN 50,000-80,000 Plant LRN-JAN87
TERPINEN-4-OL 2-28 Flower FFJ6:72
TERPINEN-4-OL 1-790 Leaf FFJ6:72

TERPINOLENE 1-20 Leaf JAD FFJ6:69
TERPINYL-ACETATE Plant FNF
THYMOL 2-3 Shoot IP35:52
TRANS-BETA-OCIMENE 6-66 Flower JAD FFJ6:69
TRANS-BETA-OCIMENE 1-140 Leaf BML JAD FFJ6:69
TRANS-HEXEN-1-OL Plant PM55:226
TRANS-NEROLIDOL Flower tr JAD FFJ6:69
TRANS-NEROLIDOL 1-80 Plant BML JAD FFJ6:69
TRANS-PINOCARVEOL 40-180 Shoot JAF42:776
TRANS-SABINENE-HYDRATE 1-6 Flower JAD FFJ6:69
TRANS-SABINENE-HYDRATE 1-40 Leaf JAD FFJ6:69
URSOLIC-ACID 4,900 Plant FNF
VERBENOL 1-78 Leaf FFJ6:72
XANTHOPHYLL 3,556 Plant FNF

Hyssopus officinalis L.
"Hyssop"
Unaggregated A

1-(1,4-DIMETHYL-3-CYCLOHEXEN-1-YL)ETHANONE 10 Shoot JAF42:776 ALLO-AROMADENDRENE 20 Shoot JAF42:776
ALPHA-CARYOPHYLLENE 180 Shoot JAF42:776
ALPHA-GURJUNENE 70 Shoot JAF42:776
ALPHA-PINENE 80 Shoot JAF42:776
ALPHA-THUJENE 30 Shoot JAF42:776
ALPHA-THUJONE 10 Shoot JAF42:776
BETA-CARYOPHYLLENE 250 Shoot JAF42:776
BETA-MYRCENE 120 Shoot JAF42:776
BETA-PHELLANDRENE 310 Shoot JAF42:776
BETA-PINENE 660 Shoot JAF42:776
BETA-SELINENOL <0.1 Shoot JAF42:776
BETA-THUJONE <0.1 Shoot JAF42:776
BICYCLOGERMACRENE 460 Shoot JAF42:776
CAMPHENE 40 Shoot JAF42:776
CIS-BETA-OCIMENE 60 Shoot JAF42:776
D-LIMONENE 50 Shoot JAF42:776
EO 10,000 Shoot ASSUMED
ESTRAGOL 40 Shoot JAF42:776
EUGENOL-METHYL-ETHER 20 Shoot JAF42:776
GERMACRENE-D 1,020 Shoot JAF42:776
HEDYCARYOL 440 Shoot JAF42:776
ISOPINOCAMPHONE 2,920 Shoot JAF42:776
LEDOL 30 Shoot JAF42:776
LINALOL 10 Shoot JAF42:776
METHYL-MYRTENATE 40 Shoot JAF42:776
MYRTENOL 190 Shoot JAF42:776
MYRTENOL-METHYL-ETHER 100 Shoot JAF42:776
N-EICOSANE Shoot JAF42:776
N-TETRADECANE Shoot JAF42:776
PINOCAMPHONE 1,800 Shoot JAF42:776
PINOCARVONE 530 Shoot JAF42:776
SABINENE 100 Shoot JAF42:776
SPATHULENOL 100 Shoot JAF42:776
T-CADINOL 20 Shoot JAF42:776
TRANS-PINOCARVEOL 60 Shoot JAF42:776

Hyssopus officinalis L.
"Hyssop"
Unaggregated B

1-(1,4-DIMETHYL-3-CYCLOHEXEN-1-YL)ETHANONE 20 Shoot JAF42:776 ALLO-AROMADENDRENE 30 Shoot JAF42:776
ALPHA-CARYOPHYLLENE 190 Shoot JAF42:776
ALPHA-GURJUNENE 50 Shoot JAF42:776
ALPHA-PINENE 20 Shoot JAF42:776
ALPHA-THUJENE 20 Shoot JAF42:776
ALPHA-THUJONE 10 Shoot JAF42:776
BETA-CARYOPHYLLENE 260 Shoot JAF42:776
BETA-MYRCENE 70 Shoot JAF42:776
BETA-PHELLANDRENE 190 Shoot JAF42:776
BETA-PINENE 520 Shoot JAF42:776
BETA-SELINENOL 20 Shoot JAF42:776
BETA-THUJONE 20 Shoot JAF42:776
BICYCLOGERMACRENE 480 Shoot JAF42:776
CAMPHENE 20 Shoot JAF42:776
CIS-BETA-OCIMENE 50 Shoot JAF42:776
D-LIMONENE 40 Shoot JAF42:776
EO 10,000 Shoot ASSUMED
ESTRAGOL 30 Shoot JAF42:776
EUGENOL-METHYL-ETHER 30 Shoot JAF42:776
GERMACRENE-D 900 Shoot JAF42:776
HEDYCARYOL 680 Shoot JAF42:776
ISOPINOCAMPHONE 2,940 Shoot JAF42:776
LEDOL 40 Shoot JAF42:776
LINALOL <0.1 Shoot JAF42:776
METHYL-MYRTENATE 20 Shoot JAF42:776
MYRTENOL 170 Shoot JAF42:776
MYRTENOL-METHYL-ETHER 110 Shoot JAF42:776
N-EICOSANE Shoot JAF42:776
N-TETRADECANE Shoot JAF42:776
PINOCAMPHONE 1,950 Shoot JAF42:776
PINOCARVONE 560 Shoot JAF42:776
SABINENE 70 Shoot JAF42:776
SPATHULENOL 130 Shoot JAF42:776
T-CADINOL 30 Shoot JAF42:776
TRANS-PINOCARVEOL 40 Shoot JAF42:776

Hyssopus officinalis L.
"Hyssop"
Unaggregated C

1-(1,4-DIMETHYL-3-CYCLOHEXEN-1-YL)ETHANONE <0.1 Shoot JAF42:776
ALLO-AROMADENDRENE 10 Shoot JAF42:776
ALPHA-CARYOPHYLLENE 160 Shoot JAF42:776
ALPHA-GURJUNENE 60 Shoot JAF42:776
ALPHA-PINENE 30 Shoot JAF42:776
ALPHA-THUJENE 30 Shoot JAF42:776
ALPHA-THUJONE 20 Shoot JAF42:776
BETA-CARYOPHYLLENE 200 Shoot JAF42:776
BETA-MYRCENE 120 Shoot JAF42:776
BETA-PHELLANDRENE 280 Shoot JAF42:776
BETA-PINENE 620 Shoot JAF42:776
BETA-SELINENOL <0.1 Shoot JAF42:776
BETA-THUJONE 20 Shoot JAF42:776
BICYCLOGERMACRENE 420 Shoot JAF42:776
CAMPHENE 20 Shoot JAF42:776
CIS-BETA-OCIMENE 110 Shoot JAF42:776
D-LIMONENE 50 Shoot JAF42:776
EO 10,000 Shoot ASSUMED
ESTRAGOL 20 Shoot JAF42:776
EUGENOL-METHYL-ETHER 30 Shoot JAF42:776
GERMACRENE-D 670 Shoot JAF42:776
HEDYCARYOL 350 Shoot JAF42:776
ISOPINOCAMPHONE 3,260 Shoot JAF42:776
LEDOL 30 Shoot JAF42:776
LINALOL 20 Shoot JAF42:776
METHYL-MYRTENATE 20 Shoot JAF42:776
MYRTENOL 190 Shoot JAF42:776
MYRTENOL-METHYL-ETHER 140 Shoot JAF42:776
N-EICOSANE Shoot JAF42:776
N-TETRADECANE Shoot JAF42:776
PINOCAMPHONE 1,970 Shoot JAF42:776
PINOCARVONE 570 Shoot JAF42:776
SABINENE 110 Shoot JAF42:776
SPATHULENOL 100 Shoot JAF42:776
T-CADINOL 30 Shoot JAF42:776
TRANS-PINOCARVEOL 40 Shoot JAF42:776

Hyssopus officinalis L.
"Hyssop"
Unaggregated D

1-(1,4-DIMETHYL-3-CYCLOHEXEN-1-YL)ETHANONE 80 Shoot JAF42:776
ALLO-AROMADENDRENE 30 Shoot JAF42:776
ALPHA-CARYOPHYLLENE 120 Shoot JAF42:776
ALPHA-GURJUNENE 30 Shoot JAF42:776
ALPHA-PINENE 80 Shoot JAF42:776
ALPHA-THUJENE 30 Shoot JAF42:776
ALPHA-THUJONE <0.1 Shoot JAF42:776
BETA-CARYOPHYLLENE 230 Shoot JAF42:776
BETA-MYRCENE 130 Shoot JAF42:776
BETA-PHELLANDRENE 380 Shoot JAF42:776
BETA-PINENE 1,750 Shoot JAF42:776
BETA-SELINENOL <0.1 Shoot JAF42:776
BETA-THUJONE 20 Shoot JAF42:776
BICYCLOGERMACRENE 510 Shoot JAF42:776
CAMPHENE 40 Shoot JAF42:776
CIS-BETA-OCIMENE 10 Shoot JAF42:776
D-LIMONENE 60 Shoot JAF42:776
EO 10,000 Shoot ASSUMED
ESTRAGOL 40 Shoot JAF42:776
EUGENOL-METHYL-ETHER 20 Shoot JAF42:776
GERMACRENE-D 990 Shoot JAF42:776
HEDYCARYOL 240 Shoot JAF42:776
ISOPINOCAMPHONE 1,380 Shoot JAF42:776
LEDOL 30 Shoot JAF42:776
LINALOL 20 Shoot JAF42:776
METHYL-MYRTENATE 20 Shoot JAF42:776
MYRTENOL 250 Shoot JAF42:776
MYRTENOL-METHYL-ETHER 110 Shoot JAF42:776
N-EICOSANE Shoot JAF42:776
N-TETRADECANE Shoot JAF42:776
PINOCAMPHONE 160 Shoot JAF42:776
PINOCARVONE 2,530 Shoot JAF42:776
SABINENE 150 Shoot JAF42:776
SPATHULENOL 160 Shoot JAF42:776
T-CADINOL <0.1 Shoot JAF42:776
TRANS-PINOCARVEOL 60 Shoot JAF42:776

Hyssopus officinalis L.
"Hyssop"
Unaggregated E

1-(1,4-DIMETHYL-3-CYCLOHEXEN-1-YL)ETHANONE 60 Shoot JAF42:776
ALLO-AROMADENDRENE 30 Shoot JAF42:776
ALPHA-CARYOPHYLLENE 200 Shoot JAF42:776
ALPHA-GURJUNENE 50 Shoot JAF42:776
ALPHA-PINENE 70 Shoot JAF42:776
ALPHA-THUJENE <0.1 Shoot JAF42:776
ALPHA-THUJONE <0.1 Shoot JAF42:776
BETA-CARYOPHYLLENE 330 Shoot JAF42:776
BETA-MYRCENE 90 Shoot JAF42:776
BETA-PHELLANDRENE 330 Shoot JAF42:776
BETA-PINENE 1,080 Shoot JAF42:776
BETA-SELINENOL <0.1 Shoot JAF42:776
BETA-THUJONE <0.1 Shoot JAF42:776
BICYCLOGERMACRENE 760 Shoot JAF42:776
CAMPHENE 40 Shoot JAF42:776
CIS-BETA-OCIMENE <0.1 Shoot JAF42:776
D-LIMONENE 40 Shoot JAF42:776
EO 10,000 Shoot ASSUMED
ESTRAGOL <0.1 Shoot JAF42:776
EUGENOL-METHYL-ETHER 40 Shoot JAF42:776
GERMACRENE-D 1,550 Shoot JAF42:776
HEDYCARYOL 410 Shoot JAF42:776
ISOPINOCAMPHONE 1,410 Shoot JAF42:776
LEDOL 50 Shoot JAF42:776
LINALOL <0.1 Shoot JAF42:776
METHYL-MYRTENATE 30 Shoot JAF42:776
MYRTENOL 200 Shoot JAF42:776
MYRTENOL-METHYL-ETHER 80 Shoot JAF42:776
N-EICOSANE Shoot JAF42:776
N-TETRADECANE Shoot JAF42:776
PINOCAMPHONE 140 Shoot JAF42:776
PINOCARVONE 2,550 Shoot JAF42:776
SABINENE 80 Shoot JAF42:776
SPATHULENOL 120 Shoot JAF42:776
T-CADINOL <0.1 Shoot JAF42:776
TRANS-PINOCARVEOL 50 Shoot JAF42:776

Hyssopus officinalis L.
"Hyssop"
Unaggregated F

1-(1,4-DIMETHYL-3-CYCLOHEXEN-1-YL)ETHANONE 40 Shoot JAF42:776
ALLO-AROMADENDRENE <0.1 Shoot JAF42:776
ALPHA-CARYOPHYLLENE 140 Shoot JAF42:776
ALPHA-GURJUNENE 40 Shoot JAF42:776
ALPHA-PINENE 40 Shoot JAF42:776
ALPHA-THUJENE 60 Shoot JAF42:776
ALPHA-THUJONE <0.1 Shoot JAF42:776
BETA-CARYOPHYLLENE 200 Shoot JAF42:776
BETA-MYRCENE 120 Shoot JAF42:776
BETA-PHELLANDRENE 290 Shoot JAF42:776
BETA-PINENE 1,050 Shoot JAF42:776
BETA-SELINENOL <0.1 Shoot JAF42:776
BETA-THUJONE 20 Shoot JAF42:776
BICYCLOGERMACRENE 460 Shoot JAF42:776
CAMPHENE 30 Shoot JAF42:776
CIS-BETA-OCIMENE 10 Shoot JAF42:776
D-LIMONENE 60 Shoot JAF42:776
EO 10,000 Shoot ASSUMED
ESTRAGOL 30 Shoot JAF42:776
EUGENOL-METHYL-ETHER 50 Shoot JAF42:776
GERMACRENE-D 1,030 Shoot JAF42:776
HEDYCARYOL 250 Shoot JAF42:776
ISOPINOCAMPHONE 2,220 Shoot JAF42:776
LEDOL 60 Shoot JAF42:776
LINALOL 80 Shoot JAF42:776
METHYL-MYRTENATE 20 Shoot JAF42:776
MYRTENOL 220 Shoot JAF42:776
MYRTENOL-METHYL-ETHER 130 Shoot JAF42:776
N-EICOSANE Shoot JAF42:776
N-TETRADECANE Shoot JAF42:776
PINOCAMPHONE 270 Shoot JAF42:776
PINOCARVONE 2,620 Shoot JAF42:776
SABINENE 110 Shoot JAF42:776
SPATHULENOL 90 Shoot JAF42:776
T-CADINOL 30 Shoot JAF42:776
TRANS-PINOCARVEOL 40 Shoot JAF42:776

Hyssopus officinalis L.
"Hyssop"
Unaggregated G

1-(1,4-DIMETHYL-3-CYCLOHEXEN-1-YL)ETHANONE 130 Shoot JAF42:776
ALLO-AROMADENDRENE 40 Shoot JAF42:776
ALPHA-CARYOPHYLLENE 260 Shoot JAF42:776
ALPHA-GURJUNENE 100 Shoot JAF42:776
ALPHA-PINENE 60 Shoot JAF42:776
ALPHA-THUJENE 50 Shoot JAF42:776
ALPHA-THUJONE 10 Shoot JAF42:776
BETA-CARYOPHYLLENE 210 Shoot JAF42:776
BETA-MYRCENE 70 Shoot JAF42:776
BETA-PHELLANDRENE 50 Shoot JAF42:776
BETA-PINENE 930 Shoot JAF42:776
BETA-SELINENOL <0.1 Shoot JAF42:776
BETA-THUJONE <0.1 Shoot JAF42:776
BICYCLOGERMACRENE 430 Shoot JAF42:776
CAMPHENE 40 Shoot JAF42:776
CIS-BETA-OCIMENE 100 Shoot JAF42:776
D-LIMONENE 40 Shoot JAF42:776
EO 10,000 Shoot ASSUMED
ESTRAGOL 40 Shoot JAF42:776
EUGENOL-METHYL-ETHER 10 Shoot JAF42:776
GERMACRENE-D 950 Shoot JAF42:776
HEDYCARYOL 560 Shoot JAF42:776
ISOPINOCAMPHONE 110 Shoot JAF42:776
LEDOL 30 Shoot JAF42:776
LINALOL 70 Shoot JAF42:776
METHYL-MYRTENATE 20 Shoot JAF42:776
MYRTENOL 210 Shoot JAF42:776
MYRTENOL-METHYL-ETHER 30 Shoot JAF42:776
N-EICOSANE Shoot JAF42:776
N-TETRADECANE Shoot JAF42:776
PINOCAMPHONE 4,870 Shoot JAF42:776
PINOCARVONE 120 Shoot JAF42:776
SABINENE 110 Shoot JAF42:776
SPATHULENOL 120 Shoot JAF42:776
T-CADINOL <0.1 Shoot JAF42:776
TRANS-PINOCARVEOL 90 Shoot JAF42:776

Hyssopus officinalis L.
"Hyssop"
Unaggregated H

1-(1,4-DIMETHYL-3-CYCLOHEXEN-1-YL)ETHANONE 70 Shoot JAF42:776
ALLO-AROMADENDRENE 70 Shoot JAF42:776
ALPHA-CARYOPHYLLENE 260 Shoot JAF42:776
ALPHA-GURJUNENE 120 Shoot JAF42:776
ALPHA-PINENE <0.1 Shoot JAF42:776
ALPHA-THUJENE <0.1 Shoot JAF42:776
ALPHA-THUJONE <0.1 Shoot JAF42:776
BETA-CARYOPHYLLENE 270 Shoot JAF42:776
BETA-MYRCENE 50 Shoot JAF42:776
BETA-PHELLANDRENE 60 Shoot JAF42:776
BETA-PINENE 790 Shoot JAF42:776
BETA-SELINENOL <0.1 Shoot JAF42:776
BETA-THUJONE <0.1 Shoot JAF42:776
BICYCLOGERMACRENE 540 Shoot JAF42:776
CAMPHENE 40 Shoot JAF42:776
CIS-BETA-OCIMENE 80 Shoot JAF42:776
D-LIMONENE 40 Shoot JAF42:776
EO 10,000 Shoot ASSUMED
ESTRAGOL <0.1 Shoot JAF42:776
EUGENOL-METHYL-ETHER 30 Shoot JAF42:776
GERMACRENE-D 970 Shoot JAF42:776
HEDYCARYOL 660 Shoot JAF42:776
ISOPINOCAMPHONE 200 Shoot JAF42:776
LEDOL 40 Shoot JAF42:776
LINALOL 50 Shoot JAF42:776
METHYL-MYRTENATE 30 Shoot JAF42:776
MYRTENOL 240 Shoot JAF42:776
MYRTENOL-METHYL-ETHER 40 Shoot JAF42:776
N-EICOSANE Shoot JAF42:776
N-TETRADECANE Shoot JAF42:776
PINOCAMPHONE 4,600 Shoot JAF42:776
PINOCARVONE 220 Shoot JAF42:776
SABINENE 80 Shoot JAF42:776
SPATHULENOL 110 Shoot JAF42:776
T-CADINOL 20 Shoot JAF42:776
TRANS-PINOCARVEOL 60 Shoot JAF42:776

Hyssopus officinalis L.
"Hyssop"
Unaggregated I

1-(1,4-DIMETHYL-3-CYCLOHEXEN-1-YL)ETHANONE 100 Shoot JAF42:776
ALLO-AROMADENDRENE 20 Shoot JAF42:776
ALPHA-CARYOPHYLLENE 290 Shoot JAF42:776
ALPHA-GURJUNENE 120 Shoot JAF42:776
ALPHA-PINENE 40 Shoot JAF42:776
ALPHA-THUJENE 20 Shoot JAF42:776
ALPHA-THUJONE 20 Shoot JAF42:776
BETA-CARYOPHYLLENE 240 Shoot JAF42:776
BETA-MYRCENE 80 Shoot JAF42:776
BETA-PHELLANDRENE 50 Shoot JAF42:776
BETA-PINENE 780 Shoot JAF42:776
BETA-SELINENOL <0.1 Shoot JAF42:776
BETA-THUJONE <0.1 Shoot JAF42:776
BICYCLOGERMACRENE 520 Shoot JAF42:776
CAMPHENE 20 Shoot JAF42:776
CIS-BETA-OCIMENE 200 Shoot JAF42:776
D-LIMONENE 40 Shoot JAF42:776
EO 10,000 Shoot ASSUMED
ESTRAGOL 40 Shoot JAF42:776
EUGENOL-METHYL-ETHER 40 Shoot JAF42:776
GERMACRENE-D 920 Shoot JAF42:776
HEDYCARYOL 530 Shoot JAF42:776
ISOPINOCAMPHONE 130 Shoot JAF42:776
LEDOL 40 Shoot JAF42:776
LINALOL 70 Shoot JAF42:776
METHYL-MYRTENATE 20 Shoot JAF42:776
MYRTENOL 260 Shoot JAF42:776
MYRTENOL-METHYL-ETHER 40 Shoot JAF42:776
N-EICOSANE Shoot JAF42:776
N-TETRADECANE Shoot JAF42:776
PINOCAMPHONE 4,620 Shoot JAF42:776
PINOCARVONE 170 Shoot JAF42:776
SABINENE 90 Shoot JAF42:776
SPATHULENOL 130 Shoot JAF42:776
T-CADINOL 30 Shoot JAF42:776
TRANS-PINOCARVEOL 70 Shoot JAF42:776

Isanthus brachiatus (L.) BSP
"False Pennyroyal"

1,8-CINEOLE Plant BML
1-OCTEN-3-OL 6 Plant BML
ALPHA-HUMULENE 17 Plant BML
ALPHA-PINENE 684 Plant BML
ALPHA-TERPINEOL 2 Plant BML
ALPHA-TERPINYL-ACETATE 1 Plant BML
BETA-PINENE 136 Plant BML
CARYOPHYLLENE 30 Plant BML
CARYOPHYLLENE-OXIDE 1 Plant BML
EO 1,000 Plant BML
GERMACRENE-D 1 Plant BML
LIMONENE 20 Plant BML
MYRCENE 58 Plant BML
TERPINEN-4-OL 2 Plant BML
TRANS-VERBENOL 4 Plant BML

Lagochilus inebrians BUNGE "Intoxicating Mint"

CALCIUM-SALTS 20,000-27,000 Plant FNF
EO 300-2,000 Plant FNF
IRON-SALTS 1,300-4,000 Plant FNF
LAGOCHILIN 10,000-30,000 Plant FNF
LAGOCHILIN-3-ACETATE Plant FNF
LAGOCHILIN-TETRAACETATE 10,000-30,000 Plant FNF
ORGANIC-ACIDS 7,000-30,000 Plant FNF
RESIN 30,000-90,000 Plant FNF
STACHYDRINE-ACETATE 2,000 Plant FNF
SUGAR 10,000-27,000 Plant FNF
TANNIN 20,000-40,000 Plant FNF
VIT-K Plant FNF

Lavandula angustifolia MILLER
"English Lavender"

1,8-CINEOLE Flower BML
1-OCTEN-3-OL Flower BML
1-OCTEN-3-YL-ACETATE Flower BML
2,6-DIMETHYL-5-ACETOXYMETHYL-HEPT-6-EN-3-ONE Flower BML
2,6-DIMETHYL-5-ACETOXYMETHYL-HEPTA-1,6-DIEN-3-ONE Flower BML
2,6-DIMETHYL-6-ACETOXY-OCT-7-EN-3-ONE Flower BML
2,6-DIMETHYL-6-ACETOXY-OCTA-1,7-DIEN-3-ONE Flower BML
2-METHYL-BUT-3-EN-1-OL Flower BML
4-BUTANOLIDE Flower 411/
5-PENTYL-5-PENTANOLIDE Flower 411/
ALPHA-NORSANTALENONE Flower BML
ALPHA-PHOTOSANTALOL-A Flower BML
ALPHA-PHOTOSANTALOL-B Flower BML
ALPHA-PINENE Plant BML
ALPHA-SANTALAL Flower BML
ALPHA-SANTALENE Flower BML
ALPHA-SANTALENIC-ACID Flower BML
AMYL-ALCOHOL Plant BML
BERGAMOTENE Plant BML
BETA-PHELLANDRENE Flower BML
BETA-PINENE Plant BML
BETA-SANTALENE Flower BML
BORNEOL Plant BML
BORNYL-ACETATE 2,250-6,750 Plant BML
CAMPHENE Plant BML
CAMPHOR Flower BML
CAPROALDEHYDE Flower BML
CAPRYLALDEHYDE Plant BML
CARYOPHYLLENE Flower BML
CARYOPHYLLENE-OXIDE Flower BML
CEDRENE Plant HHB
CITRONELLOL Flower BML
COUMARIC-ACID Plant HHB
COUMARIN 1,000-1,500 Plant HHB
CUMINALDEHYDE Flower BML
CURCURMENE Plant BML
DELTA-3-CARENE Flower BML
DELTA-CADINENE Flower BML
DIHYDROCOUMARIN Flower 411/
EO 3,000-40,000 Flower AYL HHB WOI

FARNESENE Flower BML
GAMMA-CADINENE Flower BML
GERANIAL Flower BML
GERANIOL Flower BML
GERANYL-BUTYRATE Flower BML
GERANYL-CAPROATE Flower BML
GERANYL-ISOBUTYRATE Flower BML
GERANYL-VALERATE Flower BML
HERNIARIN 20-250 Plant HHB
ISOBUTYRIC-ACID Flower BML
ISOVALERALDEHYDE Flower BML
LAVANDULOL Flower WOI
LAVANDULYL-ACETATE Flower BML
LIMONENE Flower BML
LINALOL Plant HHB
LINALOL-OXIDE Flower BML
LINALYL-ACETATE 240-7,200 Flower HHB WOI
LINALYL-BUTYRATE Flower BML
LINALYL-CAPROATE Flower BML
LINALYL-ISOBUTYRATE Flower BML
LINALYL-VALERATE Flower BML
LUTEOLIN Plant AYL
MYRCENE Flower BML
NERAL Flower BML
OCIMENE Flower BML
PLANTEOSE Seed HHB
ROSMARINIC-ACID 12,000 Plant FT62:166
SABINENE Flower BML
TANNIN 120,000 Plant HHB
UMBELLIFERONE Plant AYL
UMBELLIFERONE-METHYL-ETHER Plant HHB
URSOLIC-ACID 7,000 Leaf HHB
VALERALDEHYDE Flower BML
VALERIC-ACID Flower BML

Lavandula latifolia MEDIK.
"Aspic, Broad-Leaved Lavender, Spike Lavender"

1,8-CINEOLE 1,000-3,434 Plant AYL
3-OCTANOL Plant tr BML
3-OCTANONE 4-13 Plant BML
ALPHA-BISABOLOL 4-10 Plant BML
ALPHA-CADINOL 1-21 Plant BML
ALPHA-HUMULENE 0-51 Plant BML
ALPHA-PHELLANDRENE 2-13 Plant BML
ALPHA-PINENE 90-748 Plant BML
ALPHA-TERPINENE 4-11 Plant BML
ALPHA-TERPINEOL 35-209 Plant BML
ALPHA-TERPINYL-ACETATE 1-35 Plant BML
ALPHA-THUJENE 1-19 Plant BML
ALPHA-THUJONE 2-9 Plant BML
BETA-BISABOLENE 1-15 Plant BML
BETA-PINENE 60-374 Plant BML
BETA-THUJONE 4-20 Plant BML
BORNEOL 20-759 Plant BML JBH
BORNYL-ACETATE 1-76 Plant BML
CAMPHENE 25-231 Plant BML
CAMPHOR 480-3,300 Plant AYL BML
CARVONE 12-5 Plant BML
CARYOPHYLLENE 0-191 Plant BML
CARYOPHYLLENE-OXIDE 6-24 Plant BML
CIS-ALPHA-BISABOLENE 38-197 Plant BML
CIS-CARVEOL 1-5 Plant BML
CIS-LINALOL-EPOXIDE 2-11 Plant BML
CIS-OCIMENE 0-55 Plant BML
CITRONELLAL 1-7 Plant BML
CITRONELLOL 1-41 Plant BML
COUMARIN 0-22 Plant BML
CUMINALDEHYDE 3-42 Plant BML
EO 5,000-11,000 Plant AYL BML
EUGENOL 0-9 Plant BML
FURFURAL 8-24 Plant BML
GERANIOL 3-21 Plant BML
GERANYL-ACETATE Plant BML
GERANYL-BUTYRATE 6-28 Plant BML
HEXANOL 1-7 Plant BML
HEXYL-TIGLATE 7-87 Plant BML
ISOBORNEOL 1-43 Plant BML
LAVANDULOL 10-165 Plant BML
LAVANDULYL-ACETATE 1-33 Plant BML

LIMONENE 55-297 Plant BML
LINALOL 1,490-5,104 Plant BML
LINALYL-ACETATE 1-198 Plant AYL BML
METHYL-HEPTENONE 2-8 Plant BML
MYRCENE 39-74 Plant BML
NEROL 1-15 Plant BML
NERYL-ACETATE 6-28 Plant BML
OLEANOLIC-ACID 4,900 Leaf HHB
P-CYMENE 26-98 Plant BML
ROSMARINIC-ACID 7,000 Plant FT62:166
SABINENE 0-55 Plant BML
TERPINEN-4-OL 0-121 Plant BML
TERPINOLENE 13-28 Plant BML
TRANS-CARVEOL 1-3 Plant BML
TRANS-EPOXY-LINALYL-ACETATE Plant BML
TRANS-LINALOL-EPOXIDE 8-21 Plant BML
TRANS-OCIMENE 0-49 Plant BML
URSOLIC-ACID 10,000-19,000 Leaf HHB

Lavandula x *hybrida* BALB. EX GING. "Hybrid Lavender"

(E)-BETA-OCIMENE 22-48 Shoot JEO5:443
1,8-CINEOLE 1,750-2,465 Shoot JEO5:443
1-OCTEN-3-OL 20-31 Shoot JEO5:443
ALPHA-PINENE 300-455 Shoot JEO5:443
ALPHA-TERPINENE 1-9 Shoot JEO5:443
ALPHA-TERPINEOL 200-300 Shoot JEO5:443
BETA-BISABOLENE 20-42 Shoot JEO5:443
BETA-CARYOPHYLLENE 20-45 Shoot JEO5:443
BETA-PINENE 395-585 Shoot JEO5:443
BORNEOL 55-80 Shoot JEO5:443
CAMPHENE 55-85 Shoot JEO5:443
CAMPHOR 660-890 Shoot JEO5:443
CARYOPHYLLENE-OXIDE 14-25 Shoot JEO5:443
CUMINALDEHYDE 5-8 Shoot JEO5:443
EO 5,500-7,000 Shoot JEO5:443
GAMMA-TERPINENE 12-21 Shoot JEO5:443
GERMACRENE-D 19-45 Shoot JEO5:443
HEXYL-BUTYRATE 9-16 Shoot JEO5:443
LIMONENE 110-200 Shoot JEO5:443
LINALOL 320-580 Shoot JEO5:443
LINALYL-ACETATE 0.5-3.5 Shoot JEO5:443
MYRCENE 70-105 Shoot JEO5:443
MYRTENAL 55-120 Shoot JEO5:443
P-CYMENE 7-13 Shoot JEO5:443
PINOCARVONE 50-75 Shoot JEO5:443
SABINENE 155-235 Shoot JEO5:443
TERPINEN-4-OL 35-80 Shoot JEO5:443
TERPINOLENE 10-17 Shoot JEO5:443
TRANS-ALPHA-BERGAMOTENE 22-43 Shoot JEO5:443
TRANS-PINOCARVEOL 55-90 Shoot JEO5:443
TRANS-SABINENE-HYDRATE 47-75 Shoot JEO5:443

Lavandula x *intermedia* EMERIC ex LOIS "Dutch Lavender, Lavandin"

1,8-CINEOLE 315-1,110 Plant BML
1-OCTEN-3-OL 10-50 Plant BML
1-OCTEN-3-YL-ACETATE 45-81 Plant BML
1-OCTEN-3-YL-BUTYRATE Plant BML
2,6-DIMETHYL-3-ACETOXY-METHYL-HEPTA-1,6-DIEN-5-OL Plant BML
2,6-DIMETHYL-3-ACETOXYMETHOXYMETHYL-HEPTA-1,4-DIEN-6-OL Plant BML
2,6-DIMETHYL-3-ACETOXYMETHYL-HEPT-1-EN-5-ONE Plant BML
2,6-DIMETHYL-3-ACETOXYMETHYL-HEPTA-1,6-DIEN-5-ONE Plant BML
2,6-DIMETHYL-6-ACETOXY-OCT-7-EN-3-ONE Plant FNF
2,6-DIMETHYL-6-ACETOXY-OCTA-1,7-DIEN-3-ONE Plant BML
2-METHYL-4-BUTANOLIDE Plant BML
2-METHYL-BUTANAL Plant BML
2-METHYL-FURAN Plant BML
2-METHYL-PROPANAL Plant BML
3,4-DIHYDRO-4,7-DIMETHYL-1-(2H)-NAPHTHALENONE Plant BML
3,7-DIMETHYL-OCTA-1,7-DIEN-3,6-DIOL-6-ACETATE Plant BML
3-METHYL-BUT-3-EN-2-OL Plant BML
3-METHYL-BUTANAL Plant BML
3-METHYL-BUTANOL Plant BML
3-OCTANOL 25-45 Plant BML
3-OCTANONE 30-199 Plant BML
4,4-DIMETHYL-2-BUT-4-ENOLIDE Plant BML
4-HEXAL-4-BUTANOLIDE Plant BML
4-ISOPROPYL-4-BUTANOLIDE Plant BML
4-METHYL-4-VINYL-2-BUT-4-ENOLIDE Plant BML
4-METHYL-4-VINYL-4-BUTANOLIDE Plant BML
5,7-EPOXY-2,6-DIMETHYL-3-ACETOXYMETHYL-HEPTEN-1-ENE Plant BML
5-PENTYL-5-PENTANOLIDE Plant BML
6,7-EPOXY-1-OCTEN-3-OL-3,7-DIMETHYL-ACETATE Plant BML
6-HYDROXY-OCTA-2,7-DIENAL-2,6-DIMETHYL-ACETATE Plant BML
6-HYDROXY-OCTA-3,7-DIEN-2-ONE-6-METHYL-ACETATE Plant BML
6-METHYL-3,5-HEPTADIEN-2-ONE Plant BML
6-METHYL-3-HEPTANONE Plant BML
6-METHYL-5-HEPTEN-2-ONE Plant BML

7-HYDROXY-5-ISOPROPENYL-HEPT-2-ENAL-2-METHYL-ACETATE Plant BML
ACETIC-ACID Plant BML
ALPHA-CADINENE Plant BML
ALPHA-CAMPHOLENAL Plant BML
ALPHA-NORSANTALENONE Plant BML
ALPHA-PHOTOSANTALOL Plant BML
ALPHA-PINENE 28-120 Plant BML
ALPHA-SANTALAL Plant BML
ALPHA-SANTALENE Plant BML
ALPHA-SANTALENIC-ACID Plant BML
ALPHA-TERPINENE Plant BML
ALPHA-TERPINEOL 0-100 Plant BML
ALPHA-TERPINYL-ACETATE Plant BML
AR-CURCUMENE Plant BML
BERGAMOTENE Plant BML
BETA-PINENE 15-70 Plant BML
BETA-SANTALENE Plant BML
BORNEOL 200-330 Plant BML
BORNYL-ACETATE 190-342 Plant BML
BUTYL-ACETATE Plant BML
BUTYROLACTONE Plant BML
CAMPHENE 26-60 Plant BML
CAMPHOR 542-1,620 Plant BML
CARVONE Plant BML
CARYOPHYLLENE 147-440 Plant BML
CARYOPHYLLENE-OXIDE Plant BML
CIS-ALLO-OCIMENE Plant BML
CIS-ALPHA-BISABOLENE 60-90 Plant BML
CIS-LINALOL-OXIDE 8-98 Plant BML
CIS-OCIMENE 100-230 Plant BML
COUMARIN Plant BML
CRYPTONE Plant BML
CUMINALDEHYDE Plant BML
DECANAL Plant BML
DELTA-3-CARENE 2-8 Plant BML
DELTA-CADINENE Plant BML
DIACETYL Plant BML
DIHYDROCOUMARIN Plant BML
DIISOPROPYL-KETONE Plant BML
DIMETHYL-SULFIDE Plant BML
EO 10,000-18,000 Plant BML
EPOXY-LINALYL-ACETATE Plant BML
EUGENOL Plant BML
FURFURAL Plant BML

GAMMA-TERPINENE 2-15 Plant BML
GERANIOL 20-36 Plant BML
GERANYL-ACETATE 21-72 Plant BML
GERMACRENE-D 50-90 Plant BML
HEPTANAL Plant BML
HEPTANE Plant BML
HEXANAL 30-54 Plant BML
HEXYL-2-METHYL-PROPIONATE 25-45 Plant BML
HEXYL-3-METHYL-BUTYRATE 50-90 Plant BML
HEXYL-ACETATE Plant BML
HEXYL-BUTYRATE 35-72 Plant BML
HEXYL-ISOBUTYRATE 13-20 Plant BML
HEXYL-TIGLATE Plant FNF
ISOCARYOPHYLLENE-OXIDE Plant BML
ISOGERANIOL Plant BML
ISOVALERALDEHYDE Plant BML
L-OCTEN-3-ACETYL-ACETATE Plant BML
LAVANDULOL 30-150 Plant BML
LAVANDULYL-2-METHYL-BUTYRATE Plant BML
LAVANDULYL-2-METHYL-PROPIONATE 60-108 Plant BML
LAVANDULYL-ACETATE 120-195 Plant BML
LAVANDULYL-BENZOATE Plant BML
LAVANDULYL-BUTYRATE Plant BML
LAVANDULYL-CAPROLATE Plant BML
LIMONENE 50-140 Plant BML
LINALOL 3,098-3,710 Plant BML
LINALYL-ACETATE 2,170-3,950 Plant BML
LINALYL-BUTYRATE Plant BML
LINALYL-FORMATE Plant BML
METHYLPROPYL-KETONE Plant BML
MYRCENE 47-65 Plant BML
NEROL Plant BML
NERYL-ACETATE Plant BML
NOPINONE Plant BML
OCT-7-EN-3-ONE-6-OL-6-METHYL-ACETATE Plant BML
OCTA-1,7-DIEN-3,6-DIOL-2,6-DIMETHYL-6-ACETATE Plant BML
OCTA-1,7-DIEN-3-ONE-6-OL-6-METHYL-ACETATE Plant BML
OCTA-3,7-DIEN-2,6-DIOL-2,6-DIMETHYL-6-ACETATE Plant BML
OCTANAL Plant BML
P-CYMEN-8-OL Plant BML
P-CYMENE 32-69 Plant BML
P-ISOPROPYLACETOPHENONE Plant BML
P-MENTHA-1,4-DIEN-7-AL Plant BML
PERILLYL-ALCOHOL Plant BML
PERILLYL-ALDEHYDE Plant BML

PHELLANDRENE Plant BML
ROSMARINIC-ACID 5,000 Plant FT62:166
SABINAKETONE Plant BML
SABINENE 6-20 Plant BML
TERPINEN-4-OL 46-279 Plant BML
TERPINEOLENE 12-26 Plant BML
TRANS-2-HEXENAL Plant BML
TRANS-2-TRANS-4-HEPTADIENAL Plant BML
TRANS-ALLO-OCIMENE Plant BML
TRANS-BETA-FARNESENE Plant BML
TRANS-LINALOL-OXIDE 21-28 Plant BML
TRANS-OCIMENE 30-460 Plant BML
TRICYCLENE Plant BML
VALERALDEHYDE Plant BML

Lavandula sp.
"Lavender"

(+)-BORNEOL Plant JBH
CAMPHOR Shoot JBH
CINNAMALDEHYDE Plant JBH
FENCHONE Plant JBH
LINALOL Shoot JBH
LINALYL-ACETATE Shoot JBH

Leonotis nepetaefolia (L.) R. BR.
"Catnipleaf Lion's-Ear"

(-)-5,6-OCTADECADIENIC-ACID 44,800-60,800 Seed CRC HHB
4,6,7-TRIMETHOXY-5-METHYL-CHROMEN-2-ONE Plant ID25:488
4,6,7-TRIMETHOXY-5-METHYLCOUMARIN Plant JSG
ASH 59,000-65,000 Seed CRC
CARBOHYDRATES 336,000-368,000 Seed CRC
EO 20,000 Seed HHB
FAT 280,000-380,000 Seed CRC
FIBER 143,000-157,000 Seed CRC
GLUCOSE Seed HHB
INOSITOL Seed HHB
LABALLENIC-ACID 44,800-60,800 Seed CRC HHB
LEONOTIN Plant JSG
LEONOTININ Plant JSG
LINOLEIC-ACID 33,320-45,220 Seed CRC ID25:489
METHOXYNEPETAEFOLIN Plant JSG
MYRISTIC-ACID 3,640-4,940 Seed CRC ID25:489
NEPETAEFOLIN Leaf HHB
NEPETAEFOLININ Plant ID25:487
NEPETAEFOLINOL Plant JSG
NEPETAEFURAN Leaf HHB
NEPETAEFURANOL Leaf HHB
OLEIC-ACID 180,880-245,480 Seed CRC ID25:489
PALMITIC-ACID 33,600-45,600 Seed CRC ID25:489
PROTEIN 236,000-288,000 Seed CRC
STEARIC-ACID 3,360-4,560 Seed CRC ID25:489

Leonurus cardiaca L.
"Motherwort"

3-OCTANOL Plant FNF
8-BETA-ACETOXY-9ALPHA,13ALPHA,15,16-BIEPOXY...
LABDANOLIDE Plant 411/
AJUGOL Plant JSG
AJUGOSIDE Plant JSG
ALKALOIDS 3,500 Plant FNF
ALPHA-HUMULENE 35-175 Plant FNF
ALPHA-PINENE 6-28 Plant FNF
APIGENIN-GLUCOSIDE Plant 411/
ASCORBIC-ACID 42 Plant FNF
BENZALDEHYDE 1 Plant FNF
BETA-PINENE 1-2 Plant FNF
BUFENOLIDE Plant FNF
CAFFEIC-ACID-4-RUTINOSIDE Plant 411/
CARYOPHYLLENE 40-200 Plant FNF
CATECHIN Plant FNF
EO 100-500 Plant FNF
FAT 302,000 Seed FNF
GENKWANIN Plant 411/
GLURIDOSUDE Plant JSG
HYPEROSIDE Plant 411/
ISOQUERCITRIN Plant 411/
KAEMPFEROL-GLUCOSIDE Plant 411/
LEOCARDIN Plant 411/
LEONURIDE Plant 411/
LEONURIDINE Plant FNF
LEONURINE Plant FNF
LIMONENE 2 Plant FNF
LINALOL 1-4 Plant FNF
MALIC-ACID Plant FNF
MARRUBIIN Plant FNF
OLEANOLIC-ACID Plant FNF
PROTEIN 223,000 Seed FNF
PYROGALLIC-TANNIN Plant FNF
QUERCETIN Plant 411/
QUERCITRIN Plant 411/
QUINQUELOSIDE Plant 411/
RUTIN Plant 411/
SAPONIN Plant FNF
STACHYDRINE Plant FNF
STACHYOSE Root CRC
TANNIN 21,000-90,000 Plant FNF
URSOLIC-ACID Plant FNF
VIT.-E Plant FNF

Lepechinia calycina EPLING "Epling's Lepechinia"

1,8-CINEOLE 985 Plant BML
10-EPI-(ALPHA)-CADINOL 15 Plant BML
ALPHA-GURJUNENE 10 Plant BML
ALPHA-HUMULENE 10 Plant BML
ALPHA-PHELLANDRENE 30-50 Plant BML
ALPHA-PINENE 325 Plant BML
ALPHA-TERPINENE 115 Plant BML
ALPHA-TERPINEOL 75 Plant BML
ALPHA-THUJENE 10 Plant BML
BETA-PINENE 90 Plant BML
BORNYL-ACETATE 5 Plant BML
CAMPHENE 390 Plant BML
CAMPHOR 875 Plant BML
CARYOPHYLLENE 285 Plant BML
DELTA-3-CARENE 870 Plant BML
DELTA-CADINENE 40 Plant BML
EO 5,000 Plant BML
GAMMA-TERPINENE 70 Plant BML
GERANYL-ACETATE 5 Plant BML
ISOMENTHONE 10 Plant BML
LIMONENE 115 Plant BML
LINALOL 130 Plant BML
MENTHONE 30 Plant BML
MYRCENE 85 Plant BML
NEO-ISOPULEGOL Plant tr BML
P-CYMENE 35 Plant BML
PHENETHYL-2-METHYL-BUTYRATE Plant tr BML
PHENETHYL-BUTYRATE 30 Plant BML
PIPERITONE 40 Plant BML
TERPINEN-4-OL 35 Plant BML
TERPINOLENE 45 Plant BML
TRANS-BETA-FARNESENE 15 Plant BML
TRANS-NEROLIDOL 90 Plant BML

Lepechinia schiediana (SCHLECHT.) VATKE
"Schiede's Lepechinia"
Unaggregated A

ALPHA-PINENE 200 Plant BML
BETA-PINENE 3,000 Plant BML
CAMPHENE 40 Plant BML
CARYOPHYLLENE 200 Plant BML
EO 4,000 Plant BML

Lycopus americanus MUHL.
"American Bugle"
Unaggregated A

1-OCTEN-3-OL 179 Plant BML
3-OCTANOL 28 Plant BML
EO 500 Plant BML
GERMACRENE-D 22 Plant BML
LIANLOOL 75 Plant BML

Lycopus europeus L.
"European Bugle"

ALKALOIDS 2,400 Plant HHB
APIGENIN-7-MONOGLUCOSIDE Plant HHB
CAFFEIC-ACID Plant HHB
CARYOPHYLLENE 125 Plant BML
CARYOPHYLLENE-OXIDE 16 Plant BML
CHLOROGENIC-ACID Plant HHB
COUMARIN 1,200 Plant HHB
DELTA-CADINENE 10 Plant BML
ELLAGIC-ACID Plant HHB
EO 500-2,000 Plant BML HHB
FERULIC-ACID Plant JLS58:156
FLAVONOIDS 23,000 Plant HHB
GERMACRENE-D 22 Plant BML
GLUCOSE Plant HHB
LITHOSPERMIC-ACID Root JBH
LUTEOLIN-7-MONOGLUCOSIDE Plant HHB
RESIN 29,000 Plant HHB
ROSMARINIC-ACID 37,000 Plant FT62:166
ROSMARINIC-ACID Root JNM1:10
SINAPIC-ACID Plant HHB
TRANS-BETA-FARNESENE 85 Plant BML
URSOLIC-ACID Plant HHB

Lycopus uniflorus MICHX.
"One-Flowered Bugle"

1-OCTEN-3-OL 18 Plant BML
ALPHA-PINENE 8-22 Plant BML
ALPHA-SELINENENE 1 Plant BML
ALPHA-TERPINEOL 18 Plant BML
BETA-PHELLANDRENE 107-154 Plant BML
BETA-PINENE 7-112 Plant BML
BETA-SELINENENE 1 Plant BML
CIS-ALPHA-FARNESENE 2 Plant BML
DELTA-CADINENE 8 Plant BML
ELEMOL 1 Plant BML
EO 500 Plant BML
GAMMA-CADINENE 8 Plant BML
GAMMA-MUUROLENE 24 Plant BML
GERMACRENE-D 8-29 Plant BML
LIMONENE 111-174 Plant BML
LINALOL 5 Plant BML
MYRCENE 12-18 Plant BML
SABINENE 3-9 Plant BML
SPATHULENOL 2 Plant BML
TERPINEN-4-OL 3 Plant BML
VIRIDIFLOROL 2 Plant BML

Lycopus virginicus L.
"Bugle"

1,8-CINEOLE 7-18 Plant BML
ALPHA-PINENE 266-666 Plant BML
BETA-BOURBONENE 53-132 Plant BML
BETA-PINENE 199-498 Plant BML
CAFFEIC-ACID Plant 411/
CAMPHENE 3-8 Plant BML
CARYOPHYLLENE 30-80 Plant BML
CHLOROGENIC-ACID Plant 411/
CIS-ISOPULEGONE 2-4 Plant BML
DELTA-CADINENE 16-40 Plant BML
ELLAGIC-ACID Plant 411/
EO 800-2,000 Plant BML
GALLIC-ACID Plant BML
GAMMA-MUUROLENE 5-12 Plant BML
GERMACRENE-D 52-130 Plant BML
LIMONENE 10-26 Plant BML
LINALOL 20-50 Plant BML
LITHOSPERMIC-ACID Plant FNF
LITHOSPERMIC-ACID Root ABS JBH
LYCOPIN Plant ABS
MYRCENE 6-14 Plant BML
PULEGONE 10-24 Plant BML
ROSMARINIC-ACID Plant 411/
TANNIC-ACID Plant ABS

Marrubium vulgare L.
"Horehound"

2-(OMEGA-1)-METHYLALKANES Plant LRN-JUN88
2-METHYLALKANES Plant LRN-JUN88
3-(OMEGA-9)-METHYLALKANES Plant LRN-JUN88
3-METHYLALKANES Plant LRN-JUN88
ALPHA-PINENE Plant 411/
ALPHA-TERPINOLENE Plant 411/
APIGENIN Plant LRN-JUN88
APIGENIN-7-GLYCOSIDE Plant LRN-JUN88
BETA-SITOSTEROL-BETA-GLUCOSIDE Plant FNF
BETONICINE 3,000 Plant AYL 411/
CAFFEIC-ACID Plant FNF
CAMPHENE Plant 411/
CHOLINE 2,000 Plant AYL 411/
DIHYDROPEREGRININ Plant 411/
EO 500-600 Plant FNF
FAT 264,000-372,000 Seed FNF
FENCHENE Plant 411/
GALLIC-ACID Plant FNF
LIMONENE Plant AYL
LUTEOLIN Plant LRN-JUN88
LUTEOLIN-7-GLYCOSIDE Plant LRN-JUN88
MARRUBENOL Plant JSG
MARRUBIIN 3,000-10,000 Plant JBH WOI
MARRUBIOL Plant 411/
MUCILAGE Plant FNF
P-CYMENE Plant AYL
P-CYMOL Plant 411/
PEREGRININ Plant 411/
PEREGRINOL Plant JSG
PREMARRUBIIN Plant JBH 411/
PROTEIN 185,000-256,000 Seed FNF
QUERCETIN-3-GLYCOSIDE Plant LRN-JUN88
QUERCETIN-3-RHAMNOGLYCOSIDE Plant LRN-JUN88
SABINENE Plant AYL
SCLAREOL Plant 411/
TANNIC-ACID Plant FNF
TANNIN 65,000-70,000 Plant WOI
TURICINE Plant 411/
URSOLIC-ACID 1,200 Plant FNF
VULGAROL Plant FNF

Melissa officinalis L.
"Lemonbalm"

(+)-CITRONELLAL 39-975 Shoot FNF
1,2-HUMULENE-EPOXIDE <0.1 Shoot DAZ129:155
1-OCTEN-3-OL 1-50 Shoot FNF
10-ALPHA-CADINOL 1-8 Plant FNF
2-PHENYLETHANOL Tissue Culture JEO3:111
2Z,4E,6E-ALLOFARNESENE <0.1 Shoot DAZ129:155
3-OCTANOL 1-8 Shoot DAZ129:155
3-OCTANONE 1-15 Shoot DAZ129:155
3E,6E-ALPHA-FARNESENE 5-40 Shoot DAZ129:155
6-METHYL-5-HEPTEN-2-ONE 3-24 Shoot DAZ129:155
ALPHA-CADINENE <0.1 Shoot DAZ129:155
ALPHA-CADINOL 1-120 Shoot DAZ129:155
ALPHA-COPAENE 4-120 Shoot DAZ129:155
ALPHA-CUBEBENE 1-38 Shoot FNF
ALPHA-HUMULENE 1-80 Shoot JAR1:11 DAZ129:155
ALPHA-MUUROLENE 3-24 Shoot DAZ129:155
BENZALDEHYDE Tissue Culture JEO3:111
BETA-BOURBONENE 1-48 Shoot DAZ129:155
BETA-CARYOPHYLLENE 1-870 Shoot JAR1:11 DAZ129:155
BETA-CARYOPHYLLENE-EPOXIDE-I 1-8 Shoot DAZ129:155
BETA-CARYOPHYLLENE-EPOXIDE-II 5-40 Shoot DAZ129:155
BETA-CUBEBENE 3-24 Shoot DAZ129:155
BETA-ELEMENE 4-32 Shoot DAZ129:155
BETA-GUAIENE <0.1 Shoot DAZ129:155
BICYCLOGERMACRENE 1-8 Shoot DAZ129:155
CADINA-1,4-DIENE <0.1 Shoot DAZ129:155
CAFFEIC-ACID Plant 411/
CAMPESTEROL <0.1 Tissue Culture JEO3:111
CARYOPHYLLENE-OXIDE 0.5-225 Plant JAR1:11
CATECHINS 30,000-50,000 Plant FNF
CHLOROGENIC ACID Plant FNF
CIS-3-HEXEN-1-OL <0.1-2 Shoot DAZ129:155
CIS-OCIMENE 1-24 Shoot DAZ129:155
CIS-ROSENOXIDE 1-8 Shoot DAZ129:155
CITRAL-A Shoot FNF
CITRAL-B Shoot FNF
CITRONELLAL 1-2,900 Shoot JAR1:11 JBH DAZ129:155
CITRONELLIC-ACID 3-24 Shoot DAZ129:155
CITRONELLOL 10-80 Shoot DAZ129:155
COPAENE 4-120 Plant FNF
DELTA-CADINENE 1-145 Shoot DAZ129:155
DELTA-CADINOL 1-16 Shoot DAZ129:155

DELTA-OCTALACTONE Tissue Culture JEO3:111
E-BETA-FANESENE 3-24 Shoot JAR1:11 DAZ129:155
EO 100-2,500 Shoot GEO JAR1:11
EUGENYL-ACETATE Plant 411/
FAT 115,000 Seed FNF
FURFURAL Tissue Culture JEO3:111
GAMMA-CADINENE 1-25 Shoot DAZ129:155
GERANIAL 15-1,190 Shoot JAR1:11 DAZ129:155
GERANIC-ACID <0.1 Shoot DAZ129:155
GERANIOL 1-65 Shoot DAZ129:155
GERANYL-ACETATE 0.3-155 Shoot JAR1:11 DAZ129:155
GERMACRA-1-(10)E,5E-DIEN-4-OL 30-230 Shoot DAZ129:155
GERMACRENE-D 4-1,080 Shoot DAZ129:155
HEXANOIC-ACID Tissue Culture JEO3:111
ISOGERANIAL 2-16 Shoot DAZ129:155
ISOPULEGOL 10-70 Shoot DAZ129:155
LIMONENE <0.1 Shoot DAZ129:155
LINALOL 1-16 Shoot DAZ129:155
LUTEOLIN-7-GLUCOSIDE Shoot 411/
METHYL-CITRONELLATE 50-390 Shoot DAZ129:155
METHYL-HEPTENONE 2-215 Shoot JAR1:11
METNYL-GERANIATE 4-32 Shoot DAZ129:155
MYRCENE 1-8 Shoot DAZ129:155
N-DOCOSANE <0.1 Tissue Culture JEO3:111
N-EICOSANE <0.1 Tissue Culture JEO3:111
N-HENEICOSANE <0.1 Tissue Culture JEO3:111
N-HEPTADECANE Tissue Culture JEO3:111
N-HEXADECANE Tissue Culture JEO3:111
N-NONADECANE Tissue Culture JEO3:111
N-OCTADECANE Tissue Culture JEO3:111
N-PENTADECANE Tissue Culture JEO3:111
N-TETRADECANE Tissue Culture JEO3:111
N-TRIDECANE Tissue Culture JEO3:111
NERAL 10-900 Shoot JAR1:11 DAZ129:155
NEROL 1-24 Shoot DAZ129:155
OCTANOIC-ACID Tissue Culture JEO3:111
OCTYL-BENZOATE Shoot ABS
OLEANOLIC-ACID Shoot FNF
P-CYMOL <0.1 Shoot DAZ129:155
POMOLIC-ACID Shoot 411/
PROTEIN 293,000 Seed FNF
PROTOCATECHUIC-ACID Plant 411/
RHAMNAZIN Plant 411/
ROSMARINIC-ACID Leaf JNM1:10
ROSMARINIC-ACID 8,400-37,000 Plant JBH FT62:166

STACHYOSE Plant FNF
T-CADINOL 3-24 Shoot DAZ129:155
T-MUUROLOL 5-40 Shoot DAZ129:155
THYMOL Plant FNF
TRANS-OCIMENE 1-25,055 Plant DAZ129:155
TRANS-ROSENOXIDE 0.5-4 Shoot DAZ129:155
URSOLIC-ACID Plant 411/

Mentha aquatica L.
"Water Mint"

(+)-ISOMENTHONE 1,230-3,485 Leaf FNF
(+)-PULEGONE 510 Leaf FNF
1,8-CINEOLE Leaf FNF
ACACETIN Plant FNF
ACACETIN-7-RUTINOSIDE Plant FNF
ALPHA-PINENE 18-68 Leaf FNF
APIGENIN Plant FNF
APIGENIN-7-BETA-D-GLUCOPYRANOSIDE Plant FNF
APIGENIN-7-RUTINOSIDE Plant FNF
BETA-CARYOPHYLLENE 300-850 Leaf FNF
BETA-PINENE 15-315 Leaf FNF
CARVONE 510-1,785 Leaf FNF
CARYOPHYLLENE Leaf FNF
CINEOLE 63-1,904 Leaf FNF
EO 3,000-8,500 Leaf FNF
ERIODICTYOL-7-BETA-D-GLUCOPYRANOSIDE Plant FNF
ERIODICTYOL-7-RUTINOSIDE Plant FNF
FORMIC-ACID Leaf FNF
FURFURAL Leaf FNF
HESPERETIN Plant FNF
HESPERETIN-7-BETA-D-GLUCOPYRANOSIDE Plant FNF
ISOMENTHONE 30-170 Leaf FNF
LIMONENE 114-714 Leaf FNF
LINALOL 1,080-3,060 Leaf FNF
LINALYL-ACETATE Leaf FNF
LUTEOLIN-7-BETA-D-GLUCOPYRANOSIDE Plant FNF
LUTEOLIN-7-RUTINOSIDE Plant FNF
MELISSIC-ACID Leaf FNF
MENTHOFURAN 30-7,582 Leaf FNF
MENTHOL 3,570 Leaf FNF
MENTHONE 23-212 Leaf FNF
MENTHYL-ACETATE 1,900 Leaf FNF
MENTHYL-VALERATE Leaf FNF
MYRCENE Leaf FNF
NEOMENTHOL Leaf FNF
OCTAN-3-OL 3-162 Leaf FNF
OCTAN-3-OL-ACETATE Leaf FNF
P-CYMOL 6-51 Leaf FNF
PIPERITENONE Leaf FNF
PIPERITONE Leaf FNF
PIPERITONE-OXIDE Leaf FNF
PULEGONE 54-153 Leaf FNF

ROSMARINIC-ACID 27,000 Plant FNF
TANNIN 74,000 Leaf FNF
TERPINENE Leaf FNF
VIRIDIFLOROL 36-102 Leaf FNF

Mentha aquatica L.
"Water Mint"
Unaggregated A

1,8-CINEOLE 105 Shoot NNK67:1417
1-OCTEN-3-OL <0.1 Shoot NNK67:1417
1-OCTYL-3-YL-ACETATE 3 Shoot NNK67:1417
2,4-DECADIENAL 1 Shoot NNK67:1417
2-N-AMYLFURAN <0.1 Shoot NNK67:1417
ALLOAROMADENDRENE 1 Shoot NNK67:1417
ALPHA-COPAENE 1 Shoot NNK67:1417
ALPHA-EUDESMOL 16 Shoot NNK67:1417
ALPHA-GURJUNENE 5 Shoot NNK67:1417
ALPHA-PINENE 6 Shoot NNK67:1417
BETA-CARYOPHYLLENE 45 Shoot NNK67:1417
BETA-CUBEBENE 8 Shoot NNK67:1417
BETA-CYCLO-CITRAL <0.1 Shoot NNK67:1417
BETA-EUDESMOL 17 Shoot NNK67:1417
BETA-FARNESENE 8 Shoot NNK67:1417
BETA-MYRCENE 85 Shoot NNK67:1417
BETA-PINENE 14 Shoot NNK67:1417
BICYCLOGERMACRENE 3 Shoot NNK67:1417
BOURBONENE 5 Shoot NNK67:1417
CAMPHENE <0.1 Shoot NNK67:1417
CARYOPHYLLENE-OXIDE 4 Shoot NNK67:1417
CIS-BETA-OCIMENE 10 Shoot NNK67:1417
CIS-JASMONE 1 Shoot NNK67:1417
ELEMOL 180 Shoot NNK67:1417
EO 1,300 Shoot NNK67:1417
GAMMA-MUUROLENE <0.1 Shoot NNK67:1417
GAMMA-TERPINENE 6 Shoot NNK67:1417
GERMACRENE-D 50 Shoot NNK67:1417
ISOAMYL-ISOVALERATE <0.1 Shoot NNK67:1417
ISOMENTHONE <0.1 Shoot NNK67:1417
LIMONENE 75 Shoot NNK67:1417
MENTHOFURAN 85 Shoot NNK67:1417
MENTHOL <0.1 Shoot NNK67:1417
MENTHONE 5 Shoot NNK67:1417
METHYL-ACETATE 16 Shoot NNK67:1417
MYRTENYL-METHYL-ETHER <0.1 Shoot NNK67:1417
N-HEXANAL <0.1 Shoot NNK67:1417
NEROLIDOL 4 Shoot NNK67:1417
NONANOIC-ACID <0.1 Shoot NNK67:1417
P-CYMENE 3 Shoot NNK67:1417
PULEGONE <0.1 Shoot NNK67:1417

SABINENE 8 Shoot NNK67:1417
TERPINOLENE 1 Shoot NNK67:1417
TRANS-2-HEPTANAL <0.1 Shoot NNK67:1417
VERBENYL-ACETATE 1 Shoot NNK67:1417
VIRIDIFLOROL 110 Shoot NNK67:1417

Mentha aquatica L.
"Water Mint"
Unaggregated B

1,8-CINEOLE 110 Shoot NNK67:1417
1-OCTEN-3-OL 2 Shoot NNK67:1417
1-OCTYL-3-YL-ACETATE 3 Shoot NNK67:1417
2,4-DECADIENAL <0.1 Shoot NNK67:1417
2-N-AMYLFURAN <0.1 Shoot NNK67:1417
ALLOAROMADENDRENE 16 Shoot NNK67:1417
ALPHA-COPAENE 2 Shoot NNK67:1417
ALPHA-EUDESMOL 22 Shoot NNK67:1417
ALPHA-GURJUNENE 10 Shoot NNK67:1417
ALPHA-PINENE 9 Shoot NNK67:1417
BETA-CARYOPHYLLENE 75 Shoot NNK67:1417
BETA-CUBEBENE 9 Shoot NNK67:1417
BETA-CYCLO-CITRAL <0.1 Shoot NNK67:1417
BETA-EUDESMOL 24 Shoot NNK67:1417
BETA-FARNESENE 16 Shoot NNK67:1417
BETA-MYRCENE 36 Shoot NNK67:1417
BETA-PINENE 20 Shoot NNK67:1417
BICYCLOGERMACRENE 6 Shoot NNK67:1417
BOURBONENE 4 Shoot NNK67:1417
CAMPHENE <0.1 Shoot NNK67:1417
CARYOPHYLLENE-OXIDE 6 Shoot NNK67:1417
CIS-BETA-OCIMENE 8 Shoot NNK67:1417
CIS-JASMONE <0.1 Shoot NNK67:1417
ELEMOL 220 Shoot NNK67:1417
EO 1,500 Shoot NNK67:1417
GAMMA-MUUROLENE <0.1 Shoot NNK67:1417
GAMMA-TERPINENE 8 Shoot NNK67:1417
GERMACRENE-D 115 Shoot NNK67:1417
ISOAMYL-ISOVALERATE <0.1 Shoot NNK67:1417
ISOMENTHONE <0.1 Shoot NNK67:1417
LIMONENE 21 Shoot NNK67:1417
MENTHOFURAN 42 Shoot NNK67:1417
MENTHOL <0.1 Shoot NNK67:1417
MENTHONE <0.1 Shoot NNK67:1417
MENTHYL-ACETATE <0.1 Shoot NNK67:1417
MYRTENYL-METHYL-ETHER <0.1 Shoot NNK67:1417
N-HEXANAL <0.1 Shoot NNK67:1417
NEROLIDOL 4 Shoot NNK67:1417
NONANOIC-ACID <0.1 Shoot NNK67:1417
P-CYMENE 2 Shoot NNK67:1417
PULEGONE <0.1 Shoot NNK67:1417

SABINENE 10 Shoot NNK67:1417
TERPINOLENE 2 Shoot NNK67:1417
TRANS-2-HEPTANAL <0.1 Shoot NNK67:1417
VERBENYL-ACETATE 2 Shoot NNK67:1417
VIRIDIFLOROL 135 Shoot NNK67:1417

Mentha aquatica L.
"Water Mint"
Unaggregated C

1,8-CINEOLE 140 Shoot NNK67:1417
1-OCTEN-3-OL <0.1 Shoot NNK67:1417
1-OCTYL-3-YL-ACETATE 2 Shoot NNK67:1417
2,4-DECADIENAL <0.1 Shoot NNK67:1417
2-N-AMYLFURAN <0.1 Shoot NNK67:1417
ALLOAROMADENDRENE 2 Shoot NNK67:1417
ALPHA-COPAENE 2 Shoot NNK67:1417
ALPHA-EUDESMOL 24 Shoot NNK67:1417
ALPHA-GURJUNENE 12 Shoot NNK67:1417
ALPHA-PINENE 14 Shoot NNK67:1417
BETA-CARYOPHYLLENE 115 Shoot NNK67:1417
BETA-CUBEBENE 8 Shoot NNK67:1417
BETA-CYCLO-CITRAL <0.1 Shoot NNK67:1417
BETA-EUDESMOL 27 Shoot NNK67:1417
BETA-FARNESENE 17 Shoot NNK67:1417
BETA-MYRCENE 37 Shoot NNK67:1417
BETA-PINENE 27 Shoot NNK67:1417
BICYCLOGERMACRENE 7 Shoot NNK67:1417
BOURBONENE 3 Shoot NNK67:1417
CAMPHENE <0.1 Shoot NNK67:1417
CARYOPHYLLENE-OXIDE 3 Shoot NNK67:1417
CIS-BETA-OCIMENE 8 Shoot NNK67:1417
CIS-JASMONE <0.1 Shoot NNK67:1417
ELEMOL 275 Shoot NNK67:1417
EO 1,700 Shoot NNK67:1417
GAMMA-MUUROLENE <0.1 Shoot NNK67:1417
GAMMA-TERPINENE 5 Shoot NNK67:1417
GERMACRENE-D 130 Shoot NNK67:1417
ISOAMYL-ISOVALERATE <0.1 Shoot NNK67:1417
ISOMENTHONE <0.1 Shoot NNK67:1417
LIMONENE 60 Shoot NNK67:1417
MENTHOFURAN 180 Shoot NNK67:1417
MENTHOL 3 Shoot NNK67:1417
MENTHONE 2 Shoot NNK67:1417
MENTHYL-ACETATE 2 Shoot NNK67:1417
MYRTENYL-METHYL-ETHER 2 Shoot NNK67:1417
N-HEXANAL 2 Shoot NNK67:1417
NEROLIDOL 2 Shoot NNK67:1417
NONANOIC-ACID 2 Shoot NNK67:1417
P-CYMENE 2 Shoot NNK67:1417
PULEGONE <0.1 Shoot NNK67:1417

SABINENE 15 Shoot NNK67:1417
TERPINOLENE 2 Shoot NNK67:1417
TRANS-2-HEPTANAL <0.1 Shoot NNK67:1417
VERBENYL-ACETATE 2 Shoot NNK67:1417
VIRIDIFLOROL 195 Shoot NNK67:1417

Mentha aquatica L.
"Water Mint"
Unaggregated D

1,8-CINEOLE 115 Shoot NNK67:1417
1-OCTEN-3-OL <0.1 Shoot NNK67:1417
1-OCTYL-3-YL-ACETATE 3 Shoot NNK67:1417
2,4-DECADIENAL <0.1 Shoot NNK67:1417
2-N-AMYLFURAN <0.1 Shoot NNK67:1417
ALLOAROMADENDRENE 3 Shoot NNK67:1417
ALPHA-COPAENE 2 Shoot NNK67:1417
ALPHA-EUDESMOL 11 Shoot NNK67:1417
ALPHA-GURJUNENE 8 Shoot NNK67:1417
ALPHA-PINENE 13 Shoot NNK67:1417
BETA-CARYOPHYLLENE 75 Shoot NNK67:1417
BETA-CUBEBENE 2 Shoot NNK67:1417
BETA-CYCLO-CITRAL <0.1 Shoot NNK67:1417
BETA-EUDESMOL 13 Shoot NNK67:1417
BETA-FARNESENE 11 Shoot NNK67:1417
BETA-MYRCENE 26 Shoot NNK67:1417
BETA-PINENE 24 Shoot NNK67:1417
BICYCLOGERMACRENE 5 Shoot NNK67:1417
BOURBONENE 5 Shoot NNK67:1417
CAMPHENE <0.1 Shoot NNK67:1417
CARYOPHYLLENE-OXIDE 3 Shoot NNK67:1417
CIS-BETA-OCIMENE 5 Shoot NNK67:1417
CIS-JASMONE <0.1 Shoot NNK67:1417
ELEMOL 170 Shoot NNK67:1417
EO 1,600 Shoot NNK67:1417
GAMMA-MUUROLENE <0.1 Shoot NNK67:1417
GAMMA-TERPINENE 5 Shoot NNK67:1417
GERMACRENE-D 80 Shoot NNK67:1417
ISOAMYL-ISOVALERATE <0.1 Shoot NNK67:1417
ISOMENTHONE <0.1 Shoot NNK67:1417
LIMONENE 75 Shoot NNK67:1417
MENTHOFURAN 395 Shoot NNK67:1417
MENTHOL <0.1 Shoot NNK67:1417
MENTHONE <0.1 Shoot NNK67:1417
MENTHYL-ACETATE 2 Shoot NNK67:1417
MYRTENYL-METHYL-ETHER <0.1 Shoot NNK67:1417
N-HEXANAL <0.1 Shoot NNK67:1417
NEROLIDOL 3 Shoot NNK67:1417
NONANOIC-ACID <0.1 Shoot NNK67:1417
P-CYMENE 2 Shoot NNK67:1417
PULEGONE <0.1 Shoot NNK67:1417

SABINENE 35 Shoot NNK67:1417
TERPINOLENE 2 Shoot NNK67:1417
TRANS-2-HEPTANAL <0.1 Shoot NNK67:1417
VERBENYL-ACETATE 2 Shoot NNK67:1417
VIRIDIFLOROL 140 Shoot NNK67:1417

Mentha aquatica L.
"Water Mint"
Unaggregated E

1,8-CINEOLE 45 Shoot NNK67:1417
1-OCTEN-3-OL <0.1 Shoot NNK67:1417
1-OCTYL-3-YL-ACETATE 2 Shoot NNK67:1417
2,4-DECADIENAL 2 Shoot NNK67:1417
2-N-AMYLFURAN <0.1 Shoot NNK67:1417
ALLOAROMADENDRENE 1 Shoot NNK67:1417
ALPHA-COPAENE 2 Shoot NNK67:1417
ALPHA-EUDESMOL 7 Shoot NNK67:1417
ALPHA-GURJUNENE 4 Shoot NNK67:1417
ALPHA-PINENE 8 Shoot NNK67:1417
BETA-CARYOPHYLLENE 32 Shoot NNK67:1417
BETA-CUBEBENE 4 Shoot NNK67:1417
BETA-CYCLO-CITRAL <0.1 Shoot NNK67:1417
BETA-EUDESMOL 8 Shoot NNK67:1417
BETA-FARNESENE 6 Shoot NNK67:1417
BETA-MYRCENE 12 Shoot NNK67:1417
BETA-PINENE 13 Shoot NNK67:1417
BICYCLOGERMACRENE 2 Shoot NNK67:1417
BOURBONENE 6 Shoot NNK67:1417
CAMPHENE <0.1 Shoot NNK67:1417
CARYOPHYLLENE-OXIDE 1 Shoot NNK67:1417
CIS-BETA-OCIMENE 2 Shoot NNK67:1417
CIS-JASMONE <0.1 Shoot NNK67:1417
ELEMOL 70 Shoot NNK67:1417
EO 1,100 Shoot NNK67:1417
GAMMA-MUUROLENE 1 Shoot NNK67:1417
GAMMA-TERPINENE 2 Shoot NNK67:1417
GERMACRENE-D 25 Shoot NNK67:1417
ISOAMYL-ISOVALERATE <0.1 Shoot NNK67:1417
ISOMENTHONE <0.1 Shoot NNK67:1417
LIMONENE 60 Shoot NNK67:1417
MENTHOFURAN 500 Shoot NNK67:1417
MENTHOL 1 Shoot NNK67:1417
MENTHONE <0.1 Shoot NNK67:1417
MENTHYL-ACETATE 6 Shoot NNK67:1417
MYRTENYL-METHYL-ETHER <0.1 Shoot NNK67:1417
N-HEXANAL <0.1 Shoot NNK67:1417
NEROLIDOL <0.1 Shoot NNK67:1417
NONANOIC-ACID <0.1 Shoot NNK67:1417
P-CYMENE 1 Shoot NNK67:1417
PULEGONE <0.1 Shoot NNK67:1417

SABINENE 7 Shoot NNK67:1417
TERPINOLENE 1 Shoot NNK67:1417
TRANS-2-HEPTANAL <0.1 Shoot NNK67:1417
VERBENYL-ACETATE 3 Shoot NNK67:1417
VIRIDIFLOROL 50 Shoot NNK67:1417

Mentha arvensis L.
"Cornmint, Field Mint, Japanese Mint"

(+)-8-ACETOXY-CARVONE Essential Oil NAP
(+)-CARVONE Essential Oil NAP
(+)-ISOMENTHONE 240-7,062 Leaf GEO JAD NAP
(+)-MENTHOFURAN 32 Leaf JAD NAP
(+)-NEOMENTHOL 20-580 Leaf GEO JAD NAP
(+)-OCTAN-3-ONE 9,000-27,000 Leaf JAD NAP
(+)-PIPERITONE 90-290 Leaf JAD NAP
(+)-PULEGONE 0-25,455 Plant JAD NAP
(-)-CARVONE Essential Oil NAP
(-)-LIMONENE Leaf NAP
(-)-MENTHOL 66-25,650 Leaf HHB JAD NAP
(-)-MENTHOL-ACETATE 10-32 Leaf JAD NAP
(-)-MENTHONE 250-25,680 Leaf GEO JAD NAP
(-)-PIPERITONE 240-7,700 Leaf JAD NAP
1,8-CINEOLE Leaf NAP
1-P-MENTHEN-3-ONE Plant GEO
3',4',5,7-TETRAHYDROXY-FLAVONE-7-ALPHA-L-RHAMNOSYL-BETA-D-GLUCOSIDE Shoot NAP
3',5,7-TRIHYDROXY-4'-METHOXY-FLAVONE-O-BETA-D-GLUCOSIDE Shoot NAP
3-(5',5'-DIMETHYL-TETRAHYDROFURAN-2'-YL)-BUT-CIS-2-EN-1-OL 1.4 Plant FNF
3-METHYLPENTANOL Plant HHB
4',5,7-TRIHYDROXY-FLAVONE-7-ALPHA-L-RHAMNOSYL-BETA-D- GLUCOSIDE Shoot NAP
4',5,7-TRIHYDROXY-FLAVONE-O-BETA-D-GLUCOSIDE Shoot NAP
ACACETIN-7-O-BETA-D-GLUCOSIDE Shoot NAP
ACETIC-ACID Plant GEO
ACETYLMENTHOL Plant CCO
ALPHA,BETA-HEXENIC-ACID Plant GEO
ALPHA,GAMMA-HEXENYL-PHENYL-ACETATE Plant GEO
ALPHA-PINENE 10-300 Plant GEO JAD NAP
ALPHA-THUJENE Plant HHB
ANISALDEHYDE Shoot NAP
BETA-CAR-3-ENE Essential Oil NAP
BETA-PHELLANDRENE Plant HHB
BETA-PINENE 10-1,445 Plant JAD NAP
CALCIUM 13,500 Plant SMO
CAMPHENE Plant GEO
CARVOMENTHONE Plant HHB
CARVONE 8,100-26,000 Leaf JAD NAP

CARYOPHYLLENE Plant GEO
CHLOROPHYLL-A Tissue Culture NAP
CHLOROPHYLL-B Tissue Culture NAP
CINEOLE 40-1,785 Leaf JAD NAP
CIS-ISOPULEGONE Leaf NAP
CIS-OCIMENE Essential Oil NAP
COPPER 20 Plant SMO
D-3-OCTANOL Plant GEO
DIOSMETIN-7-O-BETA-D-GLUCOSIDE Shoot NAP
DL-ISOMENTHONE Plant GEO
EO 10,000-32,100 Plant HHB
ETHYL-AMYL-CARBINOL Plant GEO
EUGENOL Essential Oil NAP
FORMIC-ACID Plant GEO
FURFURAL 2-5 Plant GEO
GAMMA-TERPINENE Essential Oil NAP
GERMACRENE-D Essential Oil NAP
HESPERIDIN Plant HHB
IRON 400 Plant SMO
ISOMENTHOL Essential Oil NAP
ISOPULEGOL Essential Oil NAP
ISOPULEGONE 240-770 Leaf JAD NAP
ISOVALERALDEHYDE Plant GEO
ISOVALERIC-ACID 30-90 Plant GEO
KETONES 336-1,008 Plant GEO
L-MENTHOL 373-29,515 Plant GEO JAD NAP
L-MENTHYL-ACETATE 85-4,200 Plant AYL GEO
LIMONENE 20-3,520 Leaf JAD NAP
LINALOL Leaf NAP
LINALOL-ACETATE 690-2,215 Leaf JAD NAP
LUTEOLIN Shoot NAP
MAGNESIUM 2,830 Plant SMO
MANGANESE 44 Plant SMO
MENTHOFURAN 4,200-12,600 Plant AYL HHB
MENTHOFUROLACTONE Plant HHB
MENTHOL 1,000-24,385 Leaf HHB JAD NAP
MENTHOL-ACETATE 60-2,215 Leaf JAD NAP
MENTHONE 71-24,000 Plant BML JAD NAP
MENTHONONE Plant CCO
MYRCENE 10-2,485 Leaf JAD NAP
NEOISOMENTHOL Essential Oil NAP
NEOISOPULEGONE Essential Oil NAP
OCTAN-3-OL 0-10,200 Leaf JAD NAP
P-CYMENE 9-29 Leaf JAD NAP
P-CYMOL 562-1,800 Leaf JAD NAP

P-MENTHAN-TRANS-2,5-DIOL Plant HHB
PIPERITENONE Plant CCO
PIPERITENONE-OXIDE 260-7,320 Leaf JAD NAP
PIPERITONE 13-1,285 Leaf GEO JAD NAP
PIPERITONE-OXIDE 320-5,135 Leaf JAD NAP
PIPERITONE-OXIDE Plant HHB
POTASSIUM 35,100 Plant SMO
PULEGONE 100-24,960 Plant BML HHB
RAFFINOSE Plant AYL
ROSMARINIC-ACID 71-228 Leaf AYL JAD NAP
ROSMARINIC-ACID 28,000 Plant FT62:166
SABINENE-HYDRATE 30-160 Leaf JAD NAP
SANTENE Plant HHB
SODIUM 860 Plant SMO
STACHYOSE Plant AYL
THUJONE Plant HHB
TRANS-ISOPULEGONE Leaf NAP
TRANS-OCIMENE Essential Oil NAP
ZINC 28 Plant SMO

Mentha longifolia (L.) HUDS.
"Biblical Mint"

1,2-EPOXYNEOMENTHYL-ACETATE 1-3,190 Shoot BML
1,8-CINEOLE 30-10,930 Shoot BML JEO3:57
1-OCTEN-3-OL 1-150 Shoot BML
2-CARENE 10-130 Shoot PM57:A83
2-METHYLBUTYL-2-METHYLBUTYRATE 0-20 Shoot BML
2-PENTADECANONE 15-395 Shoot PM57:A83
3-CARENE 0-30 Shoot PM57:A83
3-OCTANOL 1-2,395 Shoot BML JEO3:57
3-OCTYL-ACETATE 0-40 Shoot BML
4-HYDROXYPIPERITONE 1-1,065 Shoot BML FT60:349
4-HYDROXYPIPERITONE-OXIDE 150-570 Shoot FT60:349
ACACETIN-7-RUTINOSIDE Plant HHB
ALLO-AROMADENDRENE 0-380 Shoot BML
ALPHA-CADINOL 34-250 Shoot PM57:A83
ALPHA-CARYOPHYLLENE 45-285 Shoot PM57:A83
ALPHA-FARNESENE 18-100 Shoot PM57:A83
ALPHA-GURJUNENE 0-1 Shoot BML
ALPHA-PHELLANDRENE 11-70 Shoot PM57:A83
ALPHA-PINENE 7-1,065 Shoot BML JEO3:57 PM57:A83
ALPHA-TERPINENE 0-170 Shoot BML JEO3:57 PM57:A83
ALPHA-TERPINENYL-ACETATE 0-65 Plant PM57:A83
ALPHA-TERPINEOL 1-1,175 Shoot BML JEO3:57 PM57:A83
ALPHA-YLANGENE 0-340 Shoot PM57:A83
APIGENIN-7-GLUCURONIDE Plant HHB
BETA-BOURBONENE 1-535 Shoot BML PM57:A83
BETA-CARYOPHYLLENE 30-4,155 Shoot JEO3:57 PM57:A83
BETA-COPAENE 0-1,585 Shoot BML PM57:A83
BETA-CUBEBENE 19-200 Shoot PM57:A83
BETA-ELEMENE 0-275 Shoot BML PM57:A83
BETA-PINENE 28-1,160 Shoot JEO3:57
BETA-YLANGENE 0-340 Shoot BML
BORNEOL 1-1,250 Shoot JEO3:57
BORNYL-ACETATE 1-175 Shoot JEO3:57
CAMPHENE 0-385 Shoot BML JEO3:57
CARVONE 0-29,560 Shoot BML
CARYOPHYLLENE 110-9,995 Shoot BML
CARYOPHYLLENE-OXIDE 0-835 Shoot BML
CIS-3-HEXENYL-2-METHYLBUTYRATE 0-225 Shoot BML
CIS-BETA-FARNESENE 28-325 Shoot PM57:A83
CIS-CARVEOL 10-115 Shoot BML
CIS-CARVYL-ACETATE 0-530 Shoot BML
CIS-DIHYDROCARVONE 0-5,090 Shoot BML PM57:A83

CIS-JASMONE 0-85 Shoot JEO3:57
CIS-OCIMENE 1-685 Shoot BML
CIS-PIPERITONE-OXIDE 1-9,765 Shoot BML JEO3:57 FT60:349
DELTA-CADINENE 0-140 Shoot BML PM57:A83
DIHYDROCARVEOL 1-3,150 Shoot BML
DIHYDROCARVYL-ACETATE 0-3,570 Shoot BML PM57:A83
DIOSMIN Plant HHB
EO 10,000-38,000 Plant HHB JEO3:57
ERIODICTYOL-7-RUTINOSIDE Plant HHB
GAMMA-CADINENE 1-75 Shoot BML
GAMMA-TERPINENE 0-57 Shoot BML JEO3:57 PM57:A83
GERANIOL 1-115 Shoot JEO3:57
GERANYL-ACETATE 1-45 Shoot JEO3:57
GERMACRENE-D 140-8,850 Shoot BML
HEDYOSMIN Plant HHB
HESPERIDIN Plant HHB
ISOBUTANAL 10-38 Shoot JEO3:57
ISODIHYDROCARVYL-ACETATE 35-370 Shoot PM57:A83
ISOMENTHOL 0-190 Shoot PM57:A83
ISOMENTHONE 0-16,065 Shoot BML JEO3:57 PM57:A83
ISOMENTHYL-ACETATE 0-95 Shoot PM57:A83
ISOPIPERITENONE 1-115 Shoot BML
LIMONENE 7-960 Shoot BML JEO3:57 PM57:A83
LINALOL 10-33,360 Shoot BML JEO3:57
LUTEOLIN-7-GLUCOSIDE Plant HHB
LUTEOLIN-7-GLUCURONIDE Plant HHB
LUTEOLIN-7-RUTINOSIDE Plant HHB
MENTHOFURAN 0-14,570 Shoot JEO3:57 PM57:A83
MENTHOL 0-620 Shoot BML PM57:A83
MENTHONE 0-4,485 Shoot JEO3:57 PM57:A83
MENTHYL-ACETATE 43-275 Plant PM57:A83
MYRCENE 17-550 Shoot JEO3:57 PM57:A83
MYRTENAL 1-100 Shoot JEO3:57
MYRTENOL 1-35 Shoot JEO3:57
NEODIHYDROCARVEOL 50-645 Shoot BML
NEODIHYDROCARVYL-ACETATE 0-315 Plant PM57:A83
NONANAL 1-22 Shoot JEO3:57
OCTANOL 1-34 Shoot JEO3:57
OCTANYL-ACETATE 0-125 Shoot PM57:A83
P-CYMENE 1-23 Shoot BML JEO3:57
PHENYLETHYL-2-METHYLBUTYRATE 1-60 Shoot JEO3:57
PIPERITENONE 1-9,920 Shoot BML JEO3:57
PIPERITENONE-OXIDE 1-22,380 Shoot BML JEO3:57 FT60:349
PIPERITONE 0-5,270 Shoot JEO3:57 PM57:A83
PULEGONE 0-26,600 Shoot BML JEO3:57

QUERCITRIN Plant HHB
ROTUNDIFOLINE Plant JBH
SABINENE 20-750 Shoot BML JEO3:57
SPATHULENOL 0-125 Shoot BML PM57:A83
TAU-CADINOL 0-2,700 Shoot PM57:A83
TERPINEN-4-OL 1-800 Shoot BML JEO3:57
TERPINOLENE 0-20 Shoot BML JEO3:57
THYMOL 1-15 Shoot JEO3:57
TRANS-CARVEOL 1-340 Shoot BML
TRANS-CARVYL-ACETATE 1 Shoot BML
TRANS-DIHYDROCARVONE 0-6,965 Shoot BML PM57:A83
TRANS-FARNESOL 0-85 Shoot PM57:A83
TRANS-OCIMENE 1-38 Shoot BML
TRANS-PIPERITONE-OXIDE 80-26,900 Shoot BML FT60:349
TRANS-SABINENE-HYDRATE 1-80 Shoot JEO3:57
UMBELLULONE 27-168 Shoot PM57:A83
VIRIDIFLOROL 0-85 Shoot PM57:A83

Mentha x *piperita* L.
"Peppermint"

(+)-ALPHA-PINENE 14-145 Leaf JAD NAP
(+)-ISOMINTLACTONE 0.03-0.3 Leaf JAD NAP
(+)-LIMONENE 160-1,590 Leaf JAD NAP
(+)-PULEGONE 0-440 Leaf JAD NAP
(-)-LIMONENE Leaf NAP
(-)-MINTLACTONE 0.3-3 Leaf JAD NAP
1,3-DIMETHYL-CYCLOHEXANONE Essential Oil NAP
1,4-DIMETHOXY-BENZENE 0.001-0.01 Leaf JAD NAP
1,8-CINEOLE 45-1,390 Leaf BML JAD NAP
1-MENTHYL-BETA-D-GLUCOSIDE Plant JSG
2,6-DIMETHYL-PYRIDINE Leaf NAP
2-BUTYL-ISOVALERIC-ACID-METHYL-ESTER 2-20 Leaf JAD NAP
2-ETHYL-HEXAN-1-OL 0.001-0.01 Leaf JAD NAP
2-HYDROXY-BENZALDEHYDE 0.003-0.03 Leaf JAD NAP
2-METHYL-BUT-2-EN-1-AL 0.007-0.07 Leaf JAD NAP
2-METHYL-BUTYRIC-ACID-METHYL-ESTER 0.004-0.04 Leaf JAD NAP 2-METHYL-CINNAMALDEHYDE 0.001-0.01 Leaf JAD NAP
2-PHENYLETHANOL 0.2-2 Leaf JAD NAP
2-PHENYLETHANOL-ACETATE 0.05-0.5 Leaf JAD NAP
2-PHENYLETHANOL-BUTYRATE 0.05-0.5 Leaf JAD NAP
2-PHENYLETHANOL-ISOBUTYRATE 0.05-0.5 Leaf JAD NAP
2-PHENYLETHANOL-N-VALERATE 0.4-4 Leaf JAD NAP
2-PROPYL-5-PHENYL-PYRIDINE Leaf NAP
3(5',5'-DIMETHYL-TETRAHYDROFURAN-2'-YL)-BUT-CIS-2-EN-1-OL 1.15-2.26 Essential Oil NAP
3(5,5'-DIMETHYL-TETRAHYDROFURAN-2'-YL)-BUT-CIS-2-EN-1-OL 2.94 Essential Oil NAP
3,4-DIMETHOXY-BENZALDEHYDE 0.002-0.02 Leaf JAD NAP
3,4-DIMETHOXY-SUDACHIUTIN Leaf NAP
3,4-DIMETHYL-PSEUDACHITIN Leaf NAP
3,4-DIMETHYL-SUDACHITIN Leaf NAP
3,6-DIMETHYL-6-OXO-OCTANOIC-ACID 0.001-0.01 Leaf JAD NAP
3,6-DIMETHYL-7-OXO-OCTANOIC-ACID 0.001-0.01 Leaf JAD NAP
3,6-DIMETHYL-7-OXO-OCTANOIC-ACID-ETHYL-ESTER 0.001-0.01 Leaf JAD NAP
3-METHYL-CYCLOHEXANONE 0.4-4 Leaf JAD NAP
3-PHENYL-4-PROPYL-PYRIDINE Leaf NAP
3-PHENYL-PYRIDINE Leaf NAP
4-HYDROXY-4-METHYL-CYCLOHEX-2-EN-1-ONE 0.07-0.7 Leaf JAD NAP 4-METHYL-2-PHENYL-PENT-2-EN-1-AL 0.006-0.06 Leaf JAD NAP

5,6-DIHYDROXY-3',4',7,8-TETRAMETHOXY-FLAVONE Leaf NAP
5-ETHYL-2-METHYL-PYRIDINE Leaf NAP
5-HYDROXY-3',4',6,7-TETRAMETHOXY-FLAVONE Leaf NAP
5-METHYL-2-(2'-OXO-3'-BUTYL)-PHENOL 0.05-3.85 Leaf JAD NAP
5-METHYL-2-(3'-OXO-3'-PENTYL)-PHENOL 0.01-0.04 Leaf JAD
NAP 5-METHYL-2-PHENYL-HEX-2-EN-1-AL 0.003-0.03 Leaf JAD
NAP
5-METHYL-HEPTAN-3-ONE 0.001-0.01 Leaf JAD NAP
5-O-DEMETHYL-NOBILETIN Leaf NAP
6-METHYL-HEPT-5-EN-2-ONE 0.01-0.1 Leaf JAD NAP
6-METHYL-JASMONATE 0.001-0.01 Leaf JAD NAP
ACETIC-ACID Plant WOI
ALPHA-AMORPHENE Essential Oil NAP
ALPHA-CADINENE Essential Oil NAP
ALPHA-CAROTENE Plant FNF
ALPHA-COPAENE Essential Oil NAP
ALPHA-GURJUNENE Essential Oil NAP
ALPHA-PINENE 5-230 Leaf JAD NAP
ALPHA-TERPINENE 20 Plant BML
ALPHA-TERPINEOL Essential Oil NAP
ALPHA-THUJONE Essential Oil NAP
ALUMINUM 62 Leaf PED
AMYL-ALCOHOL Plant FNF
AMYL-VALERATE 0.1-1 Leaf JAD NAP
ANETHOLE 0.001-0.01 Leaf JAD NAP
ASH 141,000 Leaf PED
AZULENE Plant [illegible]
BENZOIC-ACID 0.02-0.2 Leaf JAD NAP
BENZYL-ALCOHOL 0.01-0.1 Leaf JAD NAP
BENZYL-CYANIDE 0.002-0.02 Leaf JAD NAP
BETA-BETULENOL Plant JSG
BETA-CAROTENE 237 Leaf PED
BETA-CARYOPHYLLENE 8-80 Leaf JAD NAP
BETA-COPAENE Essential Oil NAP
BETA-IONONE 0.3-3 Leaf JAD NAP
BETA-PINENE 10-400 Leaf BML JAD NAP
BETA-THUJONE Essential Oil NAP
BETA-YLANGENE Essential Oil NAP
BETAINE Leaf PED
BICYCLOELEMENE Plant AYL
BISABOLENE Essential Oil PED
BOVOLIDE 0.001-0.01 Leaf JAD NAP
BUTAN-2-ONE 0.001-0.01 Essential Oil NAP
CADINENE Essential Oil NAP
CAFFEIC-ACID Shoot NAP

CALCIUM 16,200 Leaf PED
CAMPHENE 2-25 Leaf JAD NAP
CARBOHYDRATES 557,000 Leaf PED
CARVACROL Essential Oil NAP
CARVEOL Essential Oil NAP
CARVEOL-ACETATE Essential Oil NAP
CARVONE 4-132 Leaf USA
CARYOPHYLLENE-OXIDE Essential Oil NAP
CEDRENE Essential Oil NAP
CEDROL Essential Oil NAP
CHLOROGENIC-ACID Shoot NAP
CHROMIUM Leaf tr PED
CINEOLE 2-1,200 Leaf JAD NAP
CINEROL Essential Oil NAP
CINNAMIC-ACID-METHYL-ESTER Essential Oil NAP
CIS-PIPERITOL 0.5-5 Leaf JAD NAP
CIS-ROSEOXIDE 0.03-0.3 Plant JAD NAP
CIS-SABINOL 0.001-0.01 Leaf JAD NAP
CITRONELLIC-ACID 0.01-0.1 Leaf JAD NAP
CITRONELLOL Essential Oil NAP
COBALT 93 Leaf PED
COSMOSIIN Shoot NAP
COUMARIN 0.02-0.2 Leaf JAD NAP
CRYPTONE 0.001-0.01 Leaf JAD NAP
CUMIN-ALCOHOL 0.001-0.01 Leaf JAD NAP
CYCLOPENTANOL 0.01-0.1 Leaf JAD NAP
DELTA-DODECALACTONE 0.001-0.01 Leaf JAD NAP
DELTA-JASMINLACTONE 0.05-0.5 Leaf JAD NAP
DIHYDRO-LIMONENE-10-OL Essential Oil NAP
DIHYDRO-TERPINEOL-ACETATE 0.001-0.01 Leaf JAD NAP
DIHYDROCARVONE Leaf NAP
DIMETHYL-SULFOXIDE Essential Oil NAP
DIOSPHENOL 0.01-0.1 Leaf JAD NAP
DIPENTENE Essential Oil NAP
EO 1,000-10,000 Plant BML GEO JAD
ERIODICTYOL-7-O-RUTINOSIDE 860 Leaf NAP
EUGENOL Essential Oil NAP
EUPATORIN Essential Oil NAP
FAT 54,000 Leaf PED
FENCHENE Essential Oil NAP
FIBER 114,000 Leaf PED
GAMMA-DECALACTONE 0.001-0.01 Leaf JAD NAP
GAMMA-JASMINLACTONE 0.001-0.01 Leaf JAD NAP
GAMMA-TERPINENE 2-20 Leaf JAD NAP
GAMMA-TOCOPHEROL Plant JAD NAP

GARDENIN-B Leaf NAP
GARDENIN-D Leaf NAP
GERANIAL 0.005-0.05 Leaf JAD NAP
GERANIC-ACID 0.01-0.1 Leaf JAD NAP
GERANIOL-ACETATE 0.1-1 Leaf JAD NAP
GERMACRENE-D Essential Oil NAP
GUAIACOL 0.01-0.1 Leaf JAD NAP
HEPTAN-2-ONE 0.001-0.01 Leaf JAD NAP
HEPTAN-3-OL 0.5-5 Leaf JAD NAP
HESPERETIN Leaf PED
HEX-TRANS-2-ENOIC-ACID 0.001-0.01 Leaf JAD NAP
HYDROXY-BOVOLIDE 0.001-0.01 Leaf JAD NAP
HYMENOXIN Leaf NAP
IRON 600 Leaf PED
ISOAMYL-PHENYLACETATE 0.05-0.5 Leaf JAD NAP
ISOBUTYRIC-ACID 0.03-0.3 Leaf JAD NAP
ISOCHLOROGENIC-ACID Shoot NAP
ISOMENTHOL 19-310 Leaf BML JAD NAP
ISOMENTHOL-ACETATE 8-80 Leaf JAD NAP
ISOMENTHONE 30-1,000 Leaf JAD NAP
ISOMENTHYL-ACETATE 40 Plant FNF
ISOPULEGOL-ACETATE 0.01-0.1 Leaf JAD NAP
ISORHOIFOLIN Leaf PED
ISOVALERALDEHYDE Plant FNF
ISOVALERIC-ACID Plant WOI
ISOVALERIC-ACID-N-OCTYL-ESTER 0.005-0.05 Leaf JAD NAP
JASMONE Essential Oil NAP
KILOCALORIES 3,020/kg Leaf PED
LAVANDULOL 0.001-0.01 Leaf JAD NAP
LEDOL Plant 411/
LIMONENE 2-320 Leaf AYL JAD NAP
LIMONENE-10-OL-ACETATE Essential Oil NAP
LINALOL 12-120 Leaf JAD NAP
LITHOSPERMIC-ACID Shoot NAP
LUTEOLIN Leaf NAP
LUTEOLIN-7-O-RUTINOSIDE 120 Leaf NAP
MAGNESIUM 6,610 Leaf PED
MANGANESE 61 Leaf PED
MENTHACUBANONE Shoot NAP
MENTHOCUBANONE Shoot NAP
MENTHOKUBANONE Shoot NAP
MENTHOL 156-6,750 Leaf JAD JBH NAP
MENTHONE 20-2,730 Plant BML AYL JAD JBH NAP FT59:465
MENTHOSIDE Leaf PED
MENTHYL-ACETATE 8-5,080 Leaf JAD JBH NAP

MENTHYL-ISOVALERATE 0.01-0.1 Leaf JAD NAP
MENTHYL-VALERATE Leaf NAP
MYRCENE 1-10 Leaf BML JAD NAP
MYRTENAL 0.002-0.02 Leaf JAD NAP
MYRTENOL 0.5-5 Leaf JAD NAP
NEOISOMENTHOL-ACETATE Essential Oil NAP
NEOISOPULEGOL Essential Oil NAP
NEOMENTHOL 33-820 Leaf JAD NAP
NEOMENTHONE 70-700 Leaf JAD NAP
NEOMENTHYL-ACETATE 15-150 Leaf JAD NAP
NERAL 0.001-0.01 Leaf JAD NAP
NEROL Tissue Culture NAP
NEROLIDOL 0.05-0.5 Leaf JAD NAP
NEVADENSIN Leaf NAP
NONAN-1-OL 0.001-0.01 Leaf JAD NAP
O-CRESOL 0.001-0.01 Leaf JAD NAP
OCIMENE Essential Oil NAP
OCT-TRANS-2-EN-OL 0.001-0.01 Leaf JAD NAP
OCTAN-3-OL 4-70 Leaf JAD NAP
P-COUMARIC-ACID Shoot NAP
P-CRESOL 0.002-0.02 Leaf JAD NAP
P-CYMENE 2.5-30 Leaf BML JAD NAP
P-CYMOL 4-130 Leaf JAD NAP
P-MENTH-TRANS-2-EN-1-OL Essential Oil NAP
P-MENTHANE Essential Oil NAP
P-METHOXY-ACETOPHENONE 0.01-0.1 Leaf JAD NAP
PECTIN 30,000 Shoot NAP
PENT-CIS-2-EN-1-OL 0.5-5 Leaf JAD NAP
PENTAL-1-OL Essential Oil NAP
PERILLYL-ALCOHOL 0.05-0.5 Leaf JAD NAP
PHELLANDRENE 2-20 Leaf BML JAD NAP
PHENYLACETIC-ACID 0.001-0.01 Leaf JAD NAP
PHOSPHORUS 7,720 Leaf PED
PINENE 2-23 Leaf JAD NAP
PIPERITENONE Essential Oil NAP
PIPERITONE 0-500 Leaf JAD NAP
PIPERITONE-OXIDE Plant FNF
POTASSIUM 22,600 Leaf PED
PROTEIN 248,000 Leaf PED
PROTEIN 127,000 Plant WOI
PULEGONE 3-6,530 Leaf JAD NAP
PYRIDINE 0-0.001 Leaf JAD NAP
RIBOFLAVIN 38.9 Leaf PED
ROSMARINIC-ACID 30,000 Plant FT62:166
ROSMARINIC-ACID 700-1,800 Shoot JAD JBH NAP

RUTIN Plant 411/
SABINENE Essential Oil NAP
SABINENE-ACETATE Plant FNF
SABINENE-HYDRATE 5-110 Leaf JAD NAP
SALICYLATES 700 Leaf JAD85:950
SALVIGENIN Essential Oil NAP
SELENIUM 11 Leaf PED
SIDERITOFLAVONE Leaf NAP
SILICON Leaf tr PED
SODIUM 1,950 Leaf PED
TERPINEN-4-OL Essential Oil NAP
TERPINOLENE Leaf JAD NAP
THIAMIN 12.1 Leaf PED
THYMOL Essential Oil NAP
TIN 11 Leaf PED
TRANS-BETA-FARNESENE Essential Oil NAP
TRANS-PIPERITOL 0.5-5 Leaf JAD NAP
TRANS-ROSEOXIDE 0.01-0.1 Plant JAD NAP
VANILLIN 0.001-0.01 Leaf JAD NAP
VIRIDIFLOROL 1-15 Leaf JAD NAP
WATER 851,000 Leaf PED
XANTHOMICROL Leaf NAP
ZINC Leaf tr PED

Mentha pulegium L.
"European Pennyroyal"

(+)-CAMPHENE 106-212 Plant JAD NAP
(+)-ISOISOPULEGONE 22-76 Plant JAD NAP
(+)-ISOMENTHONE 521-4,140 Plant JAD NAP
(+)-ISOPULEGONE Plant NAP
(+)-LIMONENE 10-290 Plant JAD NAP
(+)-MENTHOL 1,624-3,248 Plant JAD NAP
(+)-MENTHONE 680-1,360 Plant JAD NAP
(+)-NEOISOMENTHOL 65-1,150 Plant JAD NAP
(+)-NEOMENTHOL 1-604 Plant JAD NAP
(+)-OCTAN-3-OL 149-416 Plant BML JAD NAP
(+)-PULEGONE 0-19,400 Plant BML JAD NAP
(-)-ISOPULEGONE 65-154 Plant JAD NAP
(-)-LIMONENE 41-148 Plant JAD NAP
(-)-MENTHOL 1-22 Plant JAD NAP
(-)-MENTHONE 50-6,160 Plant JAD NAP
(-)-OCT-3-EN-1-OL Essential Oil BML NAP
(-)-PIPERITONE 9-86 Plant JAD NAP
1,1,3-TRIMETHYL-CYCLOPENTAN-4-ONE Essential Oil NAP
1,1,3-TRIMETHYL-CYCLOPENTANONE Essential Oil NAP
1,8-CINEOLE 20 Plant BML CRC
1-METHYL-CYCLOHEX-1-EN-3-ONE Essential Oil NAP
1-METHYL-CYCLOHEX-1-ONE Essential Oil NAP
1-METHYL-CYCLOHEXAN-3-OL Essential Oil NAP
1-METHYL-CYCLOHEXAN-3-ONE 80-160 Essential Oil JAD NAP
1-METHYL-CYCLOHEXANOL Essential Oil NAP
3-METHYL-CYCLOHEXAN-1-OL Essential Oil NAP
3-METHYLCYCLOHEXANONE 70-140 Plant BML JAD
3-OCTYLACETATE 10-80 Plant BML CRC
ACETIC-ACID Essential Oil NAP
ALPHA-HUMULENE 70-140 Plant BML
ALPHA-PHELLANDRENE 30-60 Plant JAD NAP
ALPHA-PINENE 20-520 Plant BML CRC JAD NAP
ALPHA-TERPINENE 0-20 Plant JAD NAP
ALPHA-TERPINEOL 70-140 Plant FNF
ALUMINUM 1,850 Plant PED
AROMADENDRENE Plant NAP
ASH 73,000 Plant PED
BETA-BOURBONENE 15-30 Plant BML
BETA-CARYOPHYLLENE Plant NAP
BETA-FENCHENE 1-6 Plant JAD NAP
BETA-HUMULENE Plant NAP
BETA-METHYL-ADIPIC-ACID Essential Oil NAP

BETA-METHYL-DELTA-ISOBUTYLVALERIC-ACID Plant FNF
BETA-PHELLANDRENE 3-400 Plant JAD NAP
BETA-PINENE 20-288 Plant BML CRC JAD NAP
CADINENE 87-174 Plant JAD NAP
CALCIUM 9,440-16,870 Plant FNF PED
CAMPHENE 1-70 Plant JAD NAP
CARVACROL 100-400 Plant JAD NAP
CARVONE Essential Oil NAP
CARYOPHYLLENE 10-60 Plant CRC JAD NAP
CHROMIUM 17 Plant PED
CINEOLE 2-600 Plant JAD NAP
CIS-ISOPULEGONE 20-120 Plant BML JAD
CIS-PULEGONE-OXIDE Plant NAP
CITRONELLOL 90-180 Plant JAD NAP
COBALT 14 Plant PED
CYMOL 160-320 Plant JAD NAP
DECYLIC-ACID Essential Oil NAP
DEHYDROMENTHOFURAN-OXIDE Plant NAP
DEHYDROXYMENTHOFURAN-OXIDE Plant FNF
DELTA-ISOBUTYRL-VALERIC-ACID-BETA-MENTHOL Plant NAP
DELTA-MENTHENONE Essential Oil NAP
DIOSMIN Leaf NAP
DIPENTENE Plant FNF
EO 10,000-20,000 Plant BML GEO HHB WOI
EUGENOL 160-320 Plant CRC JAD NAP
FAT 45,000 Plant PED
FAT 266,000 Seed FNF
FENCHENE 70-140 Plant JAD NAP
FENCHONE 10-20 Plant JAD NAP
FIBER 123,000 Plant PED
FORMIC-ACID Essential Oil NAP
GAMMA-TERPINENE 0-50 Plant JAD NAP
GERMACRENE-D Plant FNF
GUAIAZULENE Plant NAP
HESPERIDIN Plant FNF
HEXAN-1-OL Essential Oil NAP
IRON 2,310 Plant PED
ISOHEPTYLIC-ACID Essential Oil NAP
ISOMENTHOL 1-10 Plant JAD NAP
ISOMENTHONE 0-8,420 Plant JAD NAP
ISOPIPERITENONE 0-80 Plant BML JAD NAP
ISOPIPERITONE Plant NAP
ISOPULEGOL 680-1,360 Plant NAP
ISOPULEGONE 525-2,014 Plant JAD NAP
KILOCALORIES 3,490/kg Plant PED

LIMONENE 10-2,200 Plant BML CRC JAD NAP
LINALOL 15-30 Plant BML JAD
LINALOL-ACETATE 10-820 Plant JAD NAP
LINALOL-ISOVALERATE 130-260 Plant JAD NAP
MAGNESIUM 5,500 Plant PED
MANGANESE 83 Plant PED
MENTHENE 280-560 Plant JAD NAP
MENTHOFURAN 1-160 Plant BML JAD NAP
MENTHOFURAN-OXIDE Plant NAP
MENTHOL 0-3,780 Plant JAD NAP
MENTHOL-ACETATE Plant NAP
MENTHONE 10-6,160 Plant BML JAD NAP
MYRCENE 27-600 Plant BML CRC JAD NAP
NEOISOMENTHOL 30-1,160 Plant BML JAD
NEOISOMENTHOL-ACETATE 30-286 Plant JAD NAP
NEOISOPULEGOL Plant NAP
NEOMENTHOL 1-3,880 Plant JAD NAP
NEPETALACTONE Plant NAP
NEROL 10-20 Plant JAD NAP
O-CRESOL Essential Oil NAP
OCIMENE 25-50 Plant JAD NAP
OCT-1-EN-3-OL Essential Oil BML NAP
OCTAN-1-OL Essential Oil NAP
OCTAN-1-OL-ACETATE Essential Oil NAP
OCTAN-3-OL 20-440 Plant BML JAD NAP
OCTAN-3-OL-ACETATE 12-160 Plant JAD NAP
OCTAN-3-ONE Essential Oil BML NAP
OCTEN-1-OL Essential Oil NAP
OCTOIC-ACID Essential Oil NAP
OCTYLIC-ACID Essential Oil NAP
P-CRESOL Essential Oil NAP
P-CYMENE 3-610 Plant BML CRC JAD NAP
P-CYMOL 30-60 Plant JAD NAP
P-MENTHA-1,4(8)-DIENE Plant NAP
P-MENTHANE 40-80 Plant JAD NAP
PHENOL Essential Oil NAP
PHOSPHORUS 2,520 Plant PED
PIPERITENOL 100-200 Plant JAD NAP
PIPERITENONE 30-4,800 Plant JAD NAP
PIPERITOL 10-20 Plant JAD NAP
PIPERITOL-ACETATE 100-200 Plant JAD NAP
PIPERITONE 10-11,000 Plant JAD NAP
PIPERTENONE 33-66 Plant JAD NAP
POTASSIUM 10,500-15,310 Plant FNF PED
PROTEIN 100,000-112,000 Plant FNF PED

PROTEIN 100,000-246,000 Seed FNF PED
PULEGOL 400-800 Plant JAD NAP
PULEGONE 520-18,800 Plant BML JAD JBH NAP
PULEGONE-EPOXIDE Plant NAP
ROSMARINIC-ACID 30,000 Plant FT62:166
SABINENE 8-140 Plant BML JAD NAP
SALICYLALDEHYDE Essential Oil NAP
SALICYLIC-ACID Essential Oil NAP
SALICYLIC-ACID-METHYL-ESTER Essential Oil NAP
SELENIUM 25 Plant PED
SILICON 182 Plant PED
SODIUM 2,790-5,410 Plant FNF PED
TANNIN 40,000 Plant FNF
TERPINENOL 10-20 Plant JAD NAP
TERPINOLENE 1-90 Plant JAD NAP
THUJONE 10-20 Plant JAD NAP
THYMOL Plant NAP
TIN 24 Plant PED
TRANS-BETA-OCIMENE 1-8 Plant JAD NAP
TRANS-ISOPULEGONE 70-320 Plant BML JAD
TRANS-PULEGONE-OXIDE Plant NAP
WATER 912,000 Plant PED
ZINC 56 Plant PED

Mentha sp.
"Mint"

HESPERIDIN Plant JBH
LIMONENE Plant JBH
MENTHOFURAN Plant JBH
MENTHOL Plant JBH
MENTHONE Plant JBH
MENTHYL-ACETATE Plant JBH
ROTUNDIFOLINE Plant JBH

Mentha spicata L.
"Spearmint"

(+)-DIHYDROCARVONE 550-13,750 Leaf JAD NAP
(+)-ISOMENTHONE 23-575 Leaf JAD NAP
(+)-MENTHOFURAN 4-100 Leaf JAD NAP
(+)-NEOMENTHOL 1-25 Leaf JAD NAP
(+)-PULEGONE 0-8,300 Leaf JAD NAP
(-)-CARVONE Leaf NAP
(-)-MENTHOL 8-200 Leaf JAD NAP
(-)-MENTHONE 438-10,950 Leaf JAD NAP
(-)-PIPERITONE 24-600 Leaf JAD NAP
1,8-CINEOLE 80-9,375 Plant BML GEO HHB
1-HYDROXY-2-ACETOXY-P-MENTH-8(9)-ENE Leaf NAP
2-AMINO-ADIPIC-ACID 1.6 Shoot NAP
2-AMYL-FURAN Leaf NAP
2-ETHYL-FURAN Leaf NAP
2-METHOXY-4-VINYL-PHENOL Shoot NAP
2-METHYL-2-METHYL-BUTYRATE Leaf NAP
2-METHYL-BUTAN-1-OL Leaf NAP
2-METHYL-BUTYL-2-METHYL-BUTYRATE Leaf NAP
2-METHYL-BUTYL-ISOVALERATE Leaf NAP
2-METHYL-BUTYRALDEHYDE Leaf NAP
2-METHYL-ETHYL-BUTYRATE Leaf NAP
2-METHYL-PROPAN-1-OL Leaf NAP
3-(5',5'-DIMETHYL-TETRAHYDROFURAN-2'-YL)-BUT-CIS-2-EN-1-OL 3-75 Leaf JAD NAP
3-METHYL-BUTAN-1-OL Leaf NAP
4-HYDROXY-PIPERITONE Plant NAP
5-HYDROXY-CARVONE Plant NAP
5-METHYL-HEX-2-EN-1-AL Leaf NAP
8-HYDROXY-9-ACETOXY-MENTH-1-ENE Leaf NAP
ACETIC-ACID Plant GEO
ACETOIN Leaf NAP
ACETOPHENONE Leaf NAP
ALPHA-FENCHENE Leaf NAP
ALPHA-HUMULENE Leaf NAP
ALPHA-LIMONENE-1,2-OXIDE Leaf NAP
ALPHA-METHYL-P-METHYL-STYRENE Leaf NAP
ALPHA-PINENE 5-375 Leaf BML CRC JAD NAP
ALPHA-TERPINEOL Leaf NAP
ALPHA-THUJENE Leaf NAP
APIGENIN Leaf HHB
ASH 16,000-94,000 Plant WOI
BENZALDEHYDE Leaf NAP

BENZYL-2-METHYL-BUTYRATE Leaf NAP
BENZYL-ALCOHOL Leaf NAP
BENZYL-ISOBUTYRATE Leaf NAP
BETA-BOURBONENE 2-50 Leaf BML HHB
BETA-CAROTENE 2-10 Leaf CRC WOI
BETA-CARYOPHYLLENE Leaf NAP
BETA-CARYOPHYLLENE-OXIDE Leaf NAP
BETA-ELEMENE Leaf BML
BETA-HUMULENE Leaf BML
BETA-IONONE Leaf NAP
BETA-LIMONENE-1,2-OXIDE Leaf NAP
BETA-PHENETHYL-ALCOHOL Leaf NAP
BETA-PHENETHYL-BUTYRATE Leaf NAP
BETA-PHENETHYL-ISOBUTYRATE Leaf NAP
BETA-PHENETHYL-ISOVALERATE Leaf NAP
BETA-PHENETHYL-METHYL-BUTYRATE Leaf NAP
BETA-PHENETHYL-METHYL-ETHER Leaf NAP
BETA-PINENE 9-3,425 Leaf BML CRC JAD NAP
BICYCLOELEMENE Leaf NAP
BORNEOL Leaf NAP
CADINOL Leaf BML
CALCIUM 2,000-11,760 Plant WOI
CAMPHENE Leaf BML
CAPRYLIC-ACID Plant GEO
CAR-3-ENE Leaf NAP
CARBOHYDRATES 80,000-588,300 Plant WOI
CARVACROL Leaf BML
CARVONE 495-20,000 Plant BML GEO HHB JAD JBH NAP
CARVONE-10-ACETATE Leaf NAP
CARVONE-8,9-OXIDE Leaf NAP
CARVYL-6,7-OXIDE-ACETATE Leaf NAP
CARYOPHYLLENE 3-1,950 Leaf BML HHB JAD NAP
CHLOROFORM Leaf NAP
CHLOROPHYLL-A 2,755 Leaf NAP
CHLOROPHYLL-B 1,299 Leaf NAP
CINEOLE 10-1,500 Leaf NAP
CIS-2,5-DIETHYL-TETRAHYDROFURAN Leaf NAP
CIS-BETA-COPAENE Leaf NAP
CIS-CARVEOL 4-100 Leaf BML HHB
CIS-CARVONE-OXIDE Leaf NAP
CIS-CARVYL-ACETATE 18-450 Leaf BML HHB
CIS-CARVYL-FORMATE Leaf NAP
CIS-JASMONE 1-25 Leaf BML HHB
COPAENE Leaf NAP
COPPER 1-2 Plant WOI

DECAN-1-OL Plant NAP
DECAN-3-OL Plant NAP
DEHYDROCARVACROL Leaf NAP
DELTA-CADINENE 2-38 Leaf BML HHB
DIACETYL Leaf NAP
DIHYDROCARVEOL 13-325 Leaf BML HHB
DIHYDROCARVONE 18-4,475 Leaf BML HHB JAD NAP
DIHYDROCARVYL-ACETATE 16-400 Leaf BML HHB
DIMETHYL-SULFIDE Leaf NAP
DIMETHYL-SULFOXIDE Leaf NAP
DIOSMETIN Plant JBH 411/
DIOSMETIN-7-GLUCOSIDE Plant GEO
DIOSMETIN-7-O-BETA-D-GLUCURONIDE Leaf NAP
DIOSMIN Leaf HHB
ELEMOL Leaf BML
EO 1,000-25,000 Leaf GEO HHB
EPIBULGARENE Leaf NAP
EPIMUUROLENE Leaf NAP
ETHYL-ISOBUTYRATE Leaf NAP
ETHYL-N-VALERATE Leaf NAP
EUGENOL Leaf BML
FARNESENE Leaf BML
FARNESOL Leaf BML
FAT 6,000-35,294 Plant WOI
FIBER 20,000-118,000 Plant CRC
FURFURAL Leaf NAP
GAMMA-CADINENE 2-33 Leaf BML HHB
GAMMA-TERPINENE Leaf NAP
GERANIOL Leaf BML
GERANYL-ACETONE Leaf NAP
HEPT-CIS-4-EN-1-AL Leaf NAP
HEPT-TRANS-2-EN-1-AL Leaf NAP
HEPTAN-1-AL Plant NAP
HEPTAN-1-YL-2-METHYL-BUTYRATE Plant NAP
HEPTAN-1-YL-ISOVALERATE Plant NAP
HEPTAN-3-OL Plant NAP
HEPTAN-3-ONE Plant NAP
HESPERIDIN Plant HHB
HEX-CIS-3-EN-1-OL Leaf NAP
HEX-CIS-3-EN-1-OL-2-METHYL-BUTYRATE Leaf NAP
HEX-CIS-3-EN-1-OL-ACETATE Leaf NAP
HEX-CIS-3-EN-1-OL-ISOBUTYRATE Leaf NAP
HEX-CIS-3-EN-1-OL-ISOVALERATE 2-50 Leaf BML HHB NAP
HEX-CIS-3-EN-1-OL-PHENYLACETATE Leaf NAP
HEX-TRANS-2-EN-1-OL Leaf NAP

HEX-TRANS-2-EN-1-OL-ACETATE Leaf NAP
HEX-TRANS-3-EN-1-OL Leaf NAP
HEXAHYDRO-FARNESYL-ACTONE Leaf NAP
HEXAN-1-AL Leaf NAP
HEXAN-1-OL Leaf NAP
HEXAN-1-YL-2-ISOVALERATE Leaf NAP
HEXAN-1-YL-2-METHYL-BUTYRATE Leaf NAP
HEXAN-1-YL-PHENYLACETATE Leaf NAP
IRON 156-918 Plant CRC WOI
ISOAMYL-2-METHYL-BUTYRATE Plant NAP
ISOAMYL-ISOVALERATE Plant NAP
ISOBORNYL-ACETATE 11-275 Leaf BML HHB
ISOBUTYL-2-METHYL-BUTYRATE Leaf NAP
ISOBUTYRALDEHYDE Leaf NAP
ISOCARYOPHYLLENE Leaf NAP
ISODIHYDROCARVONE 75-1,875 Leaf JAD NAP
ISOMENTHONE 50-125 Leaf BML HHB
ISOPINOCAMPHONE Leaf NAP
JASMONE Leaf NAP
L-PHELLANDRENE Plant FNF
LAVANDULYL-BUTYRATE Leaf NAP
LF Plant FNF
LIMONENE 7-9,075 Plant BML GEO HHB JAD NAP
LIMONENENE-8,9-OXIDE Leaf NAP
LINALOL 11-9,375 Plant BML GEO HHB
LUTEOLIN Leaf HHB
LUTEOLIN-7-O-BETA-D-GLUCURONIDE Leaf NAP
MENTHOL 0-375 Leaf BML HHB JAD NAP
MENTHOL-ACETATE 0-375 Leaf BML HHB JAD NAP
MENTHONE 4-1,300 Leaf BML HHB
METHANOL Leaf NAP
METHYL-HEPTENONE Leaf NAP
MYRCENE 21-1,350 Plant JAD NAP
MYRTENAL Leaf NAP
MYRTENOL Leaf NAP
MYRTENYL-ACETATE Leaf NAP
N-BUTYRALDEHYDE Leaf NAP
N-VALERALDEHYDE Leaf NAP
NEODIHYDROCARVEOL 3-75 Leaf BML HHB
NEOISOPULEGOL Plant NAP
NERYL-ACETATE 27-675 Leaf BML HHB
NIACIN 4-24 Plant WOI
NONAN-1-AL Plant NAP
NONAN-3-OL Plant NAP
OCT-1-EN-3-OL Leaf NAP

OCTAN-1-AL Leaf NAP
OCTAN-1-OL Leaf NAP
OCTAN-1-OL-2-METHYL-BUTYRATE Leaf NAP
OCTAN-1-OL-FORMATE Leaf NAP
OCTAN-3-OL 4-100 Leaf BML HHB
OCTAN-3-OL-ACETATE 2-25 Leaf BML HHB NAP
OCTAN-3-ONE Leaf NAP
OLEANOLIC-ACID Leaf NAP
OXALIC-ACID 38 Leaf WBB
P-CYMEN-8-OL Leaf NAP
P-CYMENE Leaf NAP
P-CYMOL 3-75 Leaf JAD NAP
P-MENTH-1(7)-EN-8-OL Leaf NAP
P-MENTH-1,8-DIEN-4-OL Leaf NAP
P-MENTH-CIS-1(7),8-DIEN-2-OL Leaf NAP
P-MENTH-CIS-2-EN-1-OL Leaf NAP
P-MENTH-TRANS-1(7),8-DIEN-2-OL Leaf NAP
P-MENTH-TRANS-1(7),8-DIEN-2-YL-ACETATE Leaf NAP
P-MENTH-TRANS-2,8-DIEN-1-OL Leaf NAP
PENT-1-EN-3-OL Leaf NAP
PENT-CIS-1-EN-3-OL Leaf NAP
PENTAN-1-OL Leaf NAP
PERILLALDEHYDE Leaf NAP
PERILLYL-ALCOHOL 1-25 Leaf BML HHB
PHENETHYL-ALCOHOL Leaf NAP
PHENLYALANINE Leaf NAP
PHENYL-ACETALDEHYDE Leaf NAP
PHOSPHORUS 800-4,706 Plant WOI
PIN-TRANS-3-EN-2-OL Leaf NAP
PINOCAMPHONE Leaf NAP
PINOCARVEOL Leaf NAP
PINOCARVONE Leaf NAP
PIPERITENONE Leaf NAP
PIPERITENONE-OXIDE Plant CCO
PIPERITONE Plant CCO
PIPERITONE-OXIDE 35-875 Leaf JAD NAP
POMOLIC-ACID Leaf NAP
PROPAN-1-YL-2-METHYL-BUTYRATE Leaf NAP
PROPIONALDEHYDE Leaf NAP
PROTEIN 48,000-282,240 Plant WOI
PSI LIMONENE Leaf NAP
PULEGONE 1-475 Leaf JAD NAP
RIBOFLAVIN 1-5 Plant WOI
ROSMARINIC-ACID 6,300 Leaf NAP
ROSMARINIC-ACID 43,000 Plant FT62:166

ROSMARINIC-ACID 5,900 Shoot NAP
ROTUNDIFOLINE Plant JBH
SABINENE Leaf NAP
SABINENE-HYDRATE 4-300 Leaf JAD NAP
SACCHAROPINE 1.6 Shoot NAP
SALICYLIC-ACID-METHYL-ESTER Leaf NAP
TERPINEN-4-OL Leaf NAP
TERPINOLENE Leaf NAP
THIAMIN 1-3 Plant WOI
THYMOL Leaf NAP
TIGLALDEHYDE Leaf NAP
TOLUENE Leaf NAP
TRANS-BETA-COPAENE Leaf NAP
TRANS-BETA-OCIMENE Leaf NAP
TRANS-CARVEOL 2-50 Leaf BML HHB
TRANS-CARVYL-ACETATE 3-75 Leaf BML HHB
TRANS-CARVYL-FORMATE Leaf NAP
TRANS-JASMONE Leaf NAP
TRANS-VERBENOL Leaf NAP
URSOLIC-ACID Leaf NAP
VALERIC-ACID Plant GEO
VANILLIN Leaf NAP
VERBENONE Leaf NAP
VIRIDIFLOROL Leaf NAP
WATER 0-830,000 Plant WOI

Mentha x rotundifolia (L.) HUDSON "Applemint"

(+)-PIPERITENONE-OXIDE 8-8,510 Leaf JAD NAP
(+)-PULEGONE 0-170 Leaf JAD NAP
(-)-CARVONE 240-4,000 Leaf JAD NAP
(-)-LIMONENE Leaf NAP
1,2-EPOXY-(1S,2S,3R,4S)-MENTHOL-ACETATE Shoot NAP
1,2-EPOXY-NEOMENTHOL-ACETATE 6-7,390 Leaf NAP
ALLO-CYMENE 6-100 Shoot JAD NAP
ALPHA-PHELLANDRENE 16-280 Leaf JAD NAP
ALPHA-PINENE 4-430 Leaf JAD NAP
APIGENIN Shoot NAP
BETA-PINENE 9-330 Leaf JAD NAP
BORNEOL-ACETATE 25-420 Leaf JAD NAP
CAFFEIC-ACID Shoot NAP
CALCIUM 16,300 Leaf CRC
CAMPHENE 0-20 Leaf JAD NAP
CARVONE 260-5,000 Leaf JAD NAP
CINEOLE 4-120 Leaf JAD NAP
CITRAL 9-150 Shoot JAD NAP
CYANIDIN Shoot NAP
DELPHINIDIN Shoot NAP
DIHYDROCARVONE Leaf NAP
DIOSPHENOL Leaf NAP
DIOSPHENOLENE Leaf NAP
EO 600-10,000 Leaf GEO
FAT 282,000 Seed FNF
FERULIC-ACID Shoot NAP
ISONEOPULEGOL 1-5,200 Leaf JAD NAP
ISONEOPULEGOL-ACETATE 43-720 Leaf JAD NAP
ISOPIPERITENONE 80-1,340 Leaf JAD NAP
LIMONENE 14-500 Leaf JAD NAP WOI
LUTEOLIN Shoot NAP
LUTEOLINIDIN Shoot NAP
MASLINIC-ACID Tissue Culture NAP
MENTHOFURAN Leaf NAP
MENTHOL 0-80 Leaf JAD NAP
MENTHONE Leaf tr NAP
NEOISOPULEGOL 314-5,230 Leaf JAD NAP
OCTAN-3-OL 5-822 Leaf JAD NAP
OCTAN-3-OL-ACETATE Leaf NAP
OLEANOLIC-ACID Tissue Culture NAP
P-COUMARIC-ACID Shoot NAP
P-CYMENE 14-230 Shoot JAD NAP

PELARGONIDIN Shoot NAP
PETUNIDIN Shoot NAP
PIPERITENONE 50-860 Leaf JAD NAP
PIPERITENONE-OXIDE 306-8,730 Leaf JAD NAP WOI
PIPERITONE 44-730 Leaf JAD NAP
PIPERITONE-OXIDE 40-6,020 Leaf JAD NAP
POMOLIC-ACID Tissue Culture NAP
POTASSIUM 17,100 Leaf CRC
PROTEIN 144,000 Leaf FNF
PROTEIN 196,000 Seed FNF
PULEGONE 580-9,720 Leaf JAD NAP
ROSMARINIC-ACID 3,600 Leaf NAP
ROSMARINIC-ACID 32,000 Plant FT62:166
ROTUNDIFOLONE Plant JSG
SODIUM 2,520 Leaf CRC
THUJONE 20-360 Leaf JAD NAP
URSOLIC-ACID Tissue Culture NAP

Micromeria congesta BOISS. & HAUSSKN. "Kaya Yarpuzu"

1,8-CINEOLE 11-12 Leaf JEO3:387
1-OCTEN-3-OL 30-35 Leaf JEO3:387
2,5-DIETHYLTETRAHYDROFURAN 1-35 Leaf JEO3:387
2-HEXANAL 4 Leaf JEO3:387
3-OCTANOL 12-13 Leaf JEO3:387
ALPHA-CUBEBENE 15-20 Leaf JEO3:387
ALPHA-PINENE 95-130 Leaf JEO3:387
ALPHA-TERPINENE 2 Leaf JEO3:387
ALPHA-TERPINEOL 35-40 Leaf JEO3:387
BETA-BOURBONENE 45-55 Leaf JEO3:387
BETA-PINENE 275-320 Leaf JEO3:387
CAMPHENE 1-5 Leaf JEO3:387
CARVACROL 20-30 Leaf JEO3:387
CARVONE 10-30 Leaf JEO3:387
CINEROLONE 185-305 Leaf JEO3:387
CIS-CARVEOL 15-25 Leaf JEO3:387
EO 10,000 Leaf JEO3:387 ASSUMED
ETHANOL <1-35 Leaf JEO3:387
EUGENOL 5-15 Leaf JEO3:387
GAMMA-ELEMENE 10-40 Leaf JEO3:387
GAMMA-TERPINENE 5 Leaf JEO3:387
ISOMENTHONE 80 Leaf JEO3:387
ISOPIPERITENONE 65-85 Leaf JEO3:387
LIMONENE 275-355 Leaf JEO3:387
LINALOL <1-3 Leaf JEO3:387
MENTHONE 165-170 Leaf JEO3:387
METHYL-CHAVICOL 25-30 Leaf JEO3:387
MYRCENE 40-65 Leaf JEO3:387
P-CYMEN-8-OL 35-50 Leaf JEO3:387
P-CYMENE <1-5 Leaf JEO3:387
PIPERITENONE-OXIDE 3,985-4,470 Leaf JEO3:387
PIPERITONE 60-70 Leaf JEO3:387
PIPERITONE-OXIDE-I 360-430 Leaf JEO3:387
PIPERITONE-OXIDE-II 470-520 Leaf JEO3:387
PULEGONE 975-1,185 Leaf JEO3:387
SABINENE 40-55 Leaf JEO3:387
SPATHULENOL 140-330 Leaf JEO3:387
TERPINEN-4-OL 25-30 Leaf JEO3:387
TERPINOLENE <1-3 Leaf JEO3:387
THYMOL 35-40 Leaf JEO3:387
VERBENONE 825-935 Leaf JEO3:387

Micromeria croatica

Unaggregated A

ALPHA-PINENE 90 Leaf JEO3:387
ALPHA-THUJONE 45 Leaf JEO3:387
BETA-PINENE 15 Leaf JEO3:387
BETA-THUJONE 70 Leaf JEO3:387
BORNEOL 2,240 Leaf JEO3:387
BORNYL-ACETATE 85 Leaf JEO3:387
CAMPHENE 30 Leaf JEO3:387
CAMPHOR 140 Leaf JEO3:387
CARYOPHYLLENE 300 Leaf JEO3:387
CARYOPHYLLENE-OXIDE 265 Leaf JEO3:387
ELEMENE 800 Leaf JEO3:387
EO 10,000 Leaf JEO3:387 ASSUMED
GAMMA-CADINENE 360 Leaf JEO3:387
GAMMA-MUUROLENE 100 Leaf JEO3:387
GAMMA-TERPINENE 15 Leaf JEO3:387
IOSBORNYL-ACETATE 300 Leaf JEO3:387
ISOBORNEOL 245 Leaf JEO3:387
ISOMENTHONE 600 Leaf JEO3:387
LIMONENE 30 Leaf JEO3:387
LINALYL-ACETATE 140 Leaf JEO3:387
MENTHONE 45 Leaf JEO3:387
MYRCENE 25 Leaf JEO3:387
P-CYMENE 15 Leaf JEO3:387
PIPERITENONE 210 Leaf JEO3:387
PIPERITONE 385 Leaf JEO3:387
PULEGONE 325 Leaf JEO3:387
SABINENE 1 Leaf JEO3:387
TERPINOLENE 10 Leaf JEO3:387
TRANS-CADINOL 905 Leaf JEO3:387
YLANGENE 120 Leaf JEO3:387

Micromeria dalmatica
Unaggregated A

ALPHA-PINENE 60 Leaf JEO3:387
BETA-PINENE 85 Leaf JEO3:387
CAMPHENE 6 Leaf JEO3:387
EO 10,000 Leaf JEO3:387 ASSUMED
ISOMENTHONE Leaf JEO3:387
MENTONE 10 Leaf JEO3:387
MYRCENE 35 Leaf JEO3:387
P-CYMENE 13 Leaf JEO3:387
PIPERITENONE 2,355 Leaf JEO3:387
PIPERITONE 425 Leaf JEO3:387
PULEGONE 625 Leaf JEO3:387

Micromeria fruticosa ssp. *barbata* (BOISS. & KY.) P.H. DAVIS
"Tea Hyssop, Zopha, Zuta"
Unaggregated A

(E)-BETA-OCIMENE 17 Shoot JEO3:477
(Z)-3-HEXENOL 10 Shoot JEO3:477
1,8-CINEOLE 24 Shoot JEO3:477
1-OCTEN-3-OL 210 Shoot JEO3:477
2,5-DIETHYLTETRAHYDROFURAN <0.1 Shoot JEO3:477
2,5-DIHYDROFURAN 5 Shoot JEO3:477
3-OCTANOL 120 Shoot JEO3:477
3-OCTANONE 7 Shoot JEO3:477
ACETALDEHYDE <0.1 Shoot JEO3:477
ALPHA-HUMULENE 65 Shoot JEO3:477
ALPHA-PINENE 670 Shoot JEO3:477
ALPHA-TERPINENE Shoot JEO3:477
ALPHA-TERPINEOL Shoot JEO3:477
ALPHA-THUJENE Shoot JEO3:477
ALPHA-THUJONE Shoot JEO3:477
BENZALDEHYDE 5 Shoot JEO3:477
BETA-BOURBONENE 14 Shoot JEO3:477
BETA-CADINENE 40 Shoot JEO3:477
BETA-CARYOPHYLLENE 1,025 Shoot JEO3:477
BETA-PINENE 135 Shoot JEO3:477
BETA-TERPINEOL 7 Shoot JEO3:477
BICYCLOGERMACRENE 65 Shoot JEO3:477
BORNEOL Shoot JEO3:477
CAMPHENE <0.1 Shoot JEO3:477
CARVACROL Shoot JEO3:477
CARYOPHYLLENE-OXIDE 185 Shoot JEO3:477
CIS-ISOPULEGONE 130 Shoot JEO3:477
CIS-JASMONE 65 Shoot JEO3:477
CIS-SABINENE-HYDRATE Shoot JEO3:477
DIETHYL-MALONATE 2 Shoot JEO3:477
EO 24,000 Shoot JEO3:477
ETHYL-BUTYRATE 5 Shoot JEO3:477
ETHYL-CAPROATE <0.1 Shoot JEO3:477
ETHYL-CINNAMATE Shoot JEO3:477
ETHYL-LINOLEATE 14 Shoot JEO3:477
ETHYL-LINOLENATE 31 Shoot JEO3:477
ETHYL-PALMITATE 55 Shoot JEO3:477
EUGENOL 26 Shoot JEO3:477
GAMMA-TERPINENE 5 Shoot JEO3:477

GERMACRENE-D 270 Shoot JEO3:477
HUMULENE-OXIDE 14 Shoot JEO3:477
ISOMENTHOL 1,750 Shoot JEO3:477
ISOMENTHONE 175 Shoot JEO3:477
ISOPIPERITENONE 14 Shoot JEO3:477
ISOPULEGOL 10 Shoot JEO3:477
LIMONENE 370 Shoot JEO3:477
LINALOL Shoot JEO3:477
MENTHOFURAN Shoot JEO3:477
MENTHOL 105 Shoot JEO3:477
MENTHONE 10 Shoot JEO3:477
MYRCENE 60 Shoot JEO3:477
NEOMENTHOL 85 Shoot JEO3:477
P-CYMENE 5 Shoot JEO3:477
PENTANAL Shoot JEO3:477
PIPERITENONE 480 Shoot JEO3:477
PIPERITENONE-OXIDE 705 Shoot JEO3:477
PIPERITONE 12 Shoot JEO3:477
PULEGONE 15,650 Shoot JEO3:477
SABINENE 40 Shoot JEO3:477
TERPINOLENE 10 Shoot JEO3:477
THYMOL Shoot JEO3:477
TRANS-ISOPULEGONE 100 Shoot JEO3:477
TRANS-SABINENE-HYDRATE 10 Shoot JEO3:477
VIRIDIFLORENE 15 Shoot JEO3:477

Micromeria fruticosa ssp. *barbata* (BOISS. & KY.) P.H. DAVIS
"Tea Hyssop, Zopha, Zuta"
Unaggregated B

(E)-BETA-OCIMENE 19 Shoot JEO3:477
(Z)-3-HEXENOL Shoot JEO3:477
1,8-CINEOLE 43 Shoot JEO3:477
1-OCTEN-3-OL 165 Shoot JEO3:477
2,5-DIETHYLTETRAHYDROFURAN <0.1 Shoot JEO3:477
2,5-DIHYDROFURAN Shoot JEO3:477
3-OCTANOL 120 Shoot JEO3:477
3-OCTANONE 5 Shoot JEO3:477
ACETALDEHYDE <0.1 Shoot JEO3:477
ALPHA-HUMULENE 50 Shoot JEO3:477
ALPHA-PINENE 195 Shoot JEO3:477
ALPHA-TERPINENE 7 Shoot JEO3:477
ALPHA-TERPINEOL 7 Shoot JEO3:477
ALPHA-THUJENE 7 Shoot JEO3:477
ALPHA-THUJONE Shoot JEO3:477
BENZALDEHYDE Shoot JEO3:477
BETA-BOURBONENE 10 Shoot JEO3:477
BETA-CADINENE 12 Shoot JEO3:477
BETA-CARYOPHYLLENE 645 Shoot JEO3:477
BETA-PINENE 400 Shoot JEO3:477
BETA-TERPINEOL Shoot JEO3:477
BICYCLOGERMACRENE 50 Shoot JEO3:477
BORNEOL 5 Shoot JEO3:477
CAMPHENE 7 Shoot JEO3:477
CARVACROL 17 Shoot JEO3:477
CARYOPHYLLENE-OXIDE 500 Shoot JEO3:477
CIS-ISOPULEGONE 175 Shoot JEO3:477
CIS-JASMONE 48 Shoot JEO3:477
CIS-SABINENE-HYDRATE 5 Shoot JEO3:477
DIETHYL-MALONATE Shoot JEO3:477
EO 24,000 Shoot JEO3:477
ETHYL-BUTYRATE 5 Shoot JEO3:477
ETHYL-CAPROATE Shoot JEO3:477
ETHYL-CINNAMATE 17 Shoot JEO3:477
ETHYL-LINOLEATE <0.1 Shoot JEO3:477
ETHYL-LINOLENATE <0.1 Shoot JEO3:477
ETHYL-PALMITATE Shoot JEO3:477
EUGENOL Shoot JEO3:477
GAMMA-TERPINENE 5 Shoot JEO3:477

GERMACRENE-D 105 Shoot JEO3:477
HUMULENE-OXIDE 120 Shoot JEO3:477
ISOMENTHOL 1,660 Shoot JEO3:477
ISOMENTHONE 245 Shoot JEO3:477
ISOPIPERITENONE 14 Shoot JEO3:477
ISOPULEGOL 7 Shoot JEO3:477
LIMONENE 335 Shoot JEO3:477
LINALOL 7 Shoot JEO3:477
MENTHOFURAN 7 Shoot JEO3:477
MENTHOL 85 Shoot JEO3:477
MENTHONE 10 Shoot JEO3:477
MYRCENE 120 Shoot JEO3:477
NEOMENTHOL Shoot JEO3:477
P-CYMENE 2 Shoot JEO3:477
PENTANAL 5 Shoot JEO3:477
PIPERITENONE 435 Shoot JEO3:477
PIPERITENONE-OXIDE 1,095 Shoot JEO3:477
PIPERITONE 7 Shoot JEO3:477
PULEGONE 14,890 Shoot JEO3:477
SABINENE 100 Shoot JEO3:477
TERPINOLENE 12 Shoot JEO3:477
THYMOL 19 Shoot JEO3:477
TRANS-ISOPULEGONE 60 Shoot JEO3:477
TRANS-SABINENE-HYDRATE 10 Shoot JEO3:477
VIRIDIFLORENE 17 Shoot JEO3:477

Micromeria fruticosa
"Tasnanesi"

1-OCTEN-3-OL 120 Leaf JEO3:387
3-OCTANOL <1 Leaf JEO3:387
ALPHA-GUAIENE <1 Leaf JEO3:387
ALPHA-PINENE <1-390 Leaf JEO3:387
ALPHA-TERPINEOL <1 Leaf JEO3:387
BETA-CARYOPHYLLENE 2,640 Leaf JEO3:387
BETA-PINENE 960 Leaf JEO3:387
BICYCLOELEMENE <1 Leaf JEO3:387
BORNEOL 570 Leaf JEO3:387
CAMPHENE <1-1 Leaf JEO3:387
CARVONE 90 Leaf JEO3:387
CIS-BETA-OCIMENE 120 Leaf JEO3:387
DELTA-CADINENE <1 Leaf JEO3:387
EO 30,000 Leaf JEO3:387
GAMMA-TERPINENE <1 Leaf JEO3:387
ISOMENTHONE 1,110-7,080 Leaf JEO3:387
ISOPULEGOL 180 Leaf JEO3:387
LIMONENE <1 Leaf JEO3:387
MENTHOL 3,030 Leaf JEO3:387
MENTHONE <1-6,000 Leaf JEO3:387
MYRCENE 300 Leaf JEO3:387
P-CYMENE 150 Leaf JEO3:387
PIPERITENONE 2,380 Leaf JEO3:387
PIPERITENONE-OXIDE 1,880 Leaf JEO3:387
PIPERITONE 840-7,710 Leaf JEO3:387
PULEGONE 13,710-18,780 Leaf JEO3:387
TERPINEN-4-OL 2,460 Leaf JEO3:387
TERPINOLENE <1 Leaf JEO3:387

Micromeria juliana
Unaggregated A

ALLO-OCIMENE 45 Leaf JEO3:387
ALPHA-PINENE 65 Leaf JEO3:387
ALPHA-TERPINENE 15 Leaf JEO3:387
ALPHA-TERPINEOL 155 Leaf JEO3:387
ALPHA-THUJONE 40 Leaf JEO3:387
BETA-CARYOPHYLLENE 365 Leaf JEO3:387
BETA-PINENE 50 Leaf JEO3:387
BETA-THUJONE 40 Leaf JEO3:387
CAMPHENE 25 Leaf JEO3:387
CAMPHOR 65 Leaf JEO3:387
CARVACROL 1,900 Leaf JEO3:387
CARVONE 110 Leaf JEO3:387
EO 10,000 Leaf JEO3:387
FENCHONE 30 Leaf JEO3:387
GAMMA-TERPINENE 35 Leaf JEO3:387
GERANYL-ACETATE 200 Leaf JEO3:387
GUAIACOL 75 Leaf JEO3:387
GUAIAZULENE 45 Leaf JEO3:387
ISOBORNYL-ACETATE 145 Leaf JEO3:387
ISOMENTHONE 230 Leaf JEO3:387
ISOSAFROLE 290 Leaf JEO3:387
LIMONENE 25 Leaf JEO3:387
LINALOL 150 Leaf JEO3:387
LINALYL-ACETATE 10 Leaf JEO3:387
MENTHONE 60 Leaf JEO3:387
MYRCENE 10 Leaf JEO3:387
NEROLIDOL 740 Leaf JEO3:387
P-CYMENE Leaf JEO3:387
PIPERITONE 75 Leaf JEO3:387
PULEGONE 770 Leaf JEO3:387
THYMOL 180 Leaf JEO3:387

Micromeria myrtifolia BOISS. & HOHEN "Dagcayi, Haydarotu, Topukcayi" Unaggregated A

(E)-BETA-OCIMENE 0.3 Shoot JEO4:79
1,8-CINEOLE <0.1 Shoot JEO4:79
1-OCTEN-3-OL <0.1 Shoot JEO4:79
2,5-DIETHYLTETRAHYDROFURAN Shoot JEO4:79
2-HEXANAL Shoot JEO4:79
3-OCTANOL Shoot JEO4:79
6,10,14-TRIMETHYL-2-PENTADECANOL 1 Shoot JEO4:79
6-METHYL-3-HEPTANOL <0.1 Shoot JEO4:79
ACETONE 0.1 Shoot JEO4:79
ALPHA-COPAENE 7 Shoot JEO4:79
ALPHA-CUBEBENE Shoot JEO4:79
ALPHA-HUMULENE 9 Shoot JEO4:79
ALPHA-MUUROLENE 3 Shoot JEO4:79
ALPHA-PINENE 0.3 Shoot JEO4:79
ALPHA-TERPINENE Shoot JEO4:79
ALPHA-TERPINEOL <0.1 Shoot JEO4:79
AR-CURCUMENE 2 Shoot JEO4:79
BETA-BOURBONENE 0.1 Shoot JEO4:79
BETA-CARYOPHYLLENE 130 Shoot JEO4:79
BETA-CUBEBENE 3 Shoot JEO4:79
BETA-PINENE 0.5 Shoot JEO4:79
CAMPHENE <0.1 Shoot JEO4:79
CAMPHOR 0.5 Shoot JEO4:79
CARVACROL 2 Shoot JEO4:79
CARVONE 0.3 Shoot JEO4:79
CINEROLONE Shoot JEO4:79
CIS-CARVEOL Shoot JEO4:79
CIS-GERANYL-ACETONE 0.5 Shoot JEO4:79
CIS-SABINENE-HYDRATE <0.1 Shoot JEO4:79
DECANAL 0.8 Shoot JEO4:79
DECANE <0.1 Shoot JEO4:79
DELTA-CADINENE 20 Shoot JEO4:79
EO 300 Shoot JEO4:79
EREMOPHILINE 2 Shoot JEO4:79
ETHANOL <0.1 Shoot JEO4:79
EUGENOL Shoot JEO4:79
GAMMA-ELEMENE 1 Shoot JEO4:79
GAMMA-SELINENE 7 Shoot JEO4:79
GAMMA-TERPINENE 0.2 Shoot JEO4:79
GERMACRENE-D 21 Shoot JEO4:79
ISOAMYL-ACETATE 0.1 Shoot JEO4:79

ISOMENTHONE Shoot JEO4:79
ISOPIPERITENONE Shoot JEO4:79
LIMONENE 0.9 Shoot JEO4:79
LINALOL 3 Shoot JEO4:79
LINALYL-ACETATE 0.2 Shoot JEO4:79
MENTHONE Shoot JEO4:79
METHYL-CHAVICOL Shoot JEO4:79
MYRCENE 0.1 Shoot JEO4:79
NEROLIDOL 8 Shoot JEO4:79
NONANE <0.1 Shoot JEO4:79
NONANOL 0.6 Shoot JEO4:79
OCTYL-ACETATE 0.2 Shoot JEO4:79
P-CYMEN-8-OL Shoot JEO4:79
P-CYMENE 0.1 Shoot JEO4:79
PATCHOULANE 26 Shoot JEO4:79
PIPERITENONE-OXIDE Shoot JEO4:79
PIPERITONE Shoot JEO4:79
PIPERITONE-OXIDE-I Shoot JEO4:79
PIPERITONE-OXIDE-II Shoot JEO4:79
PULEGONE Shoot JEO4:79
SABINENE <0.1 Shoot JEO4:79
SPATHULENOL 0.4 Shoot JEO4:79
TERPINEN-4-OL Shoot JEO4:79
TERPINOLENE 0.1 Shoot JEO4:79
THYMOL Shoot JEO4:79
UNDECANE <0.1 Shoot JEO4:79
VERBENONE Shoot JEO4:79

Micromeria teneriffae

Unaggregated A

1,8-CINEOLE 110 Leaf JEO3:387
ALPHA-PHELLANDRENE 20 Leaf JEO3:387
ALPHA-PINENE 2,030 Leaf JEO3:387
ALPHA-TERPINENE 90 Leaf JEO3:387
ALPHA-TERPINEOL 170 Leaf JEO3:387
BETA-PHELLANDRENE 20 Leaf JEO3:387
BETA-PINENE 110 Leaf JEO3:387
BORNEOL 1,490 Leaf JEO3:387
BORNYL-ACETATE 470 Leaf JEO3:387
CAMPHENE 470 Leaf JEO3:387
CIS-SABINENE-HYDRATE 20 Leaf JEO3:387
EO 10,000 Leaf JEO3:387
GAMMA-TERPINENE 230 Leaf JEO3:387
GERMACRENE 210 Leaf JEO3:387
LIMONENE 560 Leaf JEO3:387
LINALOL 80 Leaf JEO3:387
MYRCENE 170 Leaf JEO3:387
NEROLIDOL 1,090 Leaf JEO3:387
P-CYMENE 50 Leaf JEO3:387
SABINENE 230 Leaf JEO3:387
TERPINEN-4-OL 140 Leaf JEO3:387
TERPINOLENE 80 Leaf JEO3:387
TERPINYL-ACETATE 190 Leaf JEO3:387
TRANS-SABINENE-HYDRATE 35 Leaf JEO3:387

Micromeria thymifolia
Unaggregated A

1,8-CINEOLE 25 Leaf JEO3:387
ALPHA-PINENE 65 Leaf JEO3:387
ALPHA-TERPINENE 5 Leaf JEO3:387
ALPHA-THUJONE 1 Leaf JEO3:387
BETA-PINENE 85 Leaf JEO3:387
BETA-THUJONE 4 Leaf JEO3:387
CAMPHENE 1 Leaf JEO3:387
CAMPHOR 10 Leaf JEO3:387
CARYOPHYLLENE-OXIDE 55 Leaf JEO3:387
COPAENE 20 Leaf JEO3:387
EO 10,000 Leaf JEO3:387
GAMMA-TERPINENE 5 Leaf JEO3:387
ISOMENTHONE 695 Leaf JEO3:387
LIMONENE 175 Leaf JEO3:387
LINALYL-ACETATE 10 Leaf JEO3:387
MENTHOL 45 Leaf JEO3:387
MENTHONE 885 Leaf JEO3:387
MYRCENE 35 Leaf JEO3:387
NEOMENTHOL 105 Leaf JEO3:387
P-CYMENE Leaf JEO3:387
PIPERITENONE 270 Leaf JEO3:387
PIPERITENONE-OXIDE 280 Leaf JEO3:387
PIPERITONE 320 Leaf JEO3:387
PULEGONE 6,355 Leaf JEO3:387
TERPINOLENE 3 Leaf JEO3:387
YLANGENE 85 Leaf JEO3:387

Micromeria varia ssp. *thymoides*
"Madeiran 'Hyssop'"
Unaggregated A

ALPHA-CAMPHOLENAL 15 Shoot FFJ10(3):199
ALPHA-COPAENE Shoot FFJ10(3):199
ALPHA-CUBEBENE Shoot FFJ10(3):199
ALPHA-HUMULENE 39 Shoot FFJ10(3):199
ALPHA-PHELLANDRENE 6 Shoot FFJ10(3):199
ALPHA-PINENE 1,055 Shoot FFJ10(3):199
ALPHA-TERPINENE 9 Shoot FFJ10(3):199
ALPHA-TERPINEOL 3 Shoot FFJ10(3):199
ALPHA-THUJENE 21 Shoot FFJ10(3):199
BETA-BOURBONENE Shoot FFJ10(3):199
BETA-CARYOPHYLLENE 310 Shoot FFJ10(3):199
BETA-CARYOPHYLLENE-EPOXIDE 355 Shoot FFJ10(3):199
BETA-ELEMENE Shoot FFJ10(3):199
BETA-MYRCENE 42 Shoot FFJ10(3):199
BETA-PHELLANDRENE 9 Shoot FFJ10(3):199
BETA-PINENE 240 Shoot FFJ10(3):199
BORNEOL 24 Shoot FFJ10(3):199
CADINA-1,4-DIENE Shoot FFJ10(3):199
CAMPHENE 9 Shoot FFJ10(3):199
CIS-BETA-OCIMENE 3 Shoot FFJ10(3):199
CIS-LINALOL-OXIDE Shoot FFJ10(3):199
CIS-SABINENE-HYDRATE Shoot FFJ10(3):199
CIS-VERBENOL Shoot FFJ10(3):199
DELTA-CADINENE 3 Shoot FFJ10(3):199
EO 3,000 Shoot FFJ10(3):199
GAMMA-CADINENE 24 Shoot FFJ10(3):199
GAMMA-TERPINENE 6 Shoot FFJ10(3):199
GERANIAL 95 Shoot FFJ10(3):199
GERANYL-ACETATE Shoot FFJ10(3):199
GERMACRENE-D 75 Shoot FFJ10(3):199
LIMONENE 100 Shoot FFJ10(3):199
LINALOL 48 Shoot FFJ10(3):199
MYRTENAL 6 Shoot FFJ10(3):199
MYRTENOL 6 Shoot FFJ10(3):199
NERAL 55 Shoot FFJ10(3):199
P-CYMENE 9 Shoot FFJ10(3):199
PINOCARVONE 24 Shoot FFJ10(3):199
SABINENE 21 Shoot FFJ10(3):199
TERPINEN-4-OL 3 Shoot FFJ10(3):199
TERPINOLENE Shoot FFJ10(3):199
TRANS-CARVEOL 3 Shoot FFJ10(3):199

TRANS-HEX-2-ENAL 18 Shoot FFJ10(3):199
TRANS-LINALOL-OXIDE Shoot FFJ10(3):199
TRANS-NEROLIDOL Shoot FFJ10(3):199
TRANS-PINOCARVEOL 9 Shoot FFJ10(3):199
TRANS-SABINENE-HYDRATE Shoot FFJ10(3):199
TRANS-VERBENOL 42 Shoot FFJ10(3):199
VERBENONE 6 Shoot FFJ10(3):199

Micromeria varia ssp. *thymoides*
"Madeiran 'Hyssop'"
Unaggregated B

ALPHA-CAMPHOLENAL 15 Shoot FFJ10(3):199
ALPHA-COPAENE 45 Shoot FFJ10(3):199
ALPHA-CUBEBENE 9 Shoot FFJ10(3):199
ALPHA-HUMULENE 18 Shoot FFJ10(3):199
ALPHA-PHELLANDRENE Shoot FFJ10(3):199
ALPHA-PINENE 585 Shoot FFJ10(3):199
ALPHA-TERPINENE 3 Shoot FFJ10(3):199
ALPHA-TERPINEOL 30 Shoot FFJ10(3):199
ALPHA-THUJENE 15 Shoot FFJ10(3):199
BETA-BOURBONENE Shoot FFJ10(3):199
BETA-CARYOPHYLLENE 155 Shoot FFJ10(3):199
BETA-CARYOPHYLLENE-EPOXIDE 75 Shoot FFJ10(3):199
BETA-ELEMENE Shoot FFJ10(3):199
BETA-MYRCENE 60 Shoot FFJ10(3):199
BETA-PHELLANDRENE 15 Shoot FFJ10(3):199
BETA-PINENE 170 Shoot FFJ10(3):199
BORNEOL 24 Shoot FFJ10(3):199
CADINA-1,4-DIENE Shoot FFJ10(3):199
CAMPHENE 6 Shoot FFJ10(3):199
CIS-BETA-OCIMENE Shoot FFJ10(3):199
CIS-LINALOL-OXIDE Shoot FFJ10(3):199
CIS-SABINENE-HYDRATE 15 Shoot FFJ10(3):199
CIS-VERBENOL Shoot FFJ10(3):199
DELTA-CADINENE 21 Shoot FFJ10(3):199
EO 3,000 Shoot FFJ10(3):199
GAMMA-CADINENE 15 Shoot FFJ10(3):199
GAMMA-TERPINENE 15 Shoot FFJ10(3):199
GERANIAL 475 Shoot FFJ10(3):199
GERANYL-ACETATE 24 Shoot FFJ10(3):199
GERMACRENE-D 36 Shoot FFJ10(3):199
LIMONENE 80 Shoot FFJ10(3):199
LINALOL 27 Shoot FFJ10(3):199
MYRTENAL Shoot FFJ10(3):199
MYRTENOL Shoot FFJ10(3):199
NERAL 190 Shoot FFJ10(3):199
P-CYMENE 3 Shoot FFJ10(3):199
PINOCARVONE 12 Shoot FFJ10(3):199
SABINENE 205 Shoot FFJ10(3):199
TERPINEN-4-OL 21 Shoot FFJ10(3):199
TERPINOLENE 3 Shoot FFJ10(3):199
TRANS-CARVEOL Shoot FFJ10(3):199

TRANS-HEX-2-ENAL 6 Shoot FFJ10(3):199
TRANS-LINALOL-OXIDE Shoot FFJ10(3):199
TRANS-NEROLIDOL 440 Shoot FFJ10(3):199
TRANS-PINOCARVEOL 6 Shoot FFJ10(3):199
TRANS-SABINENE-HYDRATE 60 Shoot FFJ10(3):199
TRANS-VERBENOL Shoot FFJ10(3):199
VERBENONE Shoot FFJ10(3):199

Minthostachys mollis (KUNTH.) GRISEB "Muna, Tipo"

(E)-BETA-OCIMENE 3-18 Shoot EB48:60
(Z)-BETA-OCIMENE 4-8 Shoot EB48:60
ALPHA-HUMULENE 1-5 Shoot EB48:60
ALPHA-PINENE 2-4 Shoot EB48:60
BETA-CARYOPHYLLENE 11-32 Shoot EB48:60
BETA-PINENE 3-5 Shoot EB48:60
EO 700-900 Shoot EB48:60
ISOMENTHOL 3-9 Shoot EB48:60
ISOMENTHONE 23-44 Shoot EB48:60
ISOMENTHYL-ACETATE 95-230 Shoot EB48:60
LIMONENE 4-10 Shoot EB48:60
MENTHOL 120-215 Shoot EB48:60
MENTHONE 130-265 Shoot EB48:60
MENTHYL-ACETATE 0.5-1 Shoot EB48:60
MYRCENE 0.4-0.5 Shoot EB48:60
NEOMENTHOL 180-300 Shoot EB48:60
NEOMENTHYL-ACETATE 16-28 Shoot EB48:60
PIPERITONE 50-100 Shoot EB48:60
PULEGONE 3-10 Shoot EB48:60
SABINENE 1-2 Shoot EB48:60

Moldavica thymiflora RYDB.
"Thyme-Flowered Moldavica"
Unaggregated A

1,8-CINEOLE 134 Plant BML
ALPHA-PINENE 2 Plant BML
ALPHA-TERPINEOL 48 Plant BML
BETA-ELEMENE 148 Plant BML
BETA-PINENE 2 Plant BML
CARYOPHYLLENE 250 Plant BML
CARYOPHYLLENE-OXIDE 46 Plant BML
CIS-OCIMENE 84 Plant BML
EO 2,000 Plant BML
GERMACRENE-D 140 Plant BML
ISOMENTHONE 216 Plant BML
LIMONENE 10 Plant BML
LINALOL 32 Plant BML
MENTHONE 28 Plant BML
MYRCENE 20 Plant BML
PULEGONE 280 Plant BML
SABINENE 8 Plant BML
TRANS-OCIMENE 96 Plant BML

Monarda citriodora Cerv. ex Lagasca
"Lemon Mint"
Unaggregated A

1,8-CINEOLE 20 Flower JEO6:27
1,8-CINEOLE 40 Leaf JEO6:27
ALPHA-PINENE 115 Flower JEO6:27
ALPHA-PINENE 65 Leaf JEO6:27
ALPHA-TERPINENE 320 Flower JEO6:27
ALPHA-TERPINENE 120 Leaf JEO6:27
ALPHA-TERPINEOL 55 Flower JEO6:27
ALPHA-TERPINEOL <0.1 Leaf JEO6:27
BETA-CARYOPHYLLENE 50 Flower JEO6:27
BETA-CARYOPHYLLENE 50 Leaf JEO6:27
BETA-PINENE <0.1 Flower JEO6:27
BETA-PINENE 50 Leaf JEO6:27
CARVACROL 385 Flower JEO6:27
CARVACROL 250 Leaf JEO6:27
EO 10,000 Flower JEO6:27
EO 7,000 Leaf JEO6:27
GAMMA-TERPINENE 1,330 Flower JEO6:27
GAMMA-TERPINENE <0.1 Leaf JEO6:27
LIMONENE 40 Flower JEO6:27
LIMONENE 35 Leaf JEO6:27
MYRCENE 205 Flower JEO6:27
MYRCENE 15 Leaf JEO6:27
P-CYMENE 420 Flower JEO6:27
P-CYMENE 1,600 Leaf JEO6:27
TERPINEN-4-OL 75 Flower JEO6:27
TERPINEN-4-OL 145 Leaf JEO6:27
THYMOL 6,175 Flower JEO6:27
THYMOL 3,550 Leaf JEO6:27

Monarda citriodora Cerv. ex Lagasca
"Lemon Mint"
Aggregated

1,8-CINEOLE 85-614 Plant FNF
3-OCTANOL 124-201 Plant FNF
3-OCTYL-ACETATE 26-97 Plant FNF
ALPHA-PINENE 25-214 Plant FNF
ALPHA-TERPINEOL 11 Plant FNF
BORNYL-ACETATE 48-185 Plant FNF
CAMPHENE 6-25 Plant FNF
CARVACROL 316 Plant FNF
CITRAL Plant FNF
DIHYDROCARVEOL-ACETATE 11 Plant FNF
EO 10,000 Shoot FNF
GAMMA-TERPINENE 29 Plant FNF
ISOMENTHONE 34 Plant FNF
LIMONENE 89-90 Plant FNF
LINALOL 47 Plant FNF
LINALYL-ACETATE 24 Plant FNF
MYRCENE 14 Plant FNF
P-CYMENE 859-1,038 Plant FNF
THYMOHYDROQUINONE Plant FNF
THYMOL 6,269-8,337 Plant FNF
TRIACONTANE Shoot FNF

Monarda clinopodia L.
"Clinopod Bergamot"
Unaggregated A

1,8-CINEOLE 1,383 Plant BML
3-OCTANOL 663 Plant BML
3-OCTYL-ACETATE 333 Plant BML
ALPHA-PINENE 65 Plant BML
ALPHA-TERPINEOL 434 Plant BML
BORNYL-ACETATE 614 Plant BML
CAMPHENE 88 Plant BML
DIHYDROCARVEOL-ACETATE 348 Plant BML
EO 10,000 Plant ASSUMED
GAMMA-TERPINENE 140 Plant BML
LIMONENE 287 Plant BML
LINALOL 216 Plant BML
LINALYL-ACETATE 4,050 Plant BML
MYRCENE 10 Plant BML
P-CYMENE 218 Plant BML
PULEGONE 121 Plant BML

Monarda didyma L.
"Beebalm, Oswego Tea"

1,8-CINEOLE 22 Flower FFJ6:80
1,8-CINEOLE 30-50 Leaf FFJ6:80 HHB
1,8-CINEOLE 100-2,735 Plant BML
1-OCTEN-3-OL 30-310 Plant BML
3-OCTANOL 25-240 Plant BML
3-OCTANONE 5-40 Plant BML
3-OCTYL-ACETATE 10-110 Plant BML
ALPHA-COPAENE Flower tr FFJ6:80
ALPHA-COPAENE Leaf tr FFJ6:80
ALPHA-PINENE 18 Flower FFJ6:80
ALPHA-PINENE 50-90 Leaf FFJ6:80 HHB
ALPHA-PINENE 5-350 Plant BML
ALPHA-TERPINENE 25 Flower FFJ6:80
ALPHA-TERPINENE 10-15 Leaf FFJ6:80 HHB
ALPHA-TERPINENE 5-40 Plant BML
ALPHA-TERPINEOL 20 Flower FFJ6:80
ALPHA-TERPINEOL 25-40 Leaf FFJ6:80 HHB
ALPHA-TERPINEOL 1-420 Plant BML
ALPHA-THUJENE 2-20 Plant BML
BETA-BOURBONENE 10 Flower FFJ6:80
BETA-BOURBONENE 50-90 Leaf FFJ6:80 HHB
BETA-BOURBONENE 5-70 Plant BML
BETA-CARYOPHYLLENE 30 Flower FFJ6:80
BETA-CARYOPHYLLENE Leaf tr FFJ6:80
BETA-COPAENE 5-40 Plant BML
BETA-PINENE 7 Flower FFJ6:80
BETA-PINENE 45-80 Leaf FFJ6:80 HHB
BETA-PINENE 25-240 Plant BML
BORNEOL 7 Flower FFJ6:80
BORNEOL 40-75 Leaf FFJ6:80
BORNYL ACETATE 55 Flower FFJ6:80
BORNYL-ACETATE 380-740 Leaf FFJ6:80 HHB
BORNYL-ACETATE 15-775 Plant BML
CAMPHENE 10 Flower FFJ6:80
CAMPHENE 90-170 Leaf FFJ6:80 HHB
CAMPHENE 25-320 Plant BML
CARVACROL Plant HHB
DELTA-3-CARENE Leaf HHB
DIDYMIN 300-3,000 Plant HHB JBH
DIHYDROCARVYL-ACETATE 5-45 Plant BML
EO 3,700 Flower FFJ6:80
EO 7,000-13,000 Leaf FFJ6:80 HHB

EO 1,000-10,000 Shoot BML HHB
GAMMA-TERPINENE 195 Flower FFJ6:80
GAMMA-TERPINENE 60-115 Leaf FFJ6:80 HHB
GAMMA-TERPINENE 20-1,250 Plant BML
GENKWANIN Plant HHB
GERANIOL 4 Flower FFJ6:80
GERANIOL 14-25 Leaf FFJ6:80 HHB
GERANYL-ACETATE 4 Flower FFJ6:80
GERANYL-ACETATE 7-13 Leaf FFJ6:80 HHB
GERMACRENE-D 115 Flower FFJ6:80
GERMACRENE-D 370-690 Leaf FFJ6:80 HHB
GERMACRENE-D 20-330 Plant BML
GLUCOGENKWANIN Plant HHB
ISOMENTHONE Flower tr FFJ6:80
ISOMENTHONE 15-25 Leaf FFJ6:80 HHB
ISOSAKURANETIN Plant BML
ISOSAKURANETIN-7-RUTINOSIDE Plant JBH
ISOSAKURANETIN-GLYCOSIDE Plant BML
ISOSAKURANIN Plant BML
LIMONENE 10 Flower FFJ6:80
LIMONENE 25-40 Leaf FFJ6:80 HHB
LIMONENE 2-1,300 Plant BML
LINALOL 2,385 Flower FFJ6:80
LINALOL 5,195-9,645 Leaf FFJ6:80 HHB
LINALOL 5-5,900 Plant BML
LINALYL-ACETATE 4 Flower FFJ6:80
LINALYL-ACETATE Leaf tr FFJ6:80
LINALYL-ACETATE 50-505 Plant BML
LINARIN Plant HHB
MONARDEIN Flower HHB JBH
MYRCENE 30 Flower FFJ6:80
MYRCENE 25-40 Leaf FFJ6:80 HHB
MYRCENE 10-485 Plant BML
NARINGENIN Plant HHB
NARIRUTIN Plant HHB
NONANAL 5-50 Plant BML
OCIMENE Leaf HHB
OCT-1-ENE-3-OL 10 Flower FFJ6:80
OCT-1-ENE-3-OL 200-360 Leaf FFJ6:80 HHB
OCTAN-3-OL Flower FFJ6:80
OCTAN-3-OL Leaf tr FFJ6:80
P-CYMENE 405 Flower FFJ6:80
P-CYMENE 140-270 Leaf FFJ6:80 HHB
P-CYMENE 120-1,850 Plant BML
PELARGONIN Flower JBH

PHAEOPHYTIN Leaf HHB
PRUNIN Plant HHB
PULEGONE 10-120 Plant BML
ROSMARINIC-ACID 18,000 Flower APF48:103
SABINENE 18 Flower FFJ6:80
SABINENE 100-180 Leaf FFJ6:80 HHB
SABINENE 35-1,230 Plant BML
T-SABINENE-HYDRATE 4 Flower FFJ6:80
T-SABINENE-HYDRATE Leaf tr FFJ6:80
TERPINOLENE 4 Flower FFJ6:80
TERPINOLENE 10-15 Leaf FFJ6:80 HHB
TERPINOLENE 15-160 Plant BML
THYMOL 5-50 Plant BML
THYMOL-METHYL-ETHER 50-520 Plant BML
URSOLIC-ACID Leaf HHB

Moldavica parviflora (NUTT.) BRITT.
"Small-Flowered Moldavica"
Unaggregated A

BICYCLOGERMACRENE 15 Plant BML
CARYOPHYLLENE 50 Plant BML
EO 1,000 Plant BML
GERMACRENE-D 310 Plant BML
MYRCENE 160 Plant BML

Monarda fistulosa L.
"Wild Bergamot"

1,8-CINEOLE 3-5,006 Plant BML
1-OCTEN-3-OL 1-2,294 Plant BML
3,4-EPOXY-GAMMA-TERPINENE 3-186 Plant BML
3-OCTANOL 46-1,655 Plant BML
3-OCTANONE 1-93 Plant BML
3-OCTYL-ACETATE 21-211 Plant BML
ACETIC-ACID Essential Oil GEO
ALPHA-PINENE 3-1,823 Plant BML
ALPHA-TERPINENE 3-3,193 Plant BML
ALPHA-TERPINEOL 9-3,314 Plant BML
ALPHA-THUJENE 3-2,480 Plant BML
BETA-BOURBONENE 1-62 Plant BML
BETA-PINENE 3-93 Plant BML
BORNYL-ACETATE 1-1,085 Plant BML
CAMPHENE 1-856 Plant BML
CARVACROL 93-19,902 Plant BML
CARVACRYL-METHYL-ETHER 6-6,076 Plant BML
CIS-3-HEXENOL 1-31 Plant BML
CIS-SABINENE-HYDRATE 1-527 Plant BML
COPAENE 1-124 Plant BML
CUMINALDEHYDE 1-62 Plant BML
DELTA-3-CARENE 1-93 Plant BML
DIHYDROCARVYL-ACETATE 3-301 Plant BML
DIHYDROCUMINYL-ALCOHOL Essential Oil GEO
DIHYDROXYTHYMOQUINONE Essential Oil GEO
EO 3,000-31,000 Plant GEO
FORMALDEHYDE Essential Oil HHB
GAMMA-TERPINENE 3-11,377 Plant BML
GERANIAL 15-837 Plant BML
GERANIOL 1-29,450 Plant BML
GERMACRENE-D 9-1,891 Plant BML
ISOMENTHONE 12-146 Plant BML
ISOVALERALDEHYDE Essential Oil HHB
LIMONENE 3-589 Plant BML
LINALOL 15-341 Plant BML
MYRCENE 3-1,860 Plant BML
N-BUTYRALDEHYDE Essential Oil HHB
NERAL 24-589 Plant BML
NEROL 15-310 Plant BML
P-CYMEN-8-OL 1-124 Plant BML
P-CYMENE 3-15,810 Plant BML
PERILLYL-ALCOHOL Essential Oil HHB

PULEGONE 12-257 Plant BML
ROSMARINIC-ACID 20,000 Shoot APF48:103
SABINENE 1-93 Plant BML
THYMOHYDROQUINONE Essential Oil HHB
THYMOL 1-20,850 Plant BML
THYMOQUINONE Essential Oil HHB
TRANS-SABINENE-HYDRATE 1-217 Plant BML
VALERIC-ACID Essential Oil GEO

Monarda lindheimeri ENGL. & GRAY
"Lindheimer's Monarda"
Unaggregated A

1,8-CINEOLE 478 Plant BML
3-OCTANOL 93 Plant BML
3-OCTYL-ACETATE 10 Plant BML
ALPHA-PINENE 211 Plant BML
ALPHA-TERPINEOL 31 Plant BML
BORNYL-ACETATE 62 Plant BML
CAMPHENE 21 Plant BML
CARVACROL 7,024 Plant BML
DIHYDROCARVEOL-ACETATE 16 Plant BML
EO 10,000 Plant ASSUMED
GAMMA-TERPINENE 26 Plant BML
ISOMENTHONE 18 Plant BML
LIMONENE 94 Plant BML
MYRCENE 16 Plant BML
P-CYMENE 1,315 Plant BML

Monarda media WILLD.
"Mean Monarda"

1,8-CINEOLE 807-1,351 Plant BML
1-OCTEN-3-OL 175 Plant BML
3-OCTANOL 28-292 Plant BML
3-OCTYL-ACETATE 50 Plant BML
ALPHA-PINENE 137-434 Plant BML
ALPHA-TERPINEOL 35-196 Plant BML
BORNYL-ACETATE 161-176 Plant BML
CAMPHENE 15-231 Plant BML
CARVACROL 2,736 Plant BML
CIS-OCIMENE 63 Plant BML
DIHYDROCARVEOL-ACETATE 50 Plant BML
EO 7,000 Plant BML ASSUMED
GAMMA-TERPINENE 46 Plant BML
GERMACRENE-D 392 Plant BML
LIMONENE 236-1,484 Plant BML
LINALOL 22-63 Plant BML
LINALYL-ACETATE 69 Plant BML
MYRCENE 140 Plant BML
P-CYMENE 14-528 Plant BML
SABINENE 1,442 Plant BML
THYMOL 1,292 Plant BML
TRANS-OCIMENE 336 Plant BML

Monarda punctata L.
"Horsemint"

1,8-CINEOLE 373-6,864 Plant BML
2-METHYL-2-HEPTANONE Plant HHB
3-OCTANOL 25-801 Plant BML
3-OCTYL-ACETATE 34-825 Plant BML
ALPHA-PINENE 40-1,974 Plant BML
ALPHA-TERPINENE 550-1,740 Plant BML
ALPHA-TERPINEOL 8-4,518 Plant BML
ALPHA-THUJENE 220-720 Plant BML
BETA-PINENE 10-990 Plant BML
BORNYL-ACETATE 30-3,399 Plant BML HHB
CAMPHENE 10-1,080 Plant BML HHB
CARVACROL 4,169-12,507 Plant BML
DIHYDROCARVYL-ACETATE 32-4,518 Plant BML
EO 10,000-30,000 Plant BML HHB
FAT 313,000 Seed CRC
FORMALDEHYDE Essential Oil HHB
GAMMA-TERPINENE 11-5,610 Plant BML
GERANYL-FORMIATE 1,020-3,060 Plant HHB
HEPTANAL 480-1,440 Essential Oil HHB
ISOMENTHONE 24-243 Plant BML
ISOVALERALDEHYDE Plant BML
LIMONENE 50-2,874 Plant BML
LINALOL Plant BML
LINALYL-ACETATE 49-378 Plant BML
NERAL Plant HHB
NERYL-FORMIATE Plant HHB
NONANAL Plant HHB
NONYL-ACETATE Plant HHB
P-CYMENE 65-3,384 Plant BML
PROTEIN 213,000 Seed CRC
PULEGONE 10-267 Plant BML
SABINENE Plant BML
TERPINEN-4-OL 40-150 Plant BML
THYMOL 2,000-27,633 Plant BML JBH
THYMOL-METHYL-ETHER Essential Oil HHB
THYMOLHYDROQUINONE Essential Oil HHB
UNDECANAL Plant HHB

Monarda russeliana NUTT. ex SIMS
"Russel's Monarda"
Unaggregated A

1,8-CINEOLE 325 Plant BML
3-OCTANOL 1,498 Plant BML
3-OCTYL-ACETATE 461 Plant BML
ALPHA-PINENE 14 Plant BML
ALPHA-TERPINEOL 1,802 Plant BML
CARVACROL 528 Plant BML
EO 10,000 Plant ASSUMED
GERANIAL 470 Plant BML
ISOMENTHONE 207 Plant BML
LIMONENE 104 Plant BML
LINALOL 150 Plant BML
LINALYL-ACETATE 557 Plant BML
NERAL 54 Plant BML
P-CYMENE 923 Plant BML
PULEGONE 499 Plant BML
THYMOL 1,752 Plant BML

Nepeta cataria L.
"Catnip"

(1R,5R,8S,9S)-DEOXYLOGANIC-ACID Plant FNF
(E)-BETA-FARNESENE 16-260 Plant JEO5:159
3-HEXENYL-BENZOATE 4-39 Plant JEO5:159
3-HEXENYL-ESTER 18-155 Plant JEO5:159
5,9-DEHYDRONEPETALACTONE Plant 411/
5-EPIDEOXYLOGANIC-ACID Plant 411/
7-DEOXYLOGANIC-ACID Plant 411/
ALPHA-HUMULENE 18-170 Plant JEO5:159
ALUMINUM 250 Plant PED
BETA-ELEMENE 4-155 Plant JEO5:159
CALCIUM 6,160 Plant PED
CAMPHOR 16-80 Plant HHB
CARVACROL Plant 411/
CARYOPHYLLENE 40-3,200 Plant HHB JEO5:159
CARYOPHYLLENE-OXIDE 285-2,365 Plant JEO5:159
CHROMIUM 27 Plant PED
CIS-OCIMENE 10-50 Plant FNF
CITRAL Plant FNF
CITRONELLAL Plant 411/
COBALT 118 Plant PED
DIHYDRONEPETALACTONE 6-30 Plant HHB
DIHYDRONEPETALACTONE-N3 34-260 Plant JEO5:159
DIMETHYL-3,7-OXA-1-BICYCLO(3,3,0)OCT-2-ENE 15-90 Plant JEO5:159
EO 2,000-13,000 Plant JEO5:159
EPINEPETALACTONE 10-4,000 Plant FNF
FAT 212,000 Seed FNF
GERANIOL Plant 411/
HUMULENE-OXIDE-I 22-210 Plant JEO5:159
IRON 1,380 Plant PED
ISODIHYDRONEPETALACTONE Plant HHB
LINOLEIC-ACID 18% Seed Oil FNF
LINOLENIC-ACID 57% Seed Oil FNF
MAGNESIUM 2,070 Plant PED
MANGANESE 374 Plant PED
METHYL-NEPETALACTONE Plant HHB
MYRCENE 8-40 Plant FNF
NEPETALACTONE 25-9,900 Plant HHB JBH JEO5:159
NEPETALACTONE-N1 25-7,400 Plant JEO5:159
NEPETALIC-ACID Plant 411/
NEPETALIC-ANHYDRIDE Plant FNF
NEPETARIASIDE Plant FNF

NEPETOGLUCOSYLESTER Plant FNF
NEPETOL Plant FNF
NEROL Plant 411/
OLEIC-ACID 12% Seed Oil FNF
PHOSPHORUS 2,410 Plant PED
PIPERITONE 50-195 Plant JEO5:159
POTASSIUM 23,500 Plant PED
PROTEIN 98,000 Plant PED
PROTEIN 184,000 Seed FNF
PULEGONE Plant 411/
ROSMARINIC-ACID 2,000 Plant FT62:166
SELENIUM 123 Plant PED 1/
SILICON Plant tr PED
SODIUM Plant tr PED
THYMOL Plant 411/
THYMOL-METHYL-ETHER 8-115 Plant JEO5:159
TIN 11 Plant PED
TRANS-CIS-NEPETONIC-ACID-METHYL-ESTER Plant FNF
TRANS-OCIMENE 4-20 Plant FNF
TRANS-TRANS-NEPETONIC-ACID-METHYL-ESTER Plant FNF
ZINC Plant tr PED

Nepeta racemosa LAM.
"Catmint"
Unaggregated A

(E)-2-HEXENAL 3 Shoot JEO5:215
(Z)-3-HEXENOL Shoot JEO5:215
(Z)-BETA-OCIMENE Shoot JEO5:215
(Z)-GERANYL-ACETONE Shoot JEO5:215
1,8-CINEOLE 140 Shoot JEO5:215
1-OCTEN-E-OL Shoot JEO5:215
2-BUTANONE 0.7 Shoot JEO5:215
3-METHYL-CYCLOHEPTANONE Shoot JEO5:215
4-ALPHA,ALPHA-7-ALPHA,7-ALPHABETA-NEPETALACTONE 6,400 Shoot JEO5:215
4-ALPHA,BETA-7-ALPHA,7-ALPHABETA-NEPETALACTONE 55 Shoot JEO5:215
5-METHYL-3-HEPTANONE <0.1 Shoot JEO5:215
6-METHYL-3-HEPTANOL Shoot JEO5:215
6-METHYL-5-HEPTEN-2-ONE Shoot JEO5:215
ACETOEUGENOL Shoot JEO5:215
ACETONE 1.5 Shoot JEO5:215
ACETOPHENONE Shoot JEO5:215
ALPHA-COPAENE Shoot JEO5:215
ALPHA-HUMULENE Shoot JEO5:215
ALPHA-MUUROLENE Shoot JEO5:215
ALPHA-PHELLANDRENE Shoot JEO5:215
ALPHA-PINENE 10 Shoot JEO5:215
ALPHA-TERPINENE Shoot JEO5:215
ALPHA-TERPINEOL Shoot JEO5:215
ALPHA-TERPINYL-ACETATE Shoot JEO5:215
AMYL-FURAN Shoot JEO5:215
BETA-BISABOLENE Shoot JEO5:215
BETA-BOURBONENE Shoot JEO5:215
BETA-CARYOPHYLLENE Shoot JEO5:215
BETA-CUBEBENE Shoot JEO5:215
BETA-IONONE Shoot JEO5:215
BETA-PINENE 43 Shoot JEO5:215
BICYCLO-[4.2.0]OCTA-1,3,5-TRIENE Shoot JEO5:215
CALAMENENE Shoot JEO5:215
CAMPHENE Shoot JEO5:215
CAMPHOR Shoot JEO5:215
CARVACROL Shoot JEO5:215
CIS-SABINENE-HYDRATE Shoot JEO5:215
CUMIN-ALCOHOL Shoot JEO5:215
CUMINALDEHYDE Shoot JEO5:215

CYCLOHEXANE 3 Shoot JEO5·215
DECANAL Shoot JEO5:215
DECANE Shoot JEO5:215
DELTA-CADINENE Shoot JEO5:215
DICHLOROBENZENE Shoot JEO5:215
EO 7,000 Shoot JEO5:215
GAMMA-MUUROLENE Shoot JEO5:215
GAMMA-TERPINENE 0.7 Shoot JEO5:215
GERANIOL Shoot JEO5:215
GERANYL-ACETATE Shoot JEO5:215
GERMACRENE-D Shoot JEO5:215
HEXAHYDROFARNESYL-ACETONE Shoot JEO5:215
HEXANAL Shoot JEO5:215
ISOAMYL-ACETATE <0.1 Shoot JEO5:215
LIMONENE Shoot JEO5:215
LINALOL Shoot JEO5:215
METHYL-CYCLOHEXYL-KETONE Shoot JEO5:215
METHYL-HYDROCINNAMATE Shoot JEO5:215
METHYL-SALICYLATE Shoot JEO5:215
MYRCENE Shoot JEO5:215
MYRTENAL Shoot JEO5:215
MYRTENOL Shoot JEO5:215
NAPHTHALENE Shoot JEO5:215
NONANE Shoot JEO5:215
NONANOIC-ACID Shoot JEO5:215
OCTANE Shoot JEO5:215
P-CYMENE 14 Shoot JEO5:215
PATCHOULANE Shoot JEO5:215
PIPERITONE-OXIDE-I Shoot JEO5:215
PSEUDOLIMONENE Shoot JEO5:215
PULEGONE Shoot JEO5:215
SABINENE Shoot JEO5:215
SPATHULENOL Shoot JEO5:215
T-CADINOL Shoot JEO5:215
TERPINEN-4-OL Shoot JEO5:215
THYMOL Shoot JEO5:215
TRANS-LINALOL-OXIDE Shoot JEO5:215
TRANS-PINOCARVEOL Shoot JEO5:215
TRANS-SABINENE-HYDRATE Shoot JEO5:215
UNDECANE Shoot JEO5:215

Nepeta racemosa LAM.
"Catmint"
Unaggregated B

(E)-3-HEXENAL 8 Shoot JEO5:215
(Z)-3-HEXENOL 2 Shoot JEO5:215
(Z)-BETA-OCIMENE 33 Shoot JEO5:215
(Z)-GERANYL-ACETONE 11 Shoot JEO5:215
1,8-CINEOLE 1,150 Shoot JEO5:215
1-OCTEN-E-OL 42 Shoot JEO5:215
2-BUTANONE Shoot JEO5:215
4-ALPHA,ALPHA-7-ALPHA,7-ALPHABETA-NEPETALACTONE 3,455 Shoot JEO5:215
4-ALPHA,BETA-7-ALPHA,7-ALPHABETA-NEPETALACTONE 140 Shoot JEO5:215
5-METHYL-3-HEPTANONE 52 Shoot JEO5:215
6-METHYL-3-HEPTANOL 13 Shoot JEO5:215
6-METHYL-5-HEPTEN-2-ONE 6 Shoot JEO5:215
ACETOEUGENOL 43 Shoot JEO5:215
ACETONE 2 Shoot JEO5:215
ACETOPHENONE 22 Shoot JEO5:215
ALPHA-COPAENE 11 Shoot JEO5:215
ALPHA-HUMULENE 16 Shoot JEO5:215
ALPHA-MUUROLENE 21 Shoot JEO5:215
ALPHA-PHELLANDRENE <0.1 Shoot JEO5:215
ALPHA-PINENE 70 Shoot JEO5:215
ALPHA-TERPINENE 13 Shoot JEO5:215
ALPHA-TERPINEOL 55 Shoot JEO5:215
ALPHA-TERPINYL-ACETATE 310 Shoot JEO5:215
AMYL-FURAN 3 Shoot JEO5:215
BETA-BISABOLENE 9 Shoot JEO5:215
BETA-BOURBONENE 50 Shoot JEO5:215
BETA-CARYOPHYLLENE 240 Shoot JEO5:215
BETA-CUBEBENE 2 Shoot JEO5:215
BETA-IONONE 13 Shoot JEO5:215
BETA-PINENE 255 Shoot JEO5:215
BICYCLO-[4.2.0]OCTA-1,3,5-TRIENE 21 Shoot JEO5:215
CALAMENENE 85 Shoot JEO5:215
CAMPHENE 3 Shoot JEO5:215
CAMPHOR 2 Shoot JEO5:215
CARVACROL 170 Shoot JEO5:215
CIS-SABINENE-HYDRATE 3 Shoot JEO5:215
CUMIN-ALCOHOL 26 Shoot JEO5:215
CUMINALDEHYDE 34 Shoot JEO5:215
CYCLOHEXANE Shoot JEO5:215

DECANAL 9 Shoot JEO5:215
DECANE 2 Shoot JEO5:215
DELTA-CADINENE 80 Shoot JEO5:215
DICHLOROBENZENE 15 Shoot JEO5:215
EO 11,000 Shoot JEO5:215
ETHANOL 1 Shoot JEO5:215
GAMMA-MUUROLENE 2 Shoot JEO5:215
GAMMA-TERPINENE 44 Shoot JEO5:215
GERANIOL 70 Shoot JEO5:215
GERANYL-ACETATE 145 Shoot JEO5:215
GERMACRENE-D 100 Shoot JEO5:215
HEXAHYDROFARNESYL-ACETONE 28 Shoot JEO5:215
HEXANAL 4 Shoot JEO5:215
ISOAMYL-ACETATE 19 Shoot JEO5:215
LAURIC-ACID 8 Shoot JEO5:215
LIMONENE 90 Shoot JEO5:215
LINALOL 21 Shoot JEO5:215
METHYL-CYCLOHEPTANONE 19 Shoot JEO5:215
METHYL-CYCLOHEXYL-KETONE 11 Shoot JEO5:215
METHYL-HYDROCINNAMATE 70 Shoot JEO5:215
METHYL-SALICYLATE 6 Shoot JEO5:215
MYRCENE 24 Shoot JEO5:215
MYRTENAL 32 Shoot JEO5:215
MYRTENOL 19 Shoot JEO5:215
NAPHTHALENE 46 Shoot JEO5:215
NONANE 7 Shoot JEO5:215
NONANOIC-ACID 34 Shoot JEO5:215
OCTANE <0.1 Shoot JEO5:215
P-CYMENE 155 Shoot JEO5:215
PATCHOULANE 265 Shoot JEO5:215
PIPERITONE-OXIDE-I 30 Shoot JEO5:215
PSEUDOLIMONENE 28 Shoot JEO5:215
PULEGONE 48 Shoot JEO5:215
SABINENE 1 Shoot JEO5:215
SPATHULENOL 170 Shoot JEO5:215
T-CADINOL 60 Shoot JEO5:215
TERPINEN-4-OL 45 Shoot JEO5:215
THYMOL 305 Shoot JEO5:215
TRANS-LINALOL-OXIDE 4 Shoot JEO5:215
TRANS-PINOCARVEOL 31 Shoot JEO5:215
TRANS-SABINENE-HYDRATE 22 Shoot JEO5:215
UNDECANE 4 Shoot JEO5:215

Ocimum basilicum L.
"Basil"

(-)-LINALOL 30-300 Plant JAD NAP
(E)-BETA-FARNESENE 7 Plant DIE
(E)-BETA-OCIMENE 1-435 Plant DIE JAD JBH
(Z)-BETA-FARNESENE 16 Plant DIE
1,8-CINEOLE 0-776 Plant JAD NAP
1-EPI-BICYCLOSESQUIPHELLANDRENE Plant JSG
1-OCTEN-3-OL Plant FNF
2-EPI-ALPHA-CEDRENE Plant FNF
3-OCTANONE Plant FNF
5,10(15)-CADINEN-4-OL 1 Plant DIE
ACETIC-ACID Essential Oil NAP
AESCULETIN Leaf PMP24:193
AESCULIN Leaf NAP
ALANINE 7,470 Leaf USA
ALPHA-AMORPHENE Plant FNF
ALPHA-BERGAMOTENE 75 Plant DIE
ALPHA-BISABOLOL Essential Oil NAP
ALPHA-BISABOLOL 90 Plant DIE
ALPHA-BULNESENE 6.5 Plant DIE
ALPHA-CADINENE Plant JAD
ALPHA-CEDRENE Plant NAP
ALPHA-COPAENE 0.8-20 Plant DIE JAD
ALPHA-CUBEBENE Plant FNF
ALPHA-FARNESENE Plant FNF
ALPHA-FENCHENE Essential Oil NAP
ALPHA-GUAIENE Plant FNF
ALPHA-HUMULENE 16-313 Plant DIE JAD NAP
ALPHA-MUUROLENE Plant JAD
ALPHA-P-DIMETHYLSTYRENE Plant FNF
ALPHA-PINENE 2-180 Plant DIE JAD NAP
ALPHA-SANTALENE Essential Oil NAP
ALPHA-SELINENE Plant FNF
ALPHA-TERPINENE 1-10 Plant JAD
ALPHA-TERPINEOL 36-239 Plant JAD NAP
ALPHA-TERPINYL-ACETATE Plant FNF
ALPHA-THUJONE Plant FNF
ANETHOLE Plant NAP
APIGENIN Plant PAS
ARGININE 6,620 Leaf USA
ASCORBIC-ACID 27-612 Leaf CRC USA
ASPARTIC-ACID 16,960 Leaf USA
BENZYL-ACETATE 17-163 Leaf FNF

BENZYL-ALCOHOL Essential Oil NAP
BETA-BISABOLENE 6.5 Plant DIE
BETA-BOURBONENE Plant FNF
BETA-CADINENE Plant NAP
BETA-CADINOL Essential Oil NAP
BETA-CAROTENE 4-333 Leaf CRC USA
BETA-CARYOPHYLLENE 0-377 Plant JAD NAP
BETA-CEDRENE Plant JAD
BETA-CUBEBENE Plant FNF
BETA-CYMENE 5-36 Plant JAD NAP
BETA-ELEMENE 12 Plant DIE
BETA-MYRCENE Plant NAP
BETA-OCIMENE 1-435 Plant FNF
BETA-PINENE 3-160 Plant DIE JAD NAP
BETA-SANTALENE Essential Oil NAP
BETA-SELINENE Plant FNF
BETA-SITOSTEROL 1,051 Flower NAP
BETA-SITOSTEROL 896-1,705 Leaf NAP
BETA-SITOSTEROL 408 Root NAP
BETA-SITOSTEROL 230 Sprout Seedling NAP
BETA-SITOSTEROL 230 Stem NAP
BETA-THUJONE Plant FNF
BORNEOL 3 Plant DIE
BORNEOL-ACETATE 90-900 Plant NAP
BORON 18-31 Plant BOB
CAFFEIC-ACID 19,000 Leaf NAP
CAFFEIC-ACID-N-BUTYL-ESTER 252 Leaf NAP
CALAMENE Plant FNF
CALCIUM 20,148-22,112 Leaf CRC USA
CAMPHENE 0-400 Plant JAD
CAMPHOR 2-31 Plant JAD NAP
CARBOHYDRATES 70,000-610,000 Leaf CRC USA
CARVONE Plant PAS
CARYOPHYLLENE 18-3,196 Plant JAD
CARYOPHYLLENE-OXIDE Essential Oil NAP
CHAVICOL 3 Plant DIE
CHAVICOL-METHYL-ETHER 500-5,658 Plant JAD NAP
CINNAMIC-ACID-METHYL-ESTER Essential Oil NAP
CIS-3-HEXENOL Plant FNF
CIS-ALLOOCIMENE Plant FNF
CIS-ANETHOLE Plant FNF
CIS-CINNAMIC-ACID-METHYL-ESTER 67-900 Plant JAD NAP
CIS-LIMONENE 30-933 Plant JAD NAP
CIS-OCIMENE 3-252 Plant JAD
CIS-SABINENE-HYDRATE Plant FNF

CITRAL 560-7,000 Plant JAD NAP
CITRONELLOL 0.2-2,419 Plant DIE JAD NAP
COPPER 14 Leaf USA
CYCLOSATIVENE Plant FNF
CYSTINE 1,590 Leaf USA
D-ARABINOSE Seed RAA
D-GALACTOSE Seed RAA
D-GALACTURONIC-ACID Seed RAA
D-GLUCOSE Seed RAA
D-MANNOSE Seed RAA
D-MANNURONIC-ACID Seed RAA
DELTA-CADINENE 5 Plant DIE
DELTA-GUAIENE Plant FNF
ELEMOL Essential Oil NAP
EO 1,500-10,000 Plant CRC JAD
ERIODICTYOL Leaf PMP24:193
ERIODICTYOL-7-O-GLUCOSIDE Leaf PMP24:193
ESTRAGOLE 35-9,000 Plant JAD NAP
EUGENOL 14-8,575 Leaf DIE JAD NAP
EUGENOL-METHYL-ETHER 375-2,500 Plant JAD NAP
FARNESOL Plant FNF
FAT 35,752-43,680 Leaf CRC USA
FENCHONE 30 Plant DIE
FENCHYL-ACETATE 1-60 Plant FNF
FENCHYL-ALCOHOL 9-951 Plant CRC DIE
FIBER 145,790-154,290 Leaf CRC USA
FURFURAL Plant FNF
GAMMA-CADINENE Plant NAP
GAMMA-ELEMENE 15 Plant DIE
GAMMA-GURJUNENE Plant FNF
GAMMA-MUUROLENE Plant NAP
GAMMA-TERPINENE 1-10 Plant JAD
GERANIAL 5-3,750 Plant DIE JAD NAP
GERANIOL 1-1,000 Plant JAD NAP
GERANYL-ACETATE 11-84 Leaf JAD
GERMACRENE-D 35 Plant DIE
GLUTAMIC-ACID 15,650 Leaf USA
GLYCINE 6,900 Leaf USA
HISTIDINE 2,870 Leaf USA
HUMULENE Essential Oil NAP
HUMULENE-EPOXIDE 0.2 Plant DIE
HYDROXY-BENZOIC-ACID-4-BETA-D-GLUCOSIDE Leaf NAP
IRON 362-478 Leaf CRC USA
ISOCARYOPHYLLENE Plant FNF
ISOEUGENOL 8-95 Plant FNF

ISOEUGENOL-METHYL-ETHER Plant PAS
ISOLEUCINE 5,880 Leaf USA
ISOQUERCITRIN Leaf NAP
JUVOCIMENE-I 0.007-0.05 Plant JAD NAP
JUVOCIMENE-II 0.007-0.05 Plant JAD NAP
KAEMPFEROL Leaf NAP
KAEMPFEROL3-O-BETA-D-RUTINOSIDE Leaf NAP
L-RHAMNOSE Seed RAA
LEDENE Plant FNF
LEUCINE 10,780 Leaf USA
LIMONENE 2-934 Plant JAD NAP
LINALOL 5-8,730 Plant BML 411/
LINALYL-ACETATE 15-240 Plant JAD
LUTEOLIN Plant PAS
LYSINE 6,180 Leaf USA
MAGNESIUM 4,100-4,340 Leaf USA
MANGANESE 32 Leaf USA
MENTHOL 4-32 Plant JAD NAP
MENTHONE 1 Plant JAD NAP
METHIONINE 2,020 Leaf USA
METHYL-CHAVICOL 238-8,780 Plant CRC DIE
METHYL-CINNAMATE 1-2,800 Plant CRC DIE
METHYL-EUGENOL 13-1,400 Plant CRC DIE
METHYL-THYMOL 0.2 Plant DIE
MUCILAGE 93,000 Seed JAD NAP
MYRCENE 2-80 Leaf DIE JAD
NERAL 5 Plant DIE
NEROL 2.5-300 Plant DIE JAD NAP
NEROLIDIOL 5 Plant DIE
NEROLIDOL Essential Oil JAD
NIACIN 8-69 Leaf CRC USA
OCTANOL Plant FNF
OLEANOLIC-ACID 1,300 Flower NAP
ORIENTIN Plant PAS
P-COUMARIC-ACID 760 Leaf NAP
P-CYMENE 1-16 Plant JAD
P-METHOXYCINNAMALDEHYDE 1,000-4,000 Essential Oil NAP
PHELLANDRENE Plant 411/
PHENYLETHYL-ALCOHOL 18-136 Plant FNF
PHOSPHORUS 4,632-5,168 Leaf CRC USA
PHYTOSTEROLS 1,060 Leaf USA
PLANTEOSE Seed NAP
POTASSIUM 32,321-42,900 Leaf CRC USA
PROLINE 5,880 Leaf USA
PROPIONIC-ACID Essential Oil NAP

PROTEIN 118,550-169,850 Leaf CRC USA
QUERCETIN Leaf NAP
QUERCETIN-3-O-DIGLUCOSIDE Leaf NAP
RIBOFLAVIN 3-4 Leaf CRC USA
ROSMARINIC-ACID 1,000-19,000 Plant JAD NAP APF48:103
RUTIN Leaf NAP
SABINENE 8 Plant DIE
SAFROLE 60-400 Plant JAD JBH NAP
SALICYLIC-ACID-2-BETA-D-GLUCOSIDE Leaf NAP
SAMBULENE Leaf NAP
SERINE 5,610 Leaf USA
SESQUITHUJENE Plant FNF
SODIUM 294-386 Plant FNF
SYRINGIC-ACID-4-BETA-D-GLUCOSIDE Leaf NAP
SYRINGOYL-GLUCOSE Leaf NAP
T-CADINOL 0.5 Plant DIE
TERPINEN-4-OL 2-120 Plant JAD NAP
TERPINOLENE 1-22 Plant FNF
THIAMIN 1 Leaf USA
THREONINE 5,880 Leaf USA
THYMOL 1,415 Leaf NAP
TRANS-ALLOOCIMENE Plant FNF
TRANS-ANETHOLE 11-74 Plant JAD NAP
TRANS-CINNAMIC-ACID 350-7,000 Plant JAD NAP
TRANS-CINNAMIC-ACID-METHYL-ESTER 350-7,000 Plant JAD NAP
TRANS-OCIMENE 8-161 Plant JAD NAP
TRANS-SABINENE-HYDRATE Plant FNF
TRICYCLENE Essential Oil NAP
TRYPTOPHAN 2,210 Leaf USA
TYROSINE 4,320 Leaf USA
UNDECYLALDEHYDE Plant FNF
URSOLIC-ACID 1,740 Flower NAP
URSOLIC-ACID 413-1,143 Leaf NAP
URSOLIC-ACID 63 Sprout Seedling NAP
URSOLIC-ACID 845 Stem NAP
VALERIC-ACID Essential Oil NAP
VALINE 7,170 Leaf USA
VANILLIC-ACID-4-BETA-D-GLUCOSIDE Leaf NAP
VICENIN-2 Leaf PMP24:193
XANTHOMICROL 350 Leaf NAP
XI-BULGARENE Plant FNF
ZINC 5-6 Leaf USA

Ocimum basilicum L.
"Basil"
Unaggregated A

(3)-BETA-FARNESENE 7 Plant DIE
(E)-ALPHA-BERGAMOTENE 75 Plant DIE
(Z)-BETA-FARNESENE 16 Plant DIE
1,8-CINEOLE 135 Plant DIE
5,10(15)-CADINEN-4-OL 1 Plant DIE
ALPHA-BISABOLENE 90 Plant DIE
ALPHA-BULNESENE 6.5 Plant DIE
ALPHA-COPAENE 0.8 Plant DIE
ALPHA-HUMULENE 16 Plant DIE
BETA-BISABOLENE 6.5 Plant DIE
BETA-CARYOPHYLLENE 35 Plant DIE
BETAPELEMENE 12 Plant DIE
BORNEOL 3 Plant DIE
CHAVICOL 3 Plant DIE
CITRONELLOL 0.2 Plant DIE
DELTA-CADINENE 5 Plant DIE
EO 2,500 Plant DIE
EUGENOL 14 Plant DIE
FENCHONE 30 Plant DIE
FENCHYL-ALCOHOL 7 Plant DIE
GAMMA-ELEMENE 15 Plant DIE
GERANIAL 5 Plant DIE
GERANIOL 3 Plant DIE
GERMACRENE-D 35 Plant DIE
HUMULENE-EPOXIDE 0.2 Plant DIE
LIMONENE 20 Plant DIE
T-CADINOL 0.5 Plant DIE

Ocimum canum SIMS "Hoary Basil"

1,8-CINEOLE Shoot MPI
ALPHA-PINENE Shoot MPI
BETULINIC-ACID Shoot MPI
CAMPHENE Shoot MPI
CAMPHOR 1,900-4,550 Shoot HHB
CARYOPHYLLENE 130-270 Shoot HHB MPI
CITRAL Shoot HHB
D-GALACTOSE Seed MPI
D-GALACTURONIC-ACID Seed MPI
D-GLUCOSE Seed MPI
D-MANNOSE Seed MPI
D-MANNURONIC-ACID Seed MPI
D-XYLOSE Seed MPI
DELTA-CADINENE Shoot MPI
EO 3,800-6,500 Shoot HHB
EUGENOL Shoot MPI
EUGENYL-ACETATE Shoot MPI
FARNESENE Shoot MPI
GERANIOL Shoot MPI
GERANYL-ACETATE Shoot MPI
HUMULENE Shoot MPI
ISOBORNEOL-ACETATE Shoot MPI
L-ARABINOSE Seed MPI
L-RHAMNOSE Seed MPI
LINALOL 3,055-5,230 Shoot HHB MPI
METHYL-CHAVICOL Shoot MPI
METHYL-CINNAMATE Shoot HHB
METHYL-EUGENOL Shoot MPI
METHYL-HEPTANONE Shoot MPI
METHYL-NONYL-KETONE Shoot MPI
MYRCENE Shoot MPI
NEVADENSIN Shoot JBH MPI
OCIMIN Shoot MPI
PECTOLINARIGENIN-7-METHYL-ETHER Shoot MPI
TERPINEOL Shoot MPI
URSOLIC ACID Shoot MPI

Ocimum gratissimum L.
"Agbo, Shrubby Basil"

(Z)-ALPHA-TRANS-BERGAMOTENE 95-165 Flower JEO4:231
(Z)-ALPHA-TRANS-BERGAMOTENE 30-90 Leaf JEO4:231
(Z)-BETA-FARNESENE Flower JEO4:231
(Z)-BETA-FARNESENE Leaf JEO4:231
1,8-CINEOLE 10-65 Plant BML
2,5-DIMETHOXYBENZOIC-ACID Plant HHB
ACETIC-ACID Plant BML
ALLO-AROMADENDRENE 1-5 Plant BML
ALPHA-CARYOPHYLLENE Plant BML
ALPHA-COPAENE 100-170 Flower JEO4:231
ALPHA-COPAENE 50-95 Leaf JEO4:231
ALPHA-CUBEBENE 5-60 Plant BML
ALPHA-FENCHENE Plant BML
ALPHA-HUMULENE 1-9 Plant BML
ALPHA-P-DIMETHYL-STYRENE 15-125 Plant BML
ALPHA-PHELLANDRENE 10-125 Plant BML
ALPHA-PINENE 47-88 Flower JEO4:231
ALPHA-PINENE 23-43 Leaf JEO4:231
ALPHA-PINENE 15-155 Plant BML GEO
ALPHA-SANTALENE Plant BML
ALPHA-TERPINENE 60-370 Plant BML
ALPHA-TERPINEOL 15-145 Plant BML GEO
ALPHA-THUJENE 50-445 Plant BML
ARABINOSE Root HHB
BENZOIC-ACID Plant BML
BETA-BOURBONENE 95-145 Flower JEO4:231
BETA-BOURBONENE 40-70 Leaf JEO4:231
BETA-CARYOPHYLLENE 145-220 Flower JEO4:231
BETA-CARYOPHYLLENE 115-215 Leaf JEO4:231
BETA-CARYOPHYLLENE Plant BML
BETA-CUBEBENE Flower JEO4:231
BETA-CUBEBENE Leaf JEO4:231
BETA-ELEMENE 5-65 Plant BML
BETA-PHELLANDRENE 5-30 Plant BML
BETA-PINENE 10-37 Flower JEO4:231
BETA-PINENE 19-41 Leaf JEO4:231
BETA-PINENE 3-35 Plant BML
BETA-SANTALENE Plant BML
BETA-SELINENE 15-240 Plant BML
BORNEOL 1-9 Plant BML
CADINENE 20-600 Plant BML GEO
CAMPHENE 1-240 Plant BML

CAMPHOR 2-35 Plant BML
CARVACROL 5-190 Plant BML
CARYOPHYLLENE 5-125 Plant BML
CIS-OCIMENE 0-12 Plant BML
CITRAL Plant TRA
CLOVENE Plant BML
COPAENE 1-65 Plant BML
DELTA-3-CARENE 10-245 Plant BML
EO 13,400-16,700 Flower JEO4:231
EO 12,000-14,700 Leaf JEO4:231
EO 1,000-6,000 Plant GEO (HHB)
EO 2,700 Seed WOI
EO 1,400 Stem JEO4:231
ETHYL-CINNAMATE Plant HHB
EUGENOL <0-5,340 Plant JEO4:231 GEO
EUGENOL 1,670 Seed WOI
FORMIC-ACID Plant BML
GAMMA-MUUROLENE 190-290 Flower JEO4:231
GAMMA-MUUROLENE 240-420 Leaf JEO4:231
GERANIOL 11,010-14,245 Flower JEO4:231
GERANIOL 10,335-13,455 Leaf JEO4:231
GRATISIMENE Plant JSG
GRATISSIMIN Plant HHB
HUMULENE 5-30 Plant BML
L-LINALOL 15-115 Plant GEO
LIMONENE 158-220 Flower JEO4:231
LIMONENE 185-335 Leaf JEO4:231
LIMONENE 10-105 Plant BML
LINALOL 46-140 Flower JEO4:231
LINALOL 25-45 Leaf JEO4:231
LINALOL 2-12 Plant BML
LONGIFOLENE 30-180 Plant BML
METHYL-EUGENOL 15-100 Plant BML
METHYL-ISOEUGENOL Plant BML
MYRCENE 8-10 Flower JEO4:231
MYRCENE 24-47 Leaf JEO4:231
MYRCENE 15-480 Plant BML GEO
NERAL 466-695 Flower JEO4:231
NERAL 150-210 Leaf JEO4:231
OCIMENE 120-720 Plant GEO
OCIMENE 400 Seed WOI
OCIMIC-ACID Root HHB
OCIMOL Root HHB
P-CYMENE 160-1,945 Plant BML
PHENOLS 230-3,300 Plant GEO

PROPIONIC-ACID Plant BML
SABINENE 10-60 Plant BML
SESQUITERPENES 150-900 Plant GEO
SPATHULENOL 112-150 Flower JEO4:231
SPATHULENOL 15-105 Leaf JEO4:231
TERPENES 160-960 Plant GEO
TERPINEN-4-OL 5-50 Plant BML
THYMOL Leaf JEO4:231
THYMOL 300-2,850 Plant BML GEO
THYMOL-METHYL-ETHER 4-130 Plant BML
TRANS-SABINENE-HYDRATE 5-60 Plant BML
VALERIC-ACID Plant BML

Ocimum gratissimum L.
"Agbo, Shrubby Basil"
Unaggregated A

(E)-OCIMENE 70 Shoot ZNA45:1073
(Z)-OCIMENE 765 Shoot ZNA45:1073
1,8-CINEOLE 690 Shoot ZNA45:1073
1-OCTEN-3-OL 32 Shoot ZNA45:1073
ALLOAROMADENDRENE 295 Shoot ZNA45:1073
ALPHA-PINENE 55 Shoot ZNA45:1073
ALPHA-TERPINENE 29 Shoot ZNA45:1073
ALPHA-TERPINEOL 65 Shoot ZNA45:1073
BENZALDEHYDE <0.1 Shoot ZNA45:1073
BETA-ELEMENE 425 Shoot ZNA45:1073
BETA-PINENE 90 Shoot ZNA45:1073
CAMPHENE 29 Shoot ZNA45:1073
CARYOPHYLLENE 1,155 Shoot ZNA45:1073
DELTA-ELEMENE 36 Shoot ZNA45:1073
EO 21,000 Shoot ZNA45:1073
ESTRAGOLE <0.1 Shoot ZNA45:1073
EUGENOL 4,045 Shoot ZNA45:1073
GAMMA-TERPINENE 15 Shoot ZNA45:1073
GERMACRENE-D 210 Shoot ZNA45:1073
HUMULENE 300 Shoot ZNA45:1073
LIMONENE 42 Shoot ZNA45:1073
LINALOL 275 Shoot ZNA45:1073
METHYL-CINNAMATE <0.1 Shoot ZNA45:1073
METHYL-EUGENOL 9,835 Shoot ZNA45:1073
MYRCENE 75 Shoot ZNA45:1073
MYRCENOL 40 Shoot ZNA45:1073
OCTAN-3-OL 17 Shoot ZNA45:1073
SABINENE 17 Shoot ZNA45:1073
TERPINEN-4-OL <0.1 Shoot ZNA45:1073
TERPINOLENE 23 Shoot ZNA45:1073
THUJENE <0.1 Shoot ZNA45:1073
TRICYCLENE <0.1 Shoot ZNA45:1073

Ocimum kilimandscharicum GUERKE "African Blue Basil, Kenyan Perennial Basil"

1,8-CINEOLE 1,130-1,960 Flower JEO4:125
1,8-CINEOLE 715-6,200 Shoot JEO4:125
ALPHA-HUMULENE 0-230 Flower JEO4:125
ALPHA-HUMULENE 8-47 Leaf JEO4:125
ALPHA-PINENE 55-105 Flower JEO4:125
ALPHA-PINENE 40-85 Leaf JEO4:125
ALPHA-TERPINEOL 80-150 Flower JEO4:125
ALPHA-TERPINEOL 50-115 Leaf JEO4:125
BETA-CARYOPHYLLENE 25-945 Flower JEO4:125
BETA-CARYOPHYLLENE 90-165 Leaf JEO4:125
BETA-PINENE 205-675 Flower JEO4:125
BETA-PINENE 260-485 Leaf JEO4:125
BORNYL-ACETATE 20-70 Flower JEO4:125
BORNYL-ACETATE 45-75 Leaf JEO4:125
CAMPHENE 240-400 Flower JEO4:125
CAMPHENE 105-205 Leaf JEO4:125
CAMPHOR 2,650-5,070 Flower JEO4:125
CAMPHOR <0.1-7,000 Shoot JEO4:125
EO 19,600-28,000 Flower JEO4:125
EO 7,700-11,200 Leaf JEO4:125
EO 5,000-10,000 Shoot JEO4:125
EPI-ALPHA-CADINOL 110-565 Flower JEO4:125
EPI-ALPHA-CADINOL 80-170 Leaf JEO4:125
EUGENOL 0-35 Flower JEO4:125
EUGENOL 0-3,000 Shoot JEO4:125
GAMMA-MUUROLENE 225-2,060 Flower JEO4:125
GAMMA-MUUROLENE 190-325 Leaf JEO4:125
GERMACRENE-B 125-520 Flower JEO4:125
GERMACRENE-B 90-185 Leaf JEO4:125
LIMONENE 570-980 Flower JEO4:125
LIMONENE 360-615 Leaf JEO4:125
LINALOL 10,380-18,130 Flower JEO4:125
LINALOL 2,970-5,075 Leaf JEO4:125
MYRCENE <0.1 Flower JEO4:125
MYRCENE 8-15 Leaf JEO4:125
TERPINEN-4-OL 75-200 Flower JEO4:125
TERPINEN-4-OL 65-160 Leaf JEO4:125
TERPINOLENE Plant JBH

Ocimum sanctum L.
"Holy Basil, Tulsi"

ALDEHYDES 900-1,750 Leaf FNF WOI
ALKALOIDS Leaf WOI
ANTISTAPHYLOCOAGULASE Seed WOI
ASCORBIC-ACID 830 Leaf WOI
ASH 2,000 Seed WOI
BETA-CAROTENE 25 Leaf WOI
CARVACROL 180-210 Leaf GEO
CINEOLE Leaf GEO
EO 6,000-7,000 Leaf GEO HHB
EUGENOL 4,200-4,970 Leaf GEO
EUGENOL-METHYL-ETHER 1,200-1,400 Leaf GEO
FAT 155,000 Seed CRC
GLYCOSIDES Leaf WOI
HEXOURONIC-ACID 272,000 Seed WOI
LINALOL Leaf GEO
LINOLEIC-ACID 102,455 Seed WOI
LINOLENIC-ACID 24,355 Seed WOI
METHYL-CHAVICOL Leaf GEO
MUCILAGE Seed WOI
OLEIC-ACID 13,950 Seed WOI
PALMITIC-ACID 10,700 Seed WOI
PENTOSES 389,000 Seed WOI
PHENOLS 3,000-5,320 Leaf WOI
PROTEIN 138,000 Seed WOI
SAPONINS Plant WOI
STEARIC-ACID 3,255 Seed WOI

Ocimum suave WILLD.
"Kenyan Tree Basil"

BETA-BISABOLENE 205 Shoot JNP44:308
BETA-CARYOPHYLLENE 110 Shoot JNP44:308
BETA-CUBEBENE 205 Shoot JNP44:308
BETA-PINENE 12 Shoot JNP44:308
CIS-BETA-OCIMENE 540 Shoot JNP44:308
EO 4,000 Shoot JNP44:308
EUGENOL 110-2,860 Shoot JNP44:308
LINALOL 20 Shoot JNP44:308
LINALYL-ACETATE Shoot JNP44:308
METHYL-EUGENOL 2,240 Shoot JNP44:308
OLEANOLIC-ACID Leaf JNP44:308
PHYTOL Leaf JNP44:308

Ocimum tenuiflorum L.
"Anise-Scented Basil"
Unaggregated A

(E)-ANETHOLE <0.1 Leaf JEO:459
ALPHA-HUMULENE 360 Leaf JEO:459
ALPHA-PINENE 45 Leaf JEO:459
BETA-CARYOPHYLLENE 2,185 Leaf JEO:459
BORNEOL 240 Leaf JEO:459
CAMPHENE 65 Leaf JEO:459
CAMPHOR 1,785 Leaf JEO:459
CARYOPHYLLENE-OXIDE 75 Leaf JEO:459
EO 41,000 Leaf JEO:459
LIMONENE 145 Leaf JEO:459
LINALOL 33 Leaf JEO:459
METHYL-CHAVICOL 35,950 Leaf JEO:459
METHYL-EUGENOL 50 Leaf JEO:459
MYRCENE 8 Leaf JEO:459
TERPINOLENE 55 Leaf JEO:459

Origanum creticum L.
"Dittany Of Crete"

CAMPHOR Plant HHB
CARVACROL Plant HHB
CEDROL Plant HHB
EO 14,000-30,000 Plant HHB
LINALOL Plant HHB
P-CYMOL Plant HHB
PINENE Plant HHB

Origanum majorana L.
"Marjoram"

1-ETHOXY-P-MENTH-2-ENE Plant BML
4-ETHOXY-P-MENTH-1-ENE Plant FNF
ALLO-AROMADENDRENE Plant BML
ALPHA-COPAENE Plant BML
ALPHA-HUMULENE Plant BML
ALPHA-HYDROXYDIHYDROCAFFEIC-ACID Plant HHB
ALPHA-PHELLANDRENE 18-103 Plant BML
ALPHA-PINENE 15-86 Plant BML
ALPHA-TERPINENE 183-1,050 Plant BML JBH
ALPHA-TERPINYL-ACETATE Plant BML
ALPHA-THUJENE 15-86 Plant BML
APIGENIN-7-GLUCOSIDE Plant 411/
ARBUTIN Leaf JBH
ASCORBIC-ACID 514-555 Plant USA
ASH 96,900-123,300 Plant USA WOI
BETA-CAROTENE 48-52 Plant USA
BETA-CARYOPHYLLENE Plant HHB
BETA-FARNESENE Plant BML
BETA-PHELLANDRENE 27-155 Plant BML
BETA-PINENE 6-34 Plant BML
BICYCLOGERMACRENE 12-69 Plant BML
BORON 28-48 Plant BOB
CADINENE Plant HHB
CAFFEIC-ACID Plant HHB
CALCIUM 17,925-23,625 Plant USA
CARBOHYDRATES 605,600-654,050 Plant USA
CARVACROL 1,092-6,261 Plant WOI
CARVONE 12-69 Plant BML
CARYOPHYLLENE 52-1,307 Plant BML WOI
CHAVICOL 138-791 Plant WOI
CIS-3-ETHOXY-P-MENTH-1-ENE Plant FNF
CIS-4-ETHOXY-THUJANE Plant BML
CIS-P-MENTH-2-EN-1-OL 60-344 Plant BML
CITRAL Plant HHB
COPPER 11 Plant USA
D-ALPHA-TERPINEOL 144-826 Plant WOI
D-LINALOL 918-5,263 Plant WOI
DELTA-3-CARENE Plant HHB
DINATIN-7-GLUCOSIDE Leaf FNF
DIOSMETIN-7-GLUCOSIDE Leaf 411/
EO 3,000-17,200 Plant WOI
ESTRAGOLE Plant 411/

ETHYL-CAPRATE Plant BML
ETHYL-LAURATE Plant BML
ETHYL-LINILENATE Plant BML
ETHYL-LINOLEATE Plant BML
ETHYL-MYRISTATE Plant BML
ETHYL-OLEATE Plant BML
ETHYL-PALMITATE Plant BML
ETHYL-STEARATE Plant BML
ETHYL-TRANSHEXADECANOATE Plant BML
EUGENOL 200-1,152 Plant JBH WOI
FAT 8,000-77,950 Plant USA
FIBER 170,170-220,600 Plant USA WOI
GAMMA-ELEMENE Plant BML
GERANIOL Plant BML
GERANYL-ACETATE 12-69 Plant HHB
HYDROQUINONE Plant FNF
IRON 750-975 Plant USA
LEDENE Plant BML
LIMONENE 18-103 Plant BML
LINALOL 99-568 Plant BML
LINALYL-ACETATE 3-17 Plant BML
LINOLEIC-ACID 10,000 Seed HHB
LINOLENIC-ACID 27,500 Seed HHB
LUSITANICOSIDE Plant JBH
LUTEOLIN-7-GLUCOSIDE Plant 411/
MAGNESIUM 3,300-3,900 Plant USA
MAJARONIN Leaf HHB
MANGANESE 54 Plant USA
METHYL-CHAVICOL 96-550 Plant WOI
MYRCENE 18-103 Plant BML
N-TRIACONTANE Plant HHB
NERYL-ACETATE Plant BML
NIACIN 41-44 Plant USA
OLEANOLIC-ACID 4,700 Plant HHB
OLEIC-ACID 8,000 Seed HHB
P-CYMEN-8-OL 8-43 Plant BML
P-CYMENE 21-120 Plant BML
PENTOSANS 70,000-76,800 Plant HHB WOI
PHENOL 1,431-8,204 Plant WOI
PHOSPHORUS 2,670-3,725 Plant USA
PHYTOSTEROLS 600-650 Plant USA
PLANTEOSE Seed HHB
POTASSIUM 14,500-17,225 Plant USA
PROTEIN 121,450-143,100 Plant USA WOI
RIBOFLAVIN 3 Plant USA

ROSMARINIC-ACID 33,000 Plant FT62:166
SABINENE 75-430 Plant BML HHB
SAPONIN Plant HHB
SODIUM 675-935 Plant USA
TERPINEN-4-OL 1,365-7,826 Plant BML
TERPINEN-4-YL-ACETATE Plant BML
TERPINENE Plant HHB
TERPINENOL Plant WOI
TERPINOLENE 75-430 Plant BML
THIAMIN 3 Plant USA
TRANS-3-ETHOXY-P-MENTH-1-ENE Plant FNF
TRANS-4-ETHOXY-THUJANE Plant BML
TRANS-ANETHOLE 8-43 Plant BML
TRANS-P-MENTH-2-EN-1-OL 52-301 Plant BML
TRANS-PIPERITOL 9-52 Plant BML
TRANS-SABINENE-HYDRATE 30-172 Plant BML HHB
URSOLIC-ACID 500-2,100 Plant WOI
URSOLIC-ACID-I 4,700 Plant HHB
URSOLIC-ACID-II 1,200 Plant HHB
WATER 74,920-77,880 Plant USA
ZINC 32-43 Plant FNF

Origanum minutiflorum O. SCHWARZ & P.H. DAVIS "Small-Flowered Oregano"

(E)-BETA-OCIMENE 0-5 Shoot JEO3:445
(E)-ISOEUGENOL 2-220 Shoot JEO3:445
(Z)-3-HEXANOL 2-5 Shoot JEO3:445
(Z)-BETA-FARNESENE 20-29 Shoot JEO3:445
(Z)-ISOEUGENOL 13-37 Shoot JEO3:445
1,8-CINEOLE 60-140 Shoot JEO3:445
1-OCTEN-3-OL 49-245 Shoot JEO3:445
2-HEXANAL 2-10 Shoot JEO3:445
3-OCTANOL 9-10 Shoot JEO3:445
ALLOAROMADENDRENE 9-22 Shoot JEO3:445
ALPHA-CUBEBENE <0.1-5 Shoot JEO3:445
ALPHA-PINENE 235-310 Shoot JEO3:445
ALPHA-TERPINENE 125-145 Shoot JEO3:445
ALPHA-TERPINEOL 5-12 Shoot JEO3:445
ALPHA-THUJENE 5-7 Shoot JEO3:445
BENZYL-ALCOHOL <0.1-2 Shoot JEO3:445
BETA-CARYOPHYLLENE 165-380 Shoot JEO3:445
BETA-PINENE 31-39 Shoot JEO3:445
CAMPHENE 65-150 Shoot JEO3:445
CARVACROL 16,890-20,000 Shoot JEO3:445
CIS-PIPERITOL 5-7 Shoot JEO3:445
CIS-SABINENE-HYDRATE 60-80 Shoot JEO3:445
CUMIN-ALCOHOL <0.1-10 Shoot JEO3:445
CUMINALDEHYDE 5 Shoot JEO3:445
DELTA-3-CARENE 11-15 Shoot JEO3:445
EO 22,400-24,400 Shoot JEO3:445
ETHANOL <0.1-7 Shoot JEO3:445
EUGENOL 55-125 Shoot JEO3:445
GAMMA-TERPINENE 7-555 Shoot JEO3:445
ISOBORNEOL 285-525 Shoot JEO3:445
LIMONENE 40-60 Shoot JEO3:445
LIMONENE-OXIDE 7-12 Shoot JEO3:445
LINALOL 13-17 Shoot JEO3:445
METHANOL 5-45 Shoot JEO3:445
METHYL-ISOVALERATE 5-20 Shoot JEO3:445
METHYL-SALICYLATE 5-10 Shoot JEO3:445
MYRCENE 40-220 Shoot JEO3:445
P-CYMENE 760-2,290 Shoot JEO3:445
PATCHOULANE 60-110 Shoot JEO3:445
SABINENE 5-15 Shoot JEO3:445
TERPINEN-4-OL 155-205 Shoot JEO3:445
TERPINOLENE 20-27 Shoot JEO3:445
THYMOL 49-70 Shoot JEO3:445
TRICYCLENE 5-7 Shoot JEO3:445

Origanum onites L.
"Oregano, Pot Marjoram"

1,8-CINEOLE 23 Plant BML
ALPHA-PINENE 16 Plant BML
ALPHA-TERPINENE 47 Plant BML
ALPHA-THUJENE 16 Plant BML
BETA-BISABOLENE 246 Plant BML
BETA-PINENE 3 Plant BML
BORNEOL 315 Plant BML
BORON 28-48 Plant BOB
CAMPHENE 28 Plant BML
CARVACROL 8,310 Plant BML
CARVONE Plant tr BML
CARYOPHYLLENE 82 Plant BML
CIS-SABINENE-HYDRATE 4 Plant BML
GAMMA-TERPINENE 190 Plant BML
LIMONENE 17 Plant BML
LINALOL 39 Plant BML
MYRCENE 41 Plant BML
P-CYMENE 227 Plant BML
PENTYL-ALCOHOL Plant tr BML
TERPINEN-4-OL 144 Plant BML
THYMOL 86 Plant BML

Origanum onites L.
"Oregano, Pot Marjoram"
Unaggregated A

4-TERPINEOL 525 Shoot ZLU197:20
ALPHA-CARYOPHYLLENE Shoot ZLU197:20
ALPHA-HUMULENE <0.1 Shoot ZLU197:20
ALPHA-MUUROLENE 33 Shoot ZLU197:20
ALPHA-PHELLANDRENE 33 Shoot ZLU197:20
ALPHA-PINENE 66 Shoot ZLU197:20
ALPHA-TERPINENE 230 Shoot ZLU197:20
ALPHA-TERPINEOL 130 Shoot ZLU197:20
ALPHA-THUJENE 33 Shoot ZLU197:20
AROMADENDRENE 65 Shoot ZLU197:20
BETA-BISABOLENE 1,055 Shoot ZLU197:20
BETA-CARYOPHYLLENE 365 Shoot ZLU197:20
BETA-PINENE <0.1 Shoot ZLU197:20
BORNEOL 1,520 Shoot ZLU197:20
BORNYL-ACETATE <0.1 Shoot ZLU197:20
CAMPHENE 100 Shoot ZLU197:20
CARVACROL 24,190 Shoot ZLU197:20
CARVACROL-METHYL-ETHER 65 Shoot ZLU197:20
CARVONE <0.1 Shoot ZLU197:20
CIS-OCIMENE Shoot ZLU197:20
CIS-PINANE Shoot ZLU197:20
CIS-THUJANOL 130 Shoot ZLU197:20
DELTA-3-CARENE <0.1 Shoot ZLU197:20
EO 33,000 Shoot ZLU197:20
GAMMA-CADINENE Shoot ZLU197:20
GAMMA-GURJUNENE <0.1 Shoot ZLU197:20
GAMMA-TERPINENE 1,320 Shoot ZLU197:20
LIMONENE 200 Shoot ZLU197:20
MYRCENE 330 Shoot ZLU197:20
P-CYMENE 1,615 Shoot ZLU197:20
TERPINOLENE 65 Shoot ZLU197:20
THYMOL 230 Shoot ZLU197:20
TRANS-OCIMENE Shoot ZLU197:20
TRANS-THUJANOL 100 Shoot ZLU197:20

Origanum onites L.
"Oregano, Pot Marjoram"
Unaggregated B

1,8-CINEOLE Shoot FFJ8:331
3-CARENE 18 Shoot FFJ8:331
4,5-DIMETHYL-2-ETHYLPHENOL Shoot FFJ8:331
ALPHA,ALPHA-DIMETHYLBENZYLALCOHOL <0.1 Shoot FFJ8:331
ALPHA-BISABOLENE Shoot FFJ8:331
ALPHA-CUBEBENE <0.1 Shoot FFJ8:331
ALPHA-ELEMENOL Shoot FFJ8:331
ALPHA-ELEMOL 45 Shoot FFJ8:331
ALPHA-HUMULENE 20 Shoot FFJ8:331
ALPHA-HUMULENE-EPOXIDE Shoot FFJ8:331
ALPHA-PHELLANDRENE 45 Shoot FFJ8:331
ALPHA-PINENE 60 Shoot FFJ8:331
ALPHA-TERPINENE 350 Shoot FFJ8:331
ALPHA-TERPINEOL 28 Shoot FFJ8:331
ALPHA-THUJONE Shoot FFJ8:331
BENZALDEHYDE Shoot FFJ8:331
BETA-BERGAMOTENE 30 Shoot FFJ8:331
BETA-BISABOLENE 1,415 Shoot FFJ8:331
BETA-CUBEBENE 120 Shoot FFJ8:331
BETA-PHELLANDRENE 38 Shoot FFJ8:331
BETA-PINENE Shoot FFJ8:331
BETA-TERPINENE Shoot FFJ8:331
BETA-TERPINEOL <0.1 Shoot FFJ8:331
BETA-THUJONE Shoot FFJ8:331
BORNEOL 950 Shoot FFJ8:331
BORNYL-ACETATE Shoot FFJ8:331
CAMPHENE 370 Shoot FFJ8:331
CAMPHOR Shoot FFJ8:331
CARVACROL 15,420 Shoot FFJ8:331
CARVACROL-METHYLETHER Shoot FFJ8:331
CARVACRYL-ACETATE 40 Shoot FFJ8:331
CARVONE 38 Shoot FFJ8:331
CARYOPHYLLENE 430 Shoot FFJ8:331
CARYOPHYLLENE-OXIDE 100 Shoot FFJ8:331
CEDRANOXIDE 42 Shoot FFJ8:331
CIS-LINALOL-OXIDE Shoot FFJ8:331
CIS-OCIMENE <0.1 Shoot FFJ8:331
CIS-SABINENE-HYDRATE Shoot FFJ8:331
CUMINYL-ALCOHOL Shoot FFJ8:331
DIHYDROCARVEOL Shoot FFJ8:331
ELEMENAL 18 Shoot FFJ8:331

ELEMENE 20 Shoot FFJ8:331
EO 25,000 Shoot FFJ8:331
EUGENOL Shoot FFJ8:331
EUGENOL-METHYL-ETHER Shoot FFJ8:331
FARNESOL 45 Shoot FFJ8:331
GAMMA-CADINENE 75 Shoot FFJ8:331
GAMMA-ELEMENE <0.1 Shoot FFJ8:331
GAMMA-TERPINENE 1,550 Shoot FFJ8:331
GARANYL-ACETATE Shoot FFJ8:331
GERANIOL 48 Shoot FFJ8:331
HEPT-1-EN-3-OL Shoot FFJ8:331
HEX-2-ENAL Shoot FFJ8:331
HEX-3-EN-1-OL Shoot FFJ8:331
HEXANAL Shoot FFJ8:331
ISOPULEGOL Shoot FFJ8:331
LIMONENE 85 Shoot FFJ8:331
LINALOL 100 Shoot FFJ8:331
LINALYL-ACETATE Shoot FFJ8:331
MANOYL-OXIDE 12 Plant FFJ8:331
MUUROLENE 25 Shoot FFJ8:331
MYRCENE 170 Shoot FFJ8:331
MYRCENYL-ACETATE Shoot FFJ8:331
NERYL-ACETATE Shoot FFJ8:331
OCT-1-EN-3-OL 40 Shoot FFJ8:331
P-CYMENE 1,160 Shoot FFJ8:331
PATCHOULENE 20 Shoot FFJ8:331
SABINENE <0.1 Shoot FFJ8:331
TERPINEN-4-OL 325 Shoot FFJ8:331
TERPINOLENE 60 Shoot FFJ8:331
TERPINYL-ACETATE Shoot FFJ8:331
THUJENE 315 Shoot FFJ8:331
THYMOL 85 Shoot FFJ8:331
THYMOL-METHYLETHER Shoot FFJ8:331
THYMYL-ACETATE <0.1 Shoot FFJ8:331
TRANS-OCIMENE 20 Shoot FFJ8:331
TRANS-SABINENE-HYDRATE Shoot FFJ8:331
TRICYCLENE 15 Shoot FFJ8:331
ZINGIBERONE 105 Shoot FFJ8:331

Origanum sipyleum L.
"Bayircayi, Guveyoto"
Unaggregated A

(E)-BETA-OCIMENE 3 Shoot JEO4:139
(Z)-BETA-OCIMENE 4 Shoot JEO4:139
1,8-CINEOLE 100 Shoot JEO4:139
1-OCTEN-3-OL 5 Shoot JEO4:139
2,5-DIETHYL-FURAN 3 Shoot JEO4:139
2-BUTENYLBENZENE <0.1 Shoot JEO4:139
2-HEXENAL <0.1 Shoot JEO4:139
3-OCTANOL 5 Shoot JEO4:139
4-OCTEN-3-ONE 1.4 Shoot JEO4:139
ACETONE 2 Shoot JEO4:139
ALPHA-CUBEBENE 105 Shoot JEO4:139
ALPHA-FARNESENE 3 Shoot JEO4:139
ALPHA-PATCHOULENE 45 Shoot JEO4:139
ALPHA-PINENE 40 Shoot JEO4:139
ALPHA-TERPINENE 11 Shoot JEO4:139
ALPHA-TERPINEOL 50 Shoot JEO4:139
BETA-CUBEBENE 30 Shoot JEO4:139
BETA-EUDESMOL 13 Shoot JEO4:139
BETA-PATCHOULENE 175 Shoot JEO4:139
BETA-PINENE 14 Shoot JEO4:139
BETA-YLANGENE 2.5 Shoot JEO4:139
BORNEOL 140 Shoot JEO4:139
CAMPHENE 3 Shoot JEO4:139
CAMPHOR 2.5 Shoot JEO4:139
CARVACROL 295 Shoot JEO4:139
CARVACROL-METHYL-ETHER 365 Shoot JEO4:139
CIS-P-MENTHA-2,8-DIEN-OL 3 Shoot JEO4:139
CIS-SABINENE-HYDRATE 4 Shoot JEO4:139
DELTA-3-CARENE <0.1 Shoot JEO4:139
DELTA-CADINENE 25 Shoot JEO4:139
EO 3,600 Shoot JEO4:139
FARNESOL 8 Shoot JEO4:139
GAMMA-ELEMENE 55 Shoot JEO4:139
GAMMA-TERPINENE 390 Shoot JEO4:139
ISOEUGENOL 20 Shoot JEO4:139
LIMONENE <0.1 Shoot JEO4:139
LINALOL 30 Shoot JEO4:139
METHANOL <0.1 Shoot JEO4:139
MYRCENE 40 Shoot JEO4:139

P-CYMEN-7-OL 1.8 Shoot JEO4:139
P-CYMEN-8-OL 1.4 Shoot JEO4:139
P-CYMENE 135 Shoot JEO4:139
PERILLENE 3 Shoot JEO4:139
SABINENE 11 Shoot JEO4:139
TERPINEN-4-OL 150 Shoot JEO4:139
TERPINOLENE 4 Shoot JEO4:139
THYMOL 105 Shoot JEO4:139
THYMOL-METHYL-ETHER 715 Shoot JEO4:139
TRICYCLENE 0.7 Shoot JEO4:139

Origanum sipyleum L.
"Bayircayi, Guveyoto"
Unaggregated B

(E)-BETA-OCIMENE 1.8 Shoot JEO4:139
(Z)-BETA-OCIMENE 15 Shoot JEO4:139
1,8-CINEOLE 20 Shoot JEO4:139
1-OCTEN-3-OL 0.4 Shoot JEO4:139
2,5-DIETHYL-FURAN 0.2 Shoot JEO4:139
2-BUTENYLBENZENE 4 Shoot JEO4:139
2-HEXENAL 0.7 Shoot JEO4:139
3-OCTANOL 35 Shoot JEO4:139
4-OCTEN-3-ONE 1.3 Shoot JEO4:139
ACETONE 0.7 Shoot JEO4:139
ALPHA-CUBEBENE 25 Shoot JEO4:139
ALPHA-FARNESENE 3 Shoot JEO4:139
ALPHA PATCHOULENE 20 Shoot JEO4:139
ALPHA-PINENE 25 Shoot JEO4:139
ALPHA-TERPINENE 15 Shoot JEO4:139
ALPHA-TERPINEOL 1 Shoot JEO4:139
BETA-CUBEBENE 2 Shoot JEO4:139
BETA-EUDESMOL 6 Shoot JEO4:139
BETA-PATCHOULENE 155 Shoot JEO4:139
BETA-PINENE 13 Shoot JEO4:139
BETA-YLANGENE 20 Shoot JEO4:139
BORNEOL 50 Shoot JEO4:139
CAMPHENE 25 Shoot JEO4:139
CAMPHOR 13 Shoot JEO4:139
CARVACROL 18 Shoot JEO4:139
CARVACROL-METHYL-ETHER 9 Shoot JEO4:139
CIS-P-MENTHA-2,8-DIEN-OL <0.1 Shoot JEO4:139
CIS-SABINENE-HYDRATE 35 Shoot JEO4:139
DELTA-3-CARENE 0.2 Shoot JEO4:139
DELTA-CADINENE 3 Shoot JEO4:139
EO 2,200 Shoot JEO4:139
FARNESOL 10 Shoot JEO4:139
GAMMA-ELEMENE 4 Shoot JEO4:139
GAMMA-TERPINENE 270 Shoot JEO4:139
ISOEUGENOL 3 Shoot JEO4:139
LIMONENE 10 Shoot JEO4:139
LINALOL 18 Shoot JEO4:139
METHANOL 0.7 Shoot JEO4:139
MYRCENE 35 Shoot JEO4:139
P-CYMEN-7-OL 1.8 Shoot JEO4:139
P-CYMEN-8-OL 0.2 Shoot JEO4:139

P-CYMENE 870 Shoot JEO4:139
PERILLENE 1.3 Shoot JEO4:139
SABINENE 33 Shoot JEO4:139
TERPINEN-4-OL 200 Shoot JEO4:139
TERPINOLENE 1.8 Shoot JEO4:139
THYMOL 5 Shoot JEO4:139
THYMOL-METHYL-ETHER <0.1 Shoot JEO4:139
TRICYCLENE 20 Shoot JEO4:139

Origanum sipyleum L.
"Bayircayi, Guveyoto"
Unaggregated D

(E)-BETA-OCIMENE 17 Shoot JEO4:139
(Z)-BETA-OCIMENE 7 Shoot JEO4:139
1,8-CINEOLE 220 Shoot JEO4:139
1-OCTEN-3-OL 70 Shoot JEO4:139
2,5-DIETHYL-FURAN <0.1 Shoot JEO4:139
2-BUTENYLBENZENE 13 Shoot JEO4:139
2-HEXENAL 7 Shoot JEO4:139
3-OCTANOL 30 Shoot JEO4:139
4-OCTEN-3-ONE Shoot JEO4:139
ACETONE 1-2 Shoot JEO4:139
ALPHA-CUBEBENE 85 Shoot JEO4:139
ALPHA-FARNESENE 2 Shoot JEO4:139
ALPHA-PATCHOULENE 65 Shoot JEO4:139
ALPHA-PINENE 325 Shoot JEO4:139
ALPHA-TERPINENE 410 Shoot JEO4:139
ALPHA-TERPINEOL 15 Shoot JEO4:139
BETA-CUBEBENE 13 Shoot JEO4:139
BETA-EUDESMOL <0.1 Shoot JEO4:139
BETA-PATCHOULENE 65 Shoot JEO4:139
BETA-PINENE 55 Shoot JEO4:139
BETA-YLANGENE 185 Shoot JEO4:139
BORNEOL 200 Shoot JEO4:139
CAMPHENE 20 Shoot JEO4:139
CAMPHOR 10 Shoot JEO4:139
CARVACROL 2,050 Shoot JEO4:139
CARVACROL-METHYL-ETHER 960 Shoot JEO4:139
CIS-P-MENTHA-2,8-DIEN-OL 7 Shoot JEO4:139
CIS-SABINENE-HYDRATE 45 Shoot JEO4:139
DELTA-3-CARENE 12 Shoot JEO4:139
DELTA-CADINENE 50 Shoot JEO4:139
EO 16,800 Shoot JEO4:139
FARNESOL 3 Shoot JEO4:139
GAMMA-ELEMENE 100 Shoot JEO4:139
GAMMA-TERPINENE 4,470 Shoot JEO4:139
ISOEUGENOL <0.1 Shoot JEO4:139
LIMONENE 120 Shoot JEO4:139
LINALOL 105 Shoot JEO4:139
METHANOL <0.1 Shoot JEO4:139
MYRCENE 1,950 Shoot JEO4:139
P-CYMEN-7-OL 18 Shoot JEO4:139
P-CYMEN-8-OL Shoot JEO4:139

P-CYMENE 3,345 Shoot JEO4:139
PERILLENE 3 Shoot JEO4:139
SABINENE 100 Shoot JEO4:139
TERPINEN-4-OL 270 Shoot JEO4:139
TERPINOLENE 170 Shoot JEO4:139
THYMOL 140 Shoot JEO4:139
THYMOL-METHYL-ETHER 410 Shoot JEO4:139
TRICYCLENE <0.1 Shoot JEO4:139

Origanum syriacum L.
"Za'Atar"
Unaggregated A

1-OCTEN-3-OL 150 Shoot JEO3:121
2,5-DIETHYL-TETRAHYDROFURAN Shoot JEO3:121
3-OCTANOL 110 Shoot JEO3:121
ALPHA-GUAIENE <0.1 Shoot JEO3:121
ALPHA-HUMULENE 45 Shoot JEO3:121
ALPHA-PHELLANDRENE Shoot JEO3:121
ALPHA-PINENE Shoot JEO3:121
ALPHA-TERPINENE 840 Shoot JEO3:121
ALPHA-TERPINEOL 55 Shoot JEO3:121
ALPHA-THUJENE 320 Shoot JEO3:121
BETA-CARYOPHYLLENE 175 Shoot JEO3:121
BETA-GUAIENE <0.1 Shoot JEO3:121
BETA-PHELLANDRENE 80 Shoot JEO3:121
BETA-PINENE 30 Shoot JEO3:121
BORNEOL 55 Shoot JEO3:121
CAMPHENE 30 Shoot JEO3:121
CAMPHOR Shoot JEO3:121
CARVACROL 1,285 Shoot JEO3:121
CARVONE 15 Shoot JEO3:121
CARYOPHYLLENE-OXIDE 75 Shoot JEO3:121
CIS-CARVEOL Shoot JEO3:121
CIS-SABINENE-HYDRATE 20 Shoot JEO3:121
CITRONELLOL 12 Shoot JEO3:121
CITRONELLYL-BUTYRATE Shoot JEO3:121
DIHYDROCARVONE Shoot JEO3:121
EO 31,000 Shoot JEO3:121
ETHYL-CINNAMATE Shoot JEO3:121
GAMMA-TERPINENE 2,700 Shoot JEO3:121
GERANIOL Shoot JEO3:121
GERANYL-ACETATE Shoot JEO3:121
GERANYL-BUTYRATE Shoot JEO3:121
GERANYL-HEXANOATE Shoot JEO3:121
GERANYL-PROPIONATE Shoot JEO3:121
GERMACRENE-D Shoot JEO3:121
LIMONENE 110 Shoot JEO3:121
LINALOL 35 Shoot JEO3:121
METHYL-2-METHYL-BUTYRATE 6 Shoot JEO3:121
MYRCENE 750 Shoot JEO3:121
P-CYMENE 4,445 Shoot JEO3:121
TERPINEN-1-OL 6 Shoot JEO3:121
TERPINEN-4-OL 435 Shoot JEO3:121

TERPINOLENE Shoot JEO3:121
THYMOL 18,560 Shoot JEO3:121
TRANS-BERGAMOTTENE 9 Shoot JEO3:121
TRANS-SABINENE-HYDRATE 25 Shoot JEO3:121
TRICYCLENE Shoot JEO3:121

Origanum syriacum L.
"Za'Atar"
Unaggregated B

1-OCTEN-3-OL 65 Shoot JEO3:121
2,5-DIETHYL-TETRAHYDROFURAN 6 Shoot JEO3:121
3-OCTANOL 30 Shoot JEO3:121
ALPHA-GUAIENE 6 Shoot JEO3:121
ALPHA-HUMULENE 20 Shoot JEO3:121
ALPHA-PHELLANDRENE 40 Shoot JEO3:121
ALPHA-PINENE 325 Shoot JEO3:121
ALPHA-TERPINENE 990 Shoot JEO3:121
ALPHA-TERPINEOL 25 Shoot JEO3:121
ALPHA-THUJENE Shoot JEO3:121
BETA-CARYOPHYLLENE 185 Shoot JEO3:121
BETA-GUAIENE 3 Shoot JEO3:121
BETA-PHELLLANDRENE 70 Shoot JEO3:121
BETA-PINENE 30 Shoot JEO3:121
BORNEOL 30 Shoot JEO3:121
CAMPHENE 25 Shoot JEO3:121
CAMPHOR Shoot JEO3:121
CARVACROL 24,850 Shoot JEO3:121
CARVONE 20 Shoot JEO3:121
CARYOPHYLLENE-OXIDE 45 Shoot JEO3:121
CIS-CARVEOL 3 Shoot JEO3:121
CIS-SABINENE-HYDRATE 55 Shoot JEO3:121
CITRONELLOL Shoot JEO3:121
CITRONELLOL-BUTYRATE Shoot JEO3:121
DIHYDROCARVONE Shoot JEO3:121
EHTYL-CINNAMATE Shoot JEO3:121
EO 31,000 Shoot JEO3:121
GAMMA-TERPINENE 1,015 Shoot JEO3:121
GERANIOL Shoot JEO3:121
GERANYL-ACETATE Shoot JEO3:121
GERANYL-BUTYRATE Shoot JEO3:121
GERANYL-HEXANOATE Shoot JEO3:121
GERANYL-PROPIONATE Shoot JEO3:121
GERMACRENE-D Shoot JEO3:121
LIMONENE 55 Shoot JEO3:121
LINALOL 25 Shoot JEO3:121
METHYL-2-METHYL-BUTYRATE 12 Shoot JEO3:121
MYRCENE 565 Shoot JEO3:121
P-CYMENE 1,880 Shoot JEO3:121
TERPINEN-1-OL 15 Shoot JEO3:121
TERPINEN-4-OL 315 Shoot JEO3:121

TERPINOLENE 45 Shoot JEO3:121
THYMOL 280 Shoot JEO3:121
TRANS-BERGAMOTENE 9 Shoot JEO3:121
TRANS-SABINENE-HYDRATE 120 Shoot JEO3:121
TRICYCLENE 3 Shoot JEO3:121

Origanum syriacum L.
"Za'Atar"
Unaggregated C

1-OCTEN-3-OL 80 Shoot JEO3:121
2,5-DIETHYL-TETRAHYDROFURAN 6 Shoot JEO3:121
3-OCTANOL 12 Shoot JEO3:121
ALPHA-GUAIENE Shoot JEO3:121
ALPHA-HUMULENE 55 Shoot JEO3:121
ALPHA-PHELLANDRENE 12 Shoot JEO3:121
ALPHA-PINENE 265 Shoot JEO3:121
ALPHA-TERPINENE 990 Shoot JEO3:121
ALPHA-TERPINEOL 55 Shoot JEO3:121
ALPHA-THUJENE Shoot JEO3:121
BETA-CARYOPHYLLENE 200 Shoot JEO3:121
BETA-GUAIENE Shoot JEO3:121
BETA-PHELLANDRENE 60 Shoot JEO3:121
BETA-PINENE 560 Shoot JEO3:121
BORNEOL 15 Shoot JEO3:121
CAMPHENE 40 Shoot JEO3:121
CAMPHOR <0.1 Shoot JEO3:121
CARVACROL 23,765 Shoot JEO3:121
CARVONE <0.1 Shoot JEO3:121
CARYOPHYLLENE-OXIDE 55 Shoot JEO3:121
CIS-CARVEOL Shoot JEO3:121
CIS-SABINENE-HYDRATE Shoot JEO3:121
CITRONELLOL <0.1 Shoot JEO3:121
CITRONELLYL-BUTYRATE 15 Shoot JEO3:121
DIHYDROCARVONE 12 Shoot JEO3:121
EO 31,000 Shoot JEO3:121
ETHYL-CINNAMATE 30 Shoot JEO3:121
GAMMA-TERPINENE 240 Shoot JEO3:121
GERANIOL 65 Shoot JEO3:121
GERANYL-ACETATE 15 Shoot JEO3:121
GERANYL-BUTYRATE 20 Shoot JEO3:121
GERANYL-HEXANOATE 6 Shoot JEO3:121
GERANYL-PROPIONATE 70 Shoot JEO3:121
GERMACRENE-D 50 Shoot JEO3:121
LIMONENE 90 Shoot JEO3:121
LINALOL 55 Shoot JEO3:121
METHYL-2-METHYL-BUTYRATE 9 Shoot JEO3:121
MYRCENE <0.1 Shoot JEO3:121
P-CYMENE 3,135 Shoot JEO3:121
TERPINEN-1-OL Shoot JEO3:121
TERPINEN-4-OL 335 Shoot JEO3:121

TERPINOLENE <0.1 Shoot JEO3:121
THYMOL 385 Shoot JEO3:121
TRANS-BERGAMOTENE 9 Shoot JEO3:121
TRANS-SABINENE-HYDRATE Shoot JEO3:121
TRICYCLENE Shoot JEO3:121

Origanum vulgare L.
"Wild Oregano, Common Turkish Oregano"
Unaggregated A

(E)-2-HEXENAL Plant JEO5:425
(E)-BETA-OCIMENE 25 Plant JEO5:425
(E,Z)-2,6-NONADIENAL 0.5 Plant JEO5:425
(Z)-2-NONENAL 0.2 Plant JEO5:425
(Z)-BETA-FARNESENE 6 Plant JEO5:425
(Z)-BETA-OCIMENE 5 Plant JEO5:425
(Z)-GERANYL-ACETONE 0.7 Plant JEO5:425
1,8-CINEOLE 17 Plant JEO5:425
1-NONEN-3-OL 0.2 Plant JEO5:425
1-OCTEN-3-OL 2 Plant JEO5:425
2-ETHYL-1-HEXANOL 3 Plant JEO5:425
3-NONANONE 0.1 Plant JEO5:425
6-METHYL-3-HEPTANOL 3 Plant JEO5:425
ALPHA-BISABOLENE 7.5 Plant JEO5:425
ALPHA-CADINOL 20 Plant JEO5:425
ALPHA-FARNESENE 11 Plant JEO5:425
ALPHA-HIMACHALENE 8 Plant JEO5:425
ALPHA-HUMULENE 24 Plant JEO5:425
ALPHA-MUUROLENE 3.5 Plant JEO5:425
ALPHA-PINENE 1 Plant JEO5:425
ALPHA-TERPINENE 2 Plant JEO5:425
ALPHA-TERPINEOL 18 Plant JEO5:425
AMYL-FURAN 0.2 Plant JEO5:425
BETA-BISABOLENE 45 Plant JEO5:425
BETA-BOURBONENE 8 Plant JEO5:425
BETA-CARYOPHYLLENE 170 Plant JEO5:425
BETA-CUBEBENE 0.6 Plant JEO5:425
BETA-IONONE 3 Plant JEO5:425
BETA-PHELLANDRENE 5 Plant JEO5:425
BETA-PINENE 2.5 Plant JEO5:425
BETA-SESQUIPHELLANDRENE 1 Plant JEO5:425
BORNYL-ACETATE 11 Plant JEO5:425
CAMPHENE 0.5 Plant JEO5:425
CARVACROL 18 Plant JEO5:425
CIS-DIHYDROCARVONE Plant JEO5:425
CIS-SABINENE-HYDRATE Plant JEO5:425
DECANE <1 Plant JEO5:425
DECANOIC-ACID 10 Plant JEO5:425
DECANOL 0.4 Plant JEO5:425
DELTA-3-CARENE Plant JEO5:425
DELTA-CADINENE 11 Plant JEO5:425

DELTA-CADINOL 2.5 Plant JEO5:425
ELEMOL 1.5 Plant JEO5:425
EO 800 Plant JEO5:425
GAMMA-CADINENE Plant JEO5:425
GAMMA-ELEMENE Plant JEO5:425
GAMMA-TERPINENE 4 Plant JEO5:425
GERANIOL 1.5 Plant JEO5:425
GERANYL-ACETATE 11 Plant JEO5:425
GERMACRENE-D 140 Plant JEO5:425
HENEICOSANE 0.6 Plant JEO5:425
HEXAHYDROFARNESYL-ACETONE 1.5 Plant JEO5:425
HEXANAL <1 Plant JEO5:425
ISOBORNEOL 2.5 Plant JEO5:425
LIMONENE 5 Plant JEO5:425
LINALOL 15 Plant JEO5:425
LINALYL-ACETATE 15 Plant JEO5:425
MYRCENE 6 Plant JEO5:425
NEROL 0.6 Plant JEO5:425
NONANAL 0.1 Plant JEO5:425
NONANE <1 Plant JEO5:425
NONANOIC-ACID 0.9 Plant JEO5:425
OCTENYL-ACETATE 1 Plant JEO5:425
P-CYMEN-8-OL Plant JEO5:425
P-CYMENE 1 Plant JEO5:425
PATCHOULANE 12 Plant JEO5:425
PERILLENE 0.2 Plant JEO5:425
SABINENE 13 Plant JEO5:425
SPATHULENOL 2.5 Plant JEO5:425
TERPINEN-4-OL Plant JEO5:425
TERPINOLENE 1 Plant JEO5:425
THYMOL Plant JEO5:425
TORREYOL 6 Plant JEO5:425
TRANS-DIHYDROCARVONE Plant JEO5:425
TRANS-SABINENE-HYDRATE 2.5 Plant JEO5:425
TRICHOSANE 2.5 Plant JEO5:425
UNDECANE <1 Plant JEO5:425

Origanum vulgare L.
"Wild Oregano, Common Turkish Oregano"
Aggregated

1,8-CINEOLE 4-96 Plant BML
1-OCTEN-3-OL 106-530 Plant BML
10-ALPHA-CADINOL 20-80 Plant BML
2-CAFFEOYL-OXY-3-{2-(4-HYDROXYBENZYL)-4,5-DIHYDROXY}PHENYLPROPIONIC-ACID 21 Leaf NAP
2-UNDECANONE 6-30 Plant BML
3-OCTANOL 2-150 Plant BML
4-(3,4-DIHYDROXY-BENZOYL-OXY-METHYL)-PHENYL-BETA-D-GLUCOPYRANOSIDE Leaf NAP
4-(3,4-DIHYDROXY-BENZOYL-OXY-METHYL)-PHENYL-BETA-D-GLUCOSIDE 2,330 Leaf NAP
5-HYDROXY-3,3',4',7-TETRAMETHOXY-FLAVONE Plant NAP
ALLO-AROMADENDRENE 18-120 Plant BML
ALPHA-COPAENE 18-90 Plant BML
ALPHA-HUMULENE 6-376 Plant BML
ALPHA-LINOLENIC-ACID 4,180-4,502 Plant USA
ALPHA-MUUROLENE 3-15 Plant BML
ALPHA-PHELLANDRENE 2-244 Plant BML JAD NAP
ALPHA-PINENE 6-143 Plant BML JAD NAP
ALPHA-TERPINENE 9-178 Plant BML
ALPHA-THUJENE 6-32 Plant BML
ALPHA-THUJONE Plant BML
APIGENIN Plant NAP
APIGENIN-7-O-BETA-D-GLUCOSIDE Plant ABS
AROMADENDRIN Plant PAS
ASH 69,000-79,720 Plant USA
BETA-BISABOLENE 3-640 Plant BML
BETA-BOURBORNENE 54-430 Plant BML
BETA-CAROTENE 41-45 Plant USA
BETA-ELEMENE 6-30 Plant BML
BETA-FARNESENE Plant BML
BETA-PHELLANDRENE 12-60 Plant BML
BETA-PINENE 1-312 Plant BML
BICYCLOGERMACRENE 51-340 Plant BML
BORNEOL 32-252 Plant BML
BORON 21-41 Plant BOB
CAFFEIC-ACID 1,060 Leaf HHB
CAFFEIC-ACID 6,000 Plant HHB
CALAMENENE 24-96 Plant BML
CALCIUM 14,070-18,794 Plant BML
CAMPHENE 1-64 Plant BML

CAMPHOR 0-40 Plant BML
CAPRIC-ACID 400-431 Plant USA
CARBOHYDRATES 644,300-693,911 Plant USA
CARVACROL 6-640 Plant BML JBH
CARVACRYL-METHYL-ETHER 16-64 Plant BML
CARVONE Plant tr BML
CARYOPHYLLENE 1-920 Plant BML
CARYOPHYLLENE-OXIDE 84-336 Plant BML
CATECHOL Plant NAP
CHLOROGENIC-ACID Plant NAP
CINNAMIC-ACID Plant NAP
CIS-GAMMA-BISABOLENE 3-20 Plant BML
CIS-OCIMENE 270-1,392 Plant BML
CIS-PIPERITOL 10-40 Plant BML
CIS-SABINENE-HYDRATE 1-46 Plant BML
COPPER 9 Plant USA
COSMOSIIN Plant NAP
CUMINAL 2-10 Plant BML
DELTA-3-CARENE 2-96 Plant BML
DELTA-CADINENE 2-480 Plant BML
DIOSMETIN Plant AYL
DIPENTENE Plant HHB
EO 1,500-10,000 Plant HHB ABS
ERIODICTYOL Plant PAS
EUGENOL Shoot NAP
FAT 88,940-125,000 Plant USA
FIBER 109,900-203,870 Plant USA
GAMMA-CADINENE 50-200 Plant BML
GAMMA-ELEMENE Plant BML
GAMMA-MUUROLENE 16-88 Plant BML
GAMMA-TERPINENE 1-1,476 Plant BML JAD NAP
GERANIOL Plant BML
GERANYL-ACETATE 1-96 Plant BML
GERMACRENE-D 190-1,192 Plant BML
INDOLE-3-ACETIC-ACID Plant PAS
IRON 325-598 Plant BML
ISOVITEXIN Plant PAS
KAEMPFEROL Plant BML
LAURIC-ACID 200-215 Plant USA
LEDENE Plant BML
LEPTOSIDIN Plant NAP
LIMONENE 22-480 Plant BML
LINALOL 1-600 Plant BML
LINALYL-ACETATE 8-184 Plant BML
LINOLEIC-ACID 1,050-1,131 Plant USA

LUTEOLIN Plant NAP
LUTEOLIN-7-O-BETA-D-GLUCURONIDE Plant NAP
MAGNESIUM 2,600-3,016 Plant USA
MANGANESE 47 Plant USA
MYRCENE 8-180 Plant BML
MYRISTIC-ACID 1,700-1,831 Plant USA
NARINGENIN Plant FNF
NARINGIN Plant JBH NAP
NERYL-ACETATE Plant BML
NIACIN 62-67 Plant USA
OCIMENE 24-96 Plant BML
OLEANOLIC-ACID Plant PAS
OLEIC-ACID 5,100-5,493 Plant USA
ORIENTIN Plant PAS
P-COUMARIC-ACID Plant PAS
P-CYMENE 3-3,237 Plant BML JAD NAP
P-HYDROXYBENZOIC-ACID Plant NAP
PALMITIC-ACID 17,000-18,309 Plant USA
PALMITOLEIC-ACID 1,600-1,723 Plant USA
PENTYL-ALCOHOL Plant FNF
PEONIDIN Plant NAP
PHOSPHORUS 1,769-2,402 Plant USA
PHYTOSTEROLS 2,030-2,186 Plant USA
POTASSIUM 16,066-18,647 Plant USA
PROTEIN 96,870-132,611 Plant USA
PROTOCATECHUIC-ACID 132 Leaf NAP
QUERCETIN Plant PAS
RETUSIN Plant FNF
ROSMARINIC-ACID 790 Leaf NAP
ROSMARINIC-ACID 4,000-16,600 Plant ABS HHB NAP
ROSMARINIC-ACID 55,000 Plant FT62:166
RUTIN Plant PAS
SABINENE 50-2,096 Plant BML
SELINENE Plant HHB
SODIUM 109-205 Plant USA
SPATHULENOL 10-50 Plant BML
STACHYOSE Plant HHB
STEARIC-ACID 5,800-6,247 Plant USA
TANNINS 80,000 Plant HHB
TERPINEN-4-OL 17-390 Plant BML
TERPINOLENE 2-16 Plant BML
TERPINYL-ACETATE 54-216 Plant BML
THIAMIN 3-4 Plant USA
THYMOL 2-5,000 Plant BML HHB
THYMYL-METHYL-ETHER 12-60 Plant BML

TRANS,TRANS-ALPHA-FARNESENE 48-240 Plant BML
TRANS-2-HEXANAL 132-660 Plant BML
TRANS-ANETHOLE Plant PAS
TRANS-OCIMENE 76-824 Plant BML
URSOLIC-ACID 3,100 Plant HHB
VANILLIC-ACID Plant NAP
VITEXIN Plant PAS
WATER 68,380-74,820 Plant USA
ZINC 39-49 Plant USA

Origanum vulgare var. *gracile* (C. KOCH) IETSWAART
"Slender Turkish Oregano"
Unaggregated A

(E)-2-HEXENAL 0.5 Plant JEO5:425
(E)-BETA-OCIMENE 25 Plant JEO5:425
(Z)-BETA-OCIMENE 20 Plant JEO5:425
1,8-CINEOLE 3 Plant JEO5:425
1-NONEN-3-OL <1 Plant JEO5:425
1-OCTEN-3-OL 2.5 Plant JEO5:425
2-ETHYL-1-HEXANOL 1.5 Plant JEO5:425
6-METHYL-3-HEPTANOL 1 Plant JEO5:425
ALPHA-CADINOL 20 Plant JEO5:425
ALPHA-FARNESENE 7 Plant JEO5:425
ALPHA-HIMACHALENE 7 Plant JEO5:425
ALPHA-HUMULENE 10 Plant JEO5:425
ALPHA-MUUROLENE 2 Plant JEO5:425
ALPHA-PINENE 1 Plant JEO5:425
ALPHA-TERPINENE <1 Plant JEO5:425
ALPHA-TERPINEOL 4 Plant JEO5:425
BETA-BISABOLENE 20 Plant JEO5:425
BETA-BOURBONENE 2 Plant JEO5:425
BETA-CARYOPHYLLENE 35 Plant JEO5:425
BETA-PHELLANDRENE 1 Plant JEO5:425
BETA-PINENE 1 Plant JEO5:425
CAMPHENE <1 Plant JEO5:425
CARVACROL 6 Plant JEO5:425
CIS-DIHYDROCARVONE Plant JEO5:425
CIS-SABINENE-HYDRATE 0.5 Plant JEO5:425
DELTA-3-CARENE Plant JEO5:425
DELTA-CADINENE 7 Plant JEO5:425
DELTA-CADINOL 5 Plant JEO5:425
EO 400 Plant JEO5:425
GAMMA-CADINENE Plant JEO5:425
GAMMA-ELEMENE Plant JEO5:425
GAMMA-TERPINENE 3 Plant JEO5:425
GERANIOL 0.5 Plant JEO5:425
GERANYL-ACETATE 7 Plant JEO5:425
GERMACRENE-D 50 Plant JEO5:425
ISOBORNEOL 1 Plant JEO5:425
LIMONENE 2 Plant JEO5:425
LINALOL 6 Plant JEO5:425
MYRCENE 6 Plant JEO5:425

NONANE <1 Plant JEO5:425
NONANOL <1 Plant JEO5:425
P-CYMEN-8-OL Plant JEO5:425
P-CYMENE <1 Plant JEO5:425
PATCHOULANE 7 Plant JEO5:425
SABINENE 7 Plant JEO5:425
SPATHULENOL 2.5 Plant JEO5:425
TERPINEN-4-OL 35 Plant JEO5:425
TERPINOLENE <1 Plant JEO5:425
TORREYOL 4 Plant JEO5:425
TRANS-DIHYDROCARVONE Plant JEO5:425
TRANS-SABINENE-HYDRATE 2.5 Plant JEO5:425
UNDECANE <1 Plant JEO5:425

Origanum vulgare ssp. *hirtum* (LINK) IETSWAART
"Greek Oregano"
Unaggregated A

4-TERPINEOL 90 Shoot ZLU197:20
ALPHA-CARYOPHYLLENE Shoot ZLU197:20
ALPHA-HUMULENE Shoot ZLU197:20
ALPHA-MUUROLENE Shoot ZLU197:20
ALPHA-PHELLANDRENE 45 Shoot ZLU197:20
ALPHA-PINENE 225 Shoot ZLU197:20
ALPHA-TERPINENE 495 Shoot ZLU197:20
ALPHA-TERPINEOL Shoot ZLU197:20
ALPHA-THUJENE 315 Shoot ZLU197:20
AROMADENDRENE Shoot ZLU197:20
BETA-BISABOLENE 270 Shoot ZLU197:20
BETA-CARYOPHYLLENE 360 Shoot ZLU197:20
BETA-PINENE 45 Shoot ZLU197:20
BORNEOL 45 Shoot ZLU197:20
BORNYL-ACETATE <0.1 Shoot ZLU197:20
CAMPHENE 45 Shoot ZLU197:20
CARVACROL 26,865 Shoot ZLU197:20
CARVACROL-METHYL-ETHER Shoot ZLU197:20
CARVONE Shoot ZLU197:20
CIS-OCIMENE Shoot ZLU197:20
CIS-PINANE Shoot ZLU197:20
CIS-THUJANOL 180 Shoot ZLU197:20
DELTA-3-CARENE <0.1 Shoot ZLU197:20
EO 45,000 Shoot ZLU197:20
GAMMA-CADINENE Shoot ZLU197:20
GAMMA-GURJUNENE 270 Shoot ZLU197:20
GAMMA-TERPINENE 4,950 Shoot ZLU197:20
LIMONENE 315 Shoot ZLU197:20
MYRCENE 585 Shoot ZLU197:20
P-CYMENE 3,420 Shoot ZLU197:20
TERPINOLENE Shoot ZLU197:20
THYMOL 6,165 Shoot ZLU197:20
TRANS-OCIMENE Shoot ZLU197:20
TRANS-THUJANOL 90 Shoot ZLU197:20

Origanum vulgare var. *hirtum* (LINK) IETSWAART "Istanbul Kekigi, Turkish Oregano" Unaggregated A

(E)-2-HEXENAL 15 Plant JEO5:425
(E)-BETA-OCIMENE 40 Plant JEO5:425
(Z)-BETA-OCIMENE 7 Plant JEO5:425
1,8-CINEOLE 7 Plant JEO5:425
1-OCTEN-3-OL 105 Plant JEO5:425
6-METHYL-3-HEPTANOL <1 Plant JEO5:425
ALPHA-HUMULENE 120 Plant JEO5:425
ALPHA-PINENE 730 Plant JEO5:425
ALPHA-TERPINENE 435 Plant JEO5:425
ALPHA-TERPINEOL 65 Plant JEO5:425
BETA-BISABOLENE 200 Plant JEO5:425
BETA-CARYOPHYLLENE 720 Plant JEO5:425
BETA-PHELLANDRENE 60 Plant JEO5:425
BETA-PINENE 50 Plant JEO5:425
CAMPHENE 60 Plant JEO5:425
CARVACROL 23,960 Plant JEO5:425
CIS-DIHYDROCARVONE 3 Plant JEO5:425
CIS-SABINENE-HYDRATE 60 Plant JEO5:425
DELTA-3-CARENE 25 Plant JEO5:425
EO 34,000 Plant JEO5:425
GAMMA-CADINENE 7 Plant JEO5:425
GAMMA-ELEMENE 35 Plant JEO5:425
GAMMA-TERPINENE 2,795 Plant JEO5:425
ISOBORNEOL 65 Plant JEO5:425
LIMONENE 70 Plant JEO5:425
LINALOL 30 Plant JEO5:425
LINALYL-ACETATE 7 Plant JEO5:425
METHYL-2-METHYLBUTYRATE 14 Plant JEO5:425
METHYL-CARVACROL 30 Plant JEO5:425
MYRCENE 645 Plant JEO5:425
NONANAL <1 Plant JEO5:425
P-CYMEN-8-OL 25 Plant JEO5:425
P-CYMENE 2,600 Plant JEO5:425
PATCHOULANE 80 Plant JEO5:425
SABINENE 35 Plant JEO5:425
SPATHULENOL 7 Plant JEO5:425
TERPINEN-4-OL 95 Plant JEO5:425
TERPINOLENE 15 Plant JEO5:425
THYMOL 120 Plant JEO5:425
TRANS-DIHYDROCARVONE 20 Plant JEO5:425
TRANS-SABINENE-HYDRATE 165 Plant JEO5:425

Origanum vulgare var. *viride* (BOISS.) HAYEK
"Green Turkish Oregano"
Unaggregated A

(E)-2-HEXENOL 0.3 Plant JEO5:425
(E)-BETA-OCIMENE 50 Plant JEO5:425
(E,Z)-2,6-NONADIENAL 0.5 Plant JEO5:425
(Z)-2-NONENAL 0.6 Plant JEO5:425
(Z)-BETA-FARNESENE Plant JEO5:425
(Z)-BETA-OCIMENE 5 Plant JEO5:425
(Z)-GERANYL-ACETONE 0.5 Plant JEO5:425
1,8-CINEOLE 11 Plant JEO5:425
1-NONEN-3-OL Plant JEO5:425
1-OCTEN-3-OL 2 Plant JEO5:425
2,5-DIETHYLTETRAHYDROFURAN 0.1 Plant JEO5:425
2-ETHYL-1-HEXANOL 6 Plant JEO5:425
3-NONANONE Plant JEO5:425
6-METHYL-3-HEPTANOL 3 Plant JEO5:425
ALPHA-BISABOLENE Plant JEO5:425
ALPHA-CADINOL 50 Plant JEO5:425
ALPHA-FARNESENE 25 Plant JEO5:425
ALPHA-HIMACHALENE 14 Plant JEO5:425
ALPHA-HUMULENE 40 Plant JEO5:425
ALPHA-MUUROLENE 6 Plant JEO5:425
ALPHA-PINENE 2 Plant JEO5:425
ALPHA-TERPINENE 3 Plant JEO5:425
ALPHA-TERPINEOL 18 Plant JEO5:425
AMYL-FURAN 0.3 Plant JEO5:425
BETA-BISABOLENE 80 Plant JEO5:425
BETA-BOURBONENE 9 Plant JEO5:425
BETA-CARYOPHYLLENE 1 Plant JEO5:425
BETA-CUBEBENE Plant JEO5:425
BETA-IONONE Plant JEO5:425
BETA-PHELLANDRENE 4 Plant JEO5:425
BETA-PINENE 3.5 Plant JEO5:425
BETA-SESQUIPHELLANDRENE 1.5 Plant JEO5:425
BORNYL-ACETATE 13 Plant JEO5:425
CAMPHENE 1 Plant JEO5:425
CARVACROL 12 Plant JEO5:425
CIS-DIHYDROCARVONE Plant JEO5:425
CIS-SABINENE-HYDRATE 0.9 Plant JEO5:425
DECANE Plant JEO5:425
DECANOIC-ACID Plant JEO5:425
DECANOL 0.9 Plant JEO5:425
DELTA-3-CARENE Plant JEO5:425

DELTA-CADINENE 25 Plant JEO5:425
DELTA-CADINOL 6 Plant JEO5:425
ELEMOL Plant JEO5:425
EO 1,300 Plant JEO5:425
GAMMA-CADINENE Plant JEO5:425
GAMMA-ELEMENE Plant JEO5:425
GAMMA-TERPINENE 13 Plant JEO5:425
GERANIOL 2.5 Plant JEO5:425
GERANYL-ACETATE 25 Plant JEO5:425
GERMACRENE-D 205 Plant JEO5:425
HENEICOSANE Plant JEO5:425
HEXAHYDROFARNESYL-ACETONE 4 Plant JEO5:425
HEXANAL Plant JEO5:425
ISOBORNEOL 2.5 Plant JEO5:425
LIMONENE 17 Plant JEO5:425
LINALOL 30 Plant JEO5:425
LINALYL-ACETATE 50 Plant JEO5:425
MYRCENE 5 Plant JEO5:425
NEROL 0.9 Plant JEO5:425
NEROLIDOL 3 Plant JEO5:425
NONANAL 1.5 Plant JEO5:425
NONANE Plant JEO5:425
NONANOIC-ACID 0.8 Plant JEO5:425
OCTENYL-ACETATE 15 Plant JEO5:425
P-CYMEN-8-OL Plant JEO5:425
P-CYMENE 3 Plant JEO5:425
PATCHOULANE 21 Plant JEO5:425
PERILLENE Plant JEO5:425
PIPERITENONE-OXIDE 6 Plant JEO5:425
PIPERITONE-OXIDE <0.1 Plant JEO5:425
PIPERITONE-OXIDE-II 35 Plant JEO5:425
SABINENE 20 Plant JEO5:425
SPATHULENOL 15 Plant JEO5:425
TERPINEN-4-OL 220 Plant JEO5:425
TERPINOLENE <1 Plant JEO5:425
THYMOL Plant JEO5:425
TORREYOL 15 Plant JEO5:425
TRANS-DIHYDROCARVONE Plant JEO5:425
TRANS-SABINENE-HYDRATE 3.5 Plant JEO5:425
TRICHOSANE Plant JEO5:425
UNDECANE Plant JEO5:425

Perilla frutescens (L.) BRITTON "Perilla"

(2E,4E,63)-ALLOFARNESENE Plant JSG
(2Z,4E,63)-ALLOFARNESENE Plant JSG
ALPHA-BERGAMOTENE Plant JAF40:2328
ALPHA-PINENE Plant JSG
AMYGDALIN Leaf JBH
APIGENIN Seed CRC
ASCORBIC-ACID 460-3,286 Leaf CRC
ASCORBIC-ACID 0-200 Seed CRC
ASH 20,000-143,000 Leaf CRC
ASH 10,000-85,000 Seed CRC
BETA-CAROTENE 44-313 Leaf CRC
BETA-CAROTENE 7-66 Seed CRC
BETA-CARYOPHYLLENE Plant JAF40:2328
CAFFEIC-ACID Plant JSG
CALCIUM 1,970-14,071 Leaf CRC
CALCIUM 12,670 Plant SMO
CALCIUM 1,400-3,500 Seed CRC
CARBOHYDRATES 571,000 Leaf CRC
CARBOHYDRATES 80,000 Plant CRC
CARBOHYDRATES 69,000-585,000 Seed CRC
CITRAL Plant CCO
COPPER 17 Plant SMO
CYANIDIN-3,5-DIGLUCOSIDE Leaf CRC
DELPHINIDIN Leaf WOI
DILLAPIOL 2,200-3,400 Leaf PC25:2085
ELEMICIN 2,500-3,200 Leaf PC25:2085
ELSHOLTZIAKETONE Plant PC25:859
EO 1,000-13,000 Plant WOI
FAT 6,000 Leaf CRC
FAT 13,000-481,000 Seed CRC
FIBER 15,000 Leaf CRC
FIBER 37,000-280,000 Seed CRC
FURYLKETONE Plant CCO PC25:859
GERANYL-PYROPHOSPHATE Plant PC25:859
GLUCOSE Leaf WOI
IRON 67-479 Leaf CRC
IRON 460 Plant SMO
IRON 13-135 Seed CRC
ISOEGOMAKEKETONE Plant PC25:859
LIMONENE Plant CRC JAF40:2328 PC25:859
LINALOL Plant JAF40:2328
LUTEOLIN Leaf ABS

LUTEOLIN Seed CRC
MAGNESIUM 3,830 Plant SMO
MANGANESE 180 Plant SMO
MYRISTICIN 300-3,800 Leaf PC25:2085
NAGINATAKETONE Plant PC25:859
NIACIN 8-57 Leaf CRC
NIACIN 5-42 Seed CRC
PERILLALDEHYDE 400-825 Plant CCO JBH
PERILLANIN-CHLORIDE Plant CRC
PERILLARTINE Plant CRC
PHENYLPROPANOIDS Plant CCO
PHOSPHORUS 730-5,214 Leaf CRC
PHOSPHORUS 330-5,339 Seed CRC
POTASSIUM 6,500-46,429 Leaf CRC
POTASSIUM 26,100 Plant SMO
PROTEIN 34,000 Leaf CRC
PROTEIN 26,000-318,000 Seed CRC
PROTOCATECHUIC-ACID Leaf WOI
PRUNASIN Leaf JBH
QUERCETIN Leaf ABS
RIBOFLAVIN 3-23 Leaf CRC
RIBOFLAVIN 1-8 Seed CRC
ROSMARINIC-ACID 10,300-12,100 Leaf YAK106:1108
ROSMARINIC-ACID 2,300 Seed YAK106:1108
ROSMARINIC-ACID 2,900 Stem YAK106:1108
SCUTELLEREIN Plant JSG
SODIUM 20-1,429 Leaf CRC
SODIUM 204 Plant SMO
THIAMIN 1-5 Leaf CRC
THIAMIN 1-7 Seed CRC
WATER 860,000 Leaf CRC
WATER 63,000-178,000 Seed CRC WOI
XOI-A 11-78 Plant CPB38:1772
XOI-B 3-21 Plant CPB38:1772
ZINC 50 Plant SMO

Plectranthus amboinicus (LOUR.) SPRENGEL "Amboini Coleus, Country Borage, Cuban Oregano, French Thyme, Indian Borage, Soup Mint, Spanish Thyme"

2ALPHA,3ALPHA,19ALPHA,23-TETRAHYDROXYURSOLIC-ACID Plant JSG
2ALPHA,3ALPHA-DIHYDROXYOLEANOLIC-ACID Plant JSG
ALPHA-THUJENE Plant FNF
ASCORBIC-ACID 120-2,145 Leaf CRC
ASH 12,000-214,285 Leaf CRC
BETA-CAROTENE 10-175 Leaf CRC
CALCIUM 2,320-41,430 Leaf CRC
CARBOHYDRATES 31,000-553,570 Leaf CRC
CARVACROL 43% Plant HHB
CHRYSOERIOL Leaf JBH
CRATEGOLIC-ACID Plant JSG
EUSCAPHIC-ACID Plant JSG
FAT 4,000-71,430 Leaf CRC
FIBER 5,000-89,285 Leaf CRC
IRON 39-695 Leaf CRC
KILOCALORIES 170-3,035/kg Leaf CRC
NIACIN 2-36 Leaf CRC
OLEANOLIC-ACID Plant JSG
OXALIC-ACID Leaf HHB
PHOSPHORUS 80-1,428 Leaf CRC
POMOLIC ACID Plant JSG
PROTEIN 9,000-160,715 Leaf CRC
RIBOFLAVIN 0.7-12.5 Leaf CRC
THIAMIN 0.3-5.4 Leaf CRC
TOMENTIC-ACID Plant JSG
URSOLIC-ACID Plant JSG
WATER 944,000 Leaf CRC

Plectranthus coleoides MARGINATUS Unaggregated A

1,1-DIPHENYLPROPANE 130 Shoot JEO5:311
1,3-BIS(DIMETHYLAMINO)-BENZENE 125 Shoot JEO5:311
1,3-DIMETHYL-BENZENE 48 Shoot JEO5:311
1,3-TRIMETHYL-BENZENE 70 Shoot JEO5:311
1-PHENYLETHYL-ACETATE 12 Shoot JEO5:311
2,2-DIMETHYL-TETRADECANE 70 Shoot JEO5:311
2-ETHYL-HEXANOL 36 Shoot JEO5:311
2-METHYL-HEPTAN-4-ONE 42 Shoot JEO5:311
2-METHYL-UNDEC-2-ENE 24 Shoot JEO5:311
2-METHYLBUTANOL 24 Shoot JEO5:311
2-PROPYL-HEPT-2-ENAL 18 Shoot JEO5:311
3,4-DIMETHYL-(2H)-PYRAN-2-ONE 440 Shoot JEO5:311
3-HEXENAL 260 Shoot JEO5:311
4,6-DECADIENAL 55 Shoot JEO5:311
4-METHYL-PENT-2-ENAL 70 Shoot JEO5:311
ALPHA-HUMULENE 80 Shoot JEO5:311
BETA-CARYOPHYLLENE 150 Shoot JEO5:311
BORNEOL 120 Shoot JEO5:311
BORNYL-ACETATE 215 Shoot JEO5:311
BUTYL-ACETATE 6-42 Shoot JEO5:311
BUTYL-FORMATE 12 Shoot JEO5:311
CARVONE 55 Shoot JEO5:311
CITRONELLYL-FORMATE 70 Shoot JEO5:311
EO 6,000 Shoot JEO5:311
ETHYL-ACETATE 12 Shoot JEO5:311
FENCHONE 275 Shoot JEO5:311
GERANYL-ACETATE 65 Shoot JEO5:311
HEXADECANE 24 Shoot JEO5:311
HEXYL-CINNAMALDEHYDE 6 Shoot JEO5:311
ISOBORNYL-ACETATE 260 Shoot JEO5:311
ISOPENTYL-FORMATE 24 Shoot JEO5:311
LINALYL-FORMATE 80 Shoot JEO5:311
N,N-DIMETHYL-(1H)-PURIN-6-AMINE 490 Shoot JEO5:311
N,N-DIMETHYL-DODECYLAMINE 260 Shoot JEO5:311
N,N-DIMETHYL-FORMAMIDE Shoot JEO5:311
N,N-DIMETHYL-PENTYLAMINE 70 Shoot JEO5:311
NONYL-ACETATE 6 Shoot JEO5:311
P-MENTHA-2-EN-1-OL 95 Shoot JEO5:311
PENTADECYLAMINE 42 Shoot JEO5:311
TETRADECANE 550 Shoot JEO5:311

Plectranthus incanus LINK.
"Lal Agada"
Unaggregated A

ALPHA-CADINENE 40 Shoot JEO4:57
ALPHA-COPAENE 25 Shoot JEO4:57
ALPHA-HUMULENE 25 Shoot JEO4:57
BETA-BISABOLENE 30 Shoot JEO4:57
BETA-CARYOPHYLLENE 100 Shoot JEO4:57
BETA-CUBEBENE 20 Shoot JEO4:57
CIS-PIPERITONE-OXIDE 1,250 Shoot JEO4:57
DELTA-CADINENE 30 Shoot JEO4:57
EO 3,500 Shoot JEO4:57
FENCHONE 170 Shoot JEO4:57
HYDROXYPIPERITONES 7 Shoot JEO4:57
LIMONENE 65 Shoot JEO4:57
PIPERITENONE 65 Shoot JEO4:57
PIPERITENONE-OXIDE 1,575 Shoot JEO4:57
TRANS-PIPERITONE-OXIDE 0.4 Shoot JEO4:57

Pogostemon cablin (BLANCO) BENTH. "Patchouli"

1,10-EPOXY-ALPHABULNESENE Plant FNF
1-ALPHA,5-ALPHA-EPOXY-ALPHA-GUAIENE Plant FNF
1-BETA,5-BETA-EPOXY-ALPHA-GUAIENE Plant FNF
2-METHYLBUTYRIC-ACID Leaf FNF
2-METHYLHEXANOIC-ACID Leaf FNF
4-METHYLPENTANOIC-ACID Leaf FNF
ALPHA-BULNESENE Plant FNF
ALPHA-BULNESENE-OXIDE Plant FNF
ALPHA-BULNESONE 1,260-5,460 Plant FNF
ALPHA-GUAIENE 1,261-5,460 Plant FNF
ALPHA-GUAIENE-OXIDE Plant FNF
ALPHA-PATCHOULENE Plant FNF
ALPHA-PINENE Plant FNF
APIGENIN Plant FNF
APIGENIN-7-O-BETA-D-(-6"-P-COUMAROYL)-GLUCOSIDE Plant
FNF APIGENIN-7-O-BETA-GLUCOSIDE Plant FNF
AZULENE Plant FNF
BENZALDEHYDE Plant FNF
BETA-ELEMENE Plant FNF
BETA-PATCHOULENE 120-464 Plant FNF
BETA-PINENE Plant FNF
BULNESOL Plant FNF
CADINENE Plant FNF
CAMPHENE Plant FNF
CARYOPHYLLENE Plant FNF
CARYOPHYLLENE-OXIDE Plant FNF
CINNAMALDEHYDE Plant FNF
CIS-2-PENTYLCYCLOPROPYLCARBOXYLIC-ACID Leaf FNF
CYCLOSEYCHELLENE Plant FNF
DEHYDRACETIC-ACID Leaf FNF
DIMETHYLPHENOL Leaf FNF
EO 15,000-58,000 Plant FNF
EPIGUAIPYRIDINE Plant FNF
EPOXYCARYOPHYLLENE Plant FNF
EUGENOL Plant FNF
GAMMA-PATCHOULENE Plant FNF
GUAIACOL Leaf FNF
GUAIPYRIDINE Plant FNF
HEPTANOIC-ACID Leaf FNF
HUMULENE Plant FNF
LIMONENE Plant FNF
NONANOIC-ACID Leaf FNF

NORDEHYDROPATCHOULOL Plant FNF
NORPATCHOULENOL Plant FNF
O-CRESOL Leaf FNF
OCTANOIC-ACID Leaf FNF
OMBUINE Plant FNF
P-VINYLPHENOL Leaf FNF
PACHYPODOL Plant FNF
PATCHOULI-ALCOHOL 2,100-23,200 Plant FNF
PATCHOULIPYRIDINE Plant FNF
PENTANOIC-ACID Leaf FNF
PHENOL Leaf FNF
POGOSTOL Plant FNF
POGOSTONE Plant FNF
RHAMNETIN Plant FNF
SEYCHELLENE Plant FNF
TANNIN 17,000 Plant FNF
TRANS-2-PENTYLCYCLOPROPYLCARBOXYLIC-ACID Leaf FNF

Prunella vulgaris L.
"Self-Heal"

ARSENIC 0.37 Flower CAL
BETULINIC-ACID 738 Root NAP
BETULINIC-ACID 2.88 Shoot NAP
CALCIUM 14,500 Flower CAL
CIS-CAFFEIC-ACID Plant HHB
COPPER 8 Flower CAL
CYANIDIN Flower CCO
D-CAMPHOR Flower CCO
D-FENCHONE Flower CCO
DELPHINIDIN Flower CCO
EO 5,000 Plant HHB
HYPEROSIDE Plant CCO
IRON 640 Flower CAL
MAGNESIUM 4,560 Flower CAL
MANGANESE 96 Flower CAL
MERCURY 0.06 Flower CAL
OLEANOLIC-ACID Plant CCO
POTASIUM-CHLORIDE 23,800 Plant CCO
POTASSIUM 11,900 Flower CAL
ROSMARINIC-ACID 2,600-19,300 Flower YAK106:1108
ROSMARINIC-ACID 17,400 Leaf YAK106:1108
ROSMARINIC-ACID 17,400-61,000 Plant FT62:166
RUTIN Plant CCO
SODIUM 155 Flower CAL
TRANS-CAFFEIC-ACID Plant HHB
URSOLIC-ACID Plant JBH
ZINC 25 Flower CAL

Pycnanthemum albescens TORR. & GRAY "White Mountain Mint"

1,8-CINEOLE 80-270 Shoot BML
8-ACETOXYCARVOTANACETONE 24-81 Shoot BML
ALPHA-PINENE 16-54 Shoot BML
ALPHA-TERPINEOL 48-162 Shoot BML
ALPHA-TERPINEYL-ACETATE 2,264-7,641 Shoot BML
BETA-BOURBONENE 16-54 Shoot BML
BETA-COPAENE Shoot BML
CARVONE 128-14,310 Shoot BML
CIS-CARVEOL 496-1,674 Shoot BML
CIS-DIHYDROCARVONE 8-27 Shoot BML
ISOMENTHONE 80-270 Shoot BML
LEDENE 8-27 Shoot BML
LIMONENE 1,688-5,697 Shoot BML
LINALOL 32-108 Shoot BML
MENTHONE 680-2,295 Shoot BML
PULEGONE 24-14,283 Shoot BML
SABINENE 160-540 Shoot BML
TRANS-CARVEOL 1,112-3,753 Shoot BML
TRANS-DIHYDROCARVONE 88-297 Shoot BML

Pycnanthemum beadlei (SMALL) FERNALD
"Beadle's Mountain Mint"
Unaggregated A

1,8-CINEOLE 28 Shoot BML
2-OCTEN-3-OL 133 Shoot BML
ALPHA-HUMULENE 49 Shoot BML
ALPHA-PINENE 21 Shoot BML
BETA-PINENE 14 Shoot BML
BORNEOL 42 Shoot BML
CAMPHENE 7 Shoot BML
CARYOPHYLLENE 49 Shoot BML
CIS-ISOPULEGONE 28 Shoot BML
EO 7,000 Shoot BML
GERMACRENE-D 21 Shoot BML
ISOMENTHONE 35 Shoot BML
LIMONENE 329 Shoot BML
MENTHOL 140 Shoot BML
MENTHONE 644 Shoot BML
MYRCENE 28 Shoot BML
P-CYMENE 21 Shoot BML
PIPERITENONE 1,533 Shoot BML
PIPERITONE 168 Shoot BML
PULEGONE 3,577 Shoot BML
SABINENE 7 Shoot BML
TRANS-ISOPULEGONE 35 Shoot BML

Pycnanthemum californicum TORR. "California Mountain Mint" Unaggregated A

1,8-CINEOLE 576 Shoot BML
1-OCTEN-3-OL 252 Shoot BML
1-OCTEN-3-YL-ACETATE 144 Shoot BML
ALPHA-HUMULENE 468 Shoot BML
ALPHA-PINENE 144 Shoot BML
ALPHA-TERPINEOL Shoot tr BML
BETA-PINENE 108 Shoot BML
CARYOPHYLLENE 144 Shoot BML
CIS-ISOPULEGONE 216 Shoot BML
EO 36,000 Shoot BML
ISOMENTHONE 3,708 Shoot BML
LIMONENE 144 Shoot BML
MENTHONE Shoot tr BML
MYRCENE 108 Shoot BML
P-CYMENE Shoot tr BML
PIPERITENONE 108 Shoot BML
PULEGONE 28,836 Shoot BML
SABINENE 180 Shoot BML
TRANS-ISOPULEGONE 108 Shoot BML

Pycnanthemum clinopodioides TORR. & GRAY "Clinopod Mountain Mint"

1,8-CINEOLE 21-88 Shoot BML
1-OCTEN-3-OL 105-242 Shoot BML
3-OCTANOL 126-132 Shoot BML
3-OCTANONE 84-88 Shoot BML
ALPHA-PINENE 84-88 Shoot BML
ALPHA-TERPINEOL 42-44 Shoot BML
BETA-PINENE 63-66 Shoot BML
CAMPHOR 42-44 Shoot BML
CARVONE 42-44 Shoot BML
CARYOPHYLLENE 42-704 Shoot BML
CIS-ISOPULEGONE 63-66 Shoot BML
EO 21,000-22,000 Shoot BML
EPIBICYCLOSESQUIPHELLANDRENE 273-286 Shoot BML
GERMACRENE-D 42-374 Shoot BML
ISOMENTHONE 21-528 Shoot BML
ISOPIPERITENONE 63-66 Shoot BML
LIMONENE 1,134-10,780 Shoot BML
LINALOL 84-88 Plant FNF
MENTHOL 168-176 Shoot BML
MENTHONE 126-12,826 Shoot BML
METHYL-EUGENOL 84-88 Shoot BML
MYRCENE 84-352 Shoot BML
P-CYMENE 63-66 Shoot BML
PIPERITENONE 21-3,586 Shoot BML
PIPERITONE 42-3,586 Shoot BML
PULEGONE 105-5,984 Shoot BML
SABINENE 21-242 Shoot BML
TERPINEN-4-OL 294-308 Shoot BML
TRANS-ANETHOLE Shoot tr BML
TRANS-ISOPULEGONE 63-66 Shoot BML

Pycnanthemum curvipes (GREENE) GRANT & EPLING "Curved Mountain Mint"
Unaggregated A

ALPHA-PHELLANDRENE Shoot tr BML
ALPHA-PINENE 84 Shoot BML
BETA-PINENE 42 Shoot BML
CAMPHENE Shoot tr BML
CARVONE 42 Shoot BML
CIS-CARVEOL Shoot tr BML
CIS-DIHYDROCARVONE 16,443 Shoot BML
EO 21,000 Shoot BML
ISO-DIHYDROCARVEOL 231 Shoot BML
LIMONENE 2,323 Shoot BML
MYRCENE 63 Shoot BML
SABINENE 21 Shoot BML
TRANS-DIHYDROCARVONE 1,680 Shoot BML

Pycnanthemum incanum (L.) MICHX. "Hoary Mountain Mint"

1,8-CINEOLE 33-76 Shoot BML
1-OCTEN-3-OL 22-95 Shoot BML
3-OCTANOL 22-95 Shoot BML
3-OCTYL-3-YL-ACETATE Shoot tr BML
3-OCTYL-ACETATE Shoot tr BML
ALPHA-PINENE 55-114 Shoot BML
ALPHA-TERPINEOL Shoot tr BML
BETA-ELEMENE 22-190 Shoot BML
BETA-PINENE 44-95 Shoot BML
BICYCLOGERMACRENE Shoot tr BML
BORNEOL 33-57 Shoot BML
CARYOPHYLLENE Shoot tr BML
CIS-ISOPULEGONE 110-190 Shoot BML
EO 11,000-19,000 Shoot BML
GERMACRENE-D Shoot tr BML
ISOMENTHONE 352-931 Shoot BML
LIMONENE 209-1,216 Shoot BML
LINALOL 11-38 Shoot BML
MENTHOL 22-798 Shoot BML
MENTHONE 3,036-10,127 Shoot BML
MYRCENE 44-114 Shoot BML
NEOMENTHOL 22-95 Shoot BML
P-CYMENE 22-38 Shoot BML
PIPERITENONE 11-19 Shoot BML
PIPERITONE Shoot tr BML
PULEGONE 3,366-10,336 Shoot BML
SABINENE 11-38 Shoot BML
TRANS-ISOPULEGONE 88-152 Shoot BML

Pycnanthemum loomisii NUTT. "Loomis' Mountain Mint"

1,8-CINEOLE 17,853-26,040 Shoot BML
ALPHA-PINENE 2,046-2,646 Shoot BML
ALPHA-TERPINENE 33-168 Shoot BML
ALPHA-TERPINEOL 1,452-1,932 Shoot BML
BETA-PINENE 825-1,218 Shoot BML
BORNYL-ACETATE 33-42 Shoot BML
CAMPHENE 330-1,260 Shoot BML
CARVONE 1-42 Shoot BML
CARYOPHYLLENE 330-672 Shoot BML
CARYOPHYLLENE-OXIDE Shoot tr BML
CIS-P-MENTH-2-EN-1-OL 33-42 Shoot BML
DELTA-ELEMENE 33-42 Shoot BML
DELTA-TERPINEOL 33-84 Shoot BML
ELEMOL Shoot tr BML
EO 31,000-42,000 Shoot BML
GAMMA-TERPINENE 99-420 Shoot BML
GERMACRENE-D 66-84 Shoot BML
LIMONENE 330-2,520 Shoot BML
LINALOL 33-84 Shoot BML
MYRCENE 495-756 Shoot BML
P-CYMENE 33-84 Shoot BML
PULEGONE 33-84 Shoot BML
SABINENE 4,884-7,224 Shoot BML
TERPINEN-4-OL 396-714 Shoot BML
TERPINOLENE 33-84 Shoot BML
TRANS-NEROLIDOL Shoot tr BML
TRANS-SABINENE-HYDRATE 66-168 Shoot BML

Pycnanthemum montanum MICHX. "Montane Mountain Mint"

1,8-CINEOLE 21-720 Shoot BML
1-OCTEN-3-OL 7-152 Shoot BML
1-OCTEN-3-YL-ACETATE 7-8 Shoot BML
3-OCTANOL 35-50 Shoot BML
ALPHA-PINENE 7-272 Shoot BML
ALPHA-THUJENE 28-32 Shoot BML
BETA-BOURBONENE 91-104 Shoot BML
BETA-PINENE 7-56 Shoot BML
BICYCLOGERMACRENE 63-72 Shoot BML
BORNEOL 21-24 Shoot BML
CAMPHENE 833-952 Shoot BML
CAMPHOR 3,395-3,880 Shoot BML
CARYOPHYLLENE 42-56 Shoot BML
EO 7,000-8,000 Shoot BML
GERMACRENE-D 21-32 Shoot BML
ISOMENTHONE 42-4,848 Shoot BML
ISOPIPERITENONE 14-16 Shoot BML
LIMONENE 287-416 Shoot BML
LINALOL 105-120 Shoot BML
MENTHOL 28-32 Shoot BML
MENTHONE 371-640 Shoot BML
MYRCENE 7-16 Shoot BML
NEOMENTHOL 14-16 Shoot BML
PIPERITONE 196-224 Shoot BML
PIPERITONENE 21-24 Shoot BML
PULEGONE 518-840 Shoot BML
SABINENE 147-168 Shoot BML
TERPINEN-4-OL 14-16 Shoot BML
TRANS-ALPHA-BERGAMOTENE 14-16 Shoot BML
TRANS-ISOPULEGONE 7-8 Shoot BML

Pycnanthemum muticum (MICHX.) PERS. "Muticous Mountain Mint"

1,8-CINEOLE 1-50 Shoot BML
1-OCTEN-3-OL 29-900 Shoot BML
ALPHA-PINENE 29-200 Shoot BML
BETA-PINENE 29-150 Shoot BML
BORNEOL 58-1,350 Shoot BML
CAMPHENE 1-50 Shoot BML
CARYOPHYLLENE 29-300 Shoot BML
CIS-ISOPULEGONE 29-750 Shoot BML
CIS-OCIMENE 1-50 Shoot BML
D-MENTHOL 4,350-7,500 Shoot BML
D-MENTHONE 2,320-4,000 Shoot BML
EO 29,000-50,000 Shoot BML
GERMACRENE-D 1-350 Shoot BML
ISOMENTHONE 29-2,100 Shoot BML
LIMONENE 87-900 Shoot BML
MENTHOL 87-450 Shoot BML
MENTHONE 2,320-11,300 Shoot BML
MYRCENE 29-200 Shoot BML
NEOISOMENTHOL 1-50 Shoot BML
NEOMENTHOL 29-100 Shoot BML
P-CYMENE Shoot BML
PIPERITONE 1-200 Shoot BML
PIPERITONENE 27-950 Shoot BML
PULEGONE 17,980-40,250 Shoot BML
SABINENE 1-50 Shoot BML
TRANS-ALPHA-BERGAMOTENE 29-200 Shoot BML
TRANS-ISOPULEGONE 116-200 Shoot BML
TRANS-PIPERITOL 1-100 Shoot BML

Pycnanthemum pilosum NUTT.
"Pilose Mountain Mint"

1,8-CINEOLE 10-35 Flower BML
1,8-CINEOLE 10-35 Leaf BML
1-OCTEN-3-OL 10-35 Flower BML
1-OCTEN-3-OL 10-35 Leaf BML
3-OCTANOL Flower tr BML
3-OCTANOL Leaf tr BML
ALPHA-PINENE 38-140 Flower BML
ALPHA-PINENE 38-140 Leaf BML
ALPHA-TERPINEOL 28-105 Flower BML
ALPHA-TERPINEOL 28-105 Leaf BML
ANISALDEHYDE Flower BML
ANISALDEHYDE Leaf BML
BETA-BOURBONENE 10-35 Flower BML
BETA-BOURBONENE 10-35 Leaf BML
BETA-ELEMENE 28-105 Flower BML
BETA-ELEMENE 28-105 Leaf BML
BETA-PINENE 38-140 Flower BML
BETA-PINENE 38-140 Leaf BML
BICYCLOGERMACRENE 19-70 Flower BML
BICYCLOGERMACRENE 19-70 Leaf BML
CARYOPHYLLENE 1-350 Flower BML
CARYOPHYLLENE 1-350 Leaf BML
EO 39,000-58,000 Flower BML
EO 39,000-58,000 Leaf BML
GERMACRENE-D 28-245 Flower BML
GERMACRENE-D 28-245 Leaf BML
ISOMENTHONE 494-2,940 Flower BML
ISOMENTHONE 494-2,940 Leaf BML
ISOPIPERITENONE Flower tr BML
ISOPIPERITENONE Leaf tr BML
LIMONENE 66-1,225 Flower BML
LIMONENE 66-1,225 Leaf BML
LINALOL 1-210 Flower BML
LINALOL 1-210 Leaf BML
MENTHOL 142-700 Flower BML
MENTHOL 142-700 Leaf BML
MENTHONE 950-15,400 Flower BML
MENTHONE 950-15,400 Leaf BML
MYRCENE 10-105 Flower BML
MYRCENE 10-105 Leaf BML
NEOMENTHOL 1,482-5,460 Flower BML
NEOMENTHOL 1,482-5,460 Leaf BML

PIPERITENONE 19-105 Flower BML
PIPERITENONE 19-105 Leaf BML
PULEGONE 2,574-28,000 Flower BML
PULEGONE 2,574-28,000 Leaf BML
SABINENE 10-70 Flower BML
SABINENE 10-70 Leaf BML
SPATHULENOL Flower tr BML
SPATHULENOL Leaf tr BML
TERPINEN-4-OL 48-175 Flower BML
TERPINEN-4-OL 48-175 Leaf BML
TRANS-ALPHA-BERGAMOTENE 19-70 Flower BML
TRANS-ALPHA-BERGAMOTENE 19-70 Leaf BML
TRANS-ANETHOLE Flower BML
TRANS-ANETHOLE Leaf BML
TRANS-ISOPULEGONE 28-105 Flower BML
TRANS-ISOPULEGONE 28-105 Leaf BML

Pycnanthemum pycnanthemoides (LEAVENW.) FERNALD "Typical Mountain Mint"

1,8-CINEOLE 39-116 Shoot BML
1-OCTEN-3-OL 195-580 Shoot BML
ALPHA-PINENE 156-696 Shoot BML
ALPHA-TERPINEOL 1-696 Shoot BML
BETA-PINENE 117-290 Shoot BML
BORNEOL 78-116 Shoot BML
CAMPHOR 78-116 Shoot BML
CARYOPHYLLENE 1-1,102 Shoot BML
CIS-ISOPULEGONE 195-290 Shoot BML
EO 39,000-58,000 Shoot BML
GERANIOL 1,911-2,842 Shoot BML
GERANYL-ACETATE 1,053-1,566 Shoot BML
GERMACRENE-D 39-464 Shoot BML
ISOMENTHONE 390-580 Shoot BML
LIMONENE 234-5,800 Shoot BML
LINALOL 26,598-39,556 Shoot BML
MENTHONE 234-20,242 Shoot BML
METHYL-EUGENOL Shoot tr BML
MYRCENE 273-406 Shoot BML
P-CYMENE 39-58 Shoot BML
PIPERITENONE 312-464 Shoot BML
PULEGONE 23,088-34,336 Shoot BML
SABINENE 39-1,856 Shoot BML
TRANS-ANETHOLE Shoot tr BML
TRANS-ISOPULEGONE 234-348 Shoot BML

Pycnanthemum setosum NUTT.
"Setose Mountain Mint"

1,8-CINEOLE 20-7,099 Shoot BML
1-OCTEN-3-OL 180-279 Shoot BML
3-OCTANOL 60-93 Shoot BML
ALPHA-PINENE 20-1,209 Shoot BML
BETA-PINENE 40-527 Shoot BML
BORNEOL 160-248 Shoot BML
CAMPHENE 20-31 Shoot BML
CARVONE 120-186 Shoot BML
CARYOPHYLLENE 1,060-1,643 Shoot BML
CIS-DIHYDROCARVONE 3,954-6,129 Shoot BML
EO 20,000-31,000 Shoot BML
EUGENOL 60-93 Shoot BML
GERMACRENE-D Shoot tr BML
ISODIHYDROCARVEOL 3,660-5,693 Shoot BML
ISOMENTHONE 180-279 Shoot BML
ISOPIPERITONE 20-31 Shoot BML
LIMONENE 40-1,085 Shoot BML
LINALOL 80-124 Shoot BML
MENTHOL Shoot tr BML
MENTHONE 4,800-7,440 Shoot BML
MYRCENE 40-341 Shoot BML
NEOISODIHYDROCARVEOL 520-806 Shoot BML
NEOISODIHYDROCARVYL-ACETATE 1,080-1,674 Shoot BML
NEOMENTHOL 20-31 Shoot BML
PIPERITENONE 40-62 Shoot BML
PULEGONE 12,740-19,747 Shoot BML
SABINENE 300-465 Shoot BML
TERPINEN-4-OL 220-341 Shoot BML
TRANS-DIHYDROCARVONE 1,040-1,612 Shoot BML
TRANS-ISOPULEGONE 40-62 Shoot BML

Pycnanthemum tenuifolium SCHRAD. "Slenderleaf Mountain Mint"

1,8-CINEOLE 1-1,100 Shoot BML
1-OCTEN-3-OL 1-80 Shoot BML
1-OCTYL-3-YL-ACETATE 2-120 Shoot BML
10-(ALPHA)-CADINOL 6-60 Shoot BML
3-HEXENOL 266-2,660 Shoot BML
3-OCTANOL 2-20 Shoot BML
ALLOAROMADENDRENE 28-440 Shoot BML
ALPHA-HUMULENE 8-640 Shoot BML
ALPHA-PHELLANDRENE 2-420 Shoot BML
ALPHA-PINENE 1-1,020 Shoot BML
ALPHA-TERPINEOL 28-280 Shoot BML
BETA-BOURBONENE 16-400 Shoot BML
BETA-ELEMENE 16-1,600 Shoot BML
BETA-PINENE 1-100 Shoot BML
BORNEOL 6-220 Shoot BML
BORNYL-ACETATE 14-880 Shoot BML
CAMPHENE 2-420 Shoot BML
CAMPHOR 4-40 Shoot BML
CARVEOL 80-800 Shoot BML
CARVONE 460-1,320 Shoot BML
CARYOPHYLLENE 28-2,460 Shoot BML
CARYOPHYLLENE-OXIDE 2-320 Shoot BML
CIS-ISOPULEGONE 14-140 Shoot BML
DELTA-CADINENE 1-620 Shoot BML
EO 2,000-20,000 Shoot BML
GAMMA-TERPINENE 1-260 Shoot BML
GERMACRENE-D 114-9,840 Shoot BML
ISOMENTHONE 1-10,580 Shoot BML
ISOPULEGONE Shoot tr BML
LIMONENE 2-8,520 Shoot BML
LINALOL 46-700 Shoot BML
MENTHOL 62-620 Shoot BML
MENTHONE 2-3,660 Shoot BML
MENTHYL-ACETATE 4-40 Shoot BML
MYRCENE 16-2,560 Shoot BML
P-CYMENE 1-140 Shoot BML
PALUSTROL 4-180 Shoot BML
PULEGONE 4-13,440 Shoot BML
SABINENE 1-360 Shoot BML
TERPINEN-4-OL 2-40 Shoot BML
TERPINOLENE 1-20 Shoot BML
THUJONE 19-190 Shoot BML

TRANS-ISOPULEGONE 4-40 Shoot BML
TRANS-METHYL-MYRCENATE 1,086-10,860 Shoot BML
TRANS-OCIMENE 30-300 Shoot BML

Pycnanthemum torreyi BENTH. "Torrey's Mountain Mint" Unaggregated A

1,8-CINEOLE 10 Shoot BML
1-OCTEN-3-OL 70 Shoot BML
ALPHA-PINENE 20 Shoot BML
BETA-PINENE 20 Shoot BML
CAMPHENE 10 Shoot BML
CARYOPHYLLENE 140 Shoot BML
CIS-ISOPULEGONE 20 Shoot BML
EO 10,000 Shoot BML
GERMACRENE-D 90 Shoot BML
ISOMENTHONE 310 Shoot BML
LIMONENE 60 Shoot BML
LINALOL 10 Shoot BML
MENTHONE 8,100 Shoot BML
MYRCENE 90 Shoot BML
P-CYMENE 30 Shoot BML
PIPERITENONE 40-62 Shoot BML
PIPERITONE 30 Shoot BML
PULEGONE 820 Shoot BML
SABINENE 10 Shoot BML
TRANS-ISOPULEGONE 70 Shoot BML

Pycnanthemum verticillatum (MICHX.) PERS.
"Whorled Mountain Mint"
Unaggregated A

1,8-CINEOLE 42 Shoot BML
1-OCTEN-3-OL 63 Shoot BML
ALPHA-PINENE 63 Shoot BML
BETA-PINENE 63 Shoot BML
CARYOPHYLLENE 315 Shoot BML
EO 21,000 Shoot BML
GERMACRENE-D 63 Shoot BML
ISOMENTHONE 147 Shoot BML
LIMONENE 336 Shoot BML
MENTHONE 4,557 Shoot BML
MYRCENE 126 Shoot BML
P-CYMENE 42 Shoot BML
PIPERITENONE Shoot tr BML
PULEGONE 14,595 Shoot BML
SABINENE 21 Shoot BML

Pycnanthemum virginianum (L.) DURAND & JACKSON "Virginia Mountain Mint"

1,8-CINEOLE 6-58 Shoot BML
1-OCTEN-3-OL 18-609 Shoot BML
1-OCTEN-3-YL-ACETATE 6-203 Shoot BML
10-(ALPHA)-CADINOL 18-87 Shoot BML
ALLO-AROMADENDRENE 24-116 Shoot BML
ALPHA-PINENE 6-145 Shoot BML
ALPHA-TERPINEOL 18-493 Shoot BML
BETA-BOURBONENE 6-232 Shoot BML
BETA-ELEMENE 240-1,160 Shoot BML
BETA-PINENE 6-145 Shoot BML
BORNYL-ACETATE 90-435 Shoot BML
CAMPHENE 6-87 Shoot BML
CARVACROL Shoot BML
CARYOPHYLLENE 72-1,015 Shoot BML
CIS-BETA-TERPINEOL 30-145 Shoot BML
EO 6,000-29,000 Shoot BML
GERANIOL Shoot BML
GERMACRENE-D 18-6,032 Shoot BML
ISOMENTHONE 102-22,533 Shoot BML
LIMONENE 12-14,674 Shoot BML
MENTHOL 18-3,016 Shoot BML
MENTHONE 25-14,819 Shoot BML
MYRCENE 6-261 Shoot BML
NEOMENTHOL 48-232 Shoot BML
P-CYMENE 6-290 Shoot BML
PIPERITENONE 6-203 Shoot BML
PIPERITONE 6-203 Shoot BML
PULEGONE 102-19,024 Shoot BML
SABINENE 1-6 Shoot BML
THYMOL 66-377 Shoot BML

Rosmarinus eriocalyx JORDAN & FOURR.
"Rosemary"
Aggregated A

(E)-BETA-OCIMENE 45-95 Shoot JEO5:243
(Z)-BETA-OCIMENE Shoot JEO5:243
1,8-CINEOLE 405-1,175 Shoot JEO5:243
ALPHA-HUMULENE <0.1-80 Shoot JEO5:243
ALPHA-PHELLANDRENE 44-165 Shoot JEO5:243
ALPHA-PINENE 895-1,505 Shoot JEO5:243
ALPHA-TERPINENE 60-155 Shoot JEO5:243
ALPHA-TERPINEOL 55-120 Shoot JEO5:243
ALPHA-THUJENE 38-70 Shoot JEO5:243
BETA-CARYOPHYLLENE 80-280 Shoot JEO5:243
BETA-PINENE 95-185 Shoot JEO5:243
BORNEOL 205-355 Shoot JEO5:243
BORNYL-ACETATE 45-235 Shoot JEO5:243
CAMPHENE 845-1,420 Shoot JEO5:243
CAMPHOR 1,470-2,600 Shoot JEO5:243
CARYOPHYLLENE-OXIDE 16-35 Shoot JEO5:243
DELTA-3-CARENE <0.1 Shoot JEO5:243
DELTA-CADINENE <0.1 Shoot JEO5:243
EO 5,500-8,700 Shoot JEO5:243
GAMMA-TERPINENE 5-190 Shoot JEO5:243
LIMONENE 155-280 Shoot JEO5:243
LINALOL <0.1-9 Shoot JEO5:243
MYRCENE 33-60 Shoot JEO5:243
P-CYMENE 160-305 Shoot JEO5:243
SABINENE <0.1 Shoot JEO5:243
TERPINEN-4-OL 60-95 Shoot JEO5:243
TERPINOLENE 5-26 Shoot JEO5:243
VERBENONE <0.1 Shoot JEO5:243

Rosmarinus eriocalyx JORDAN & FOURR.
"Rosemary"
Aggregated B

(E)-BETA-OCIMENE 60-185 Shoot JEO5:243
(Z)-BETA-OCIMENE <0.1 Shoot JEO5:243
1,8-CINEOLE 380-705 Shoot JEO5:243
ALPHA-HUMULENE 22-80 Shoot JEO5:243
ALPHA-PHELLANDRENE 44-150 Shoot JEO5:243
ALPHA-PINENE 680-1,225 Shoot JEO5:243
ALPHA-TERPINENE 150-550 Shoot JEO5:243
ALPHA-TERPINEOL 60-95 Shoot JEO5:243
ALPHA-THUJENE 33-52 Shoot JEO5:243
BETA-CARYOPHYLLENE 50-260 Shoot JEO5:243
BETA-PINENE 33-105 Shoot JEO5:243
BORNEOL 105-245 Shoot JEO5:243
BORNYL-ACETATE 105-260 Shoot JEO5:243
CAMPHENE 735-1,375 Shoot JEO5:243
CAMPHOR 1,590-2,670 Shoot JEO5:243
CARYOPHYLLENE-OXIDE 16-35 Shoot JEO5:243
DELTA-3-CARENE <0.1 Shoot JEO5:243
DELTA-CADINENE <0.1 Shoot JEO5:243
EO 5,500-8,700 Shoot JEO5:243
GAMMA-TERPINENE 16-95 Shoot JEO5:243
LIMONENE 145-330 Shoot JEO5:243
LINALOL <0.1 Shoot JEO5:243
MYRCENE 38-70 Shoot JEO5:243
P-CYMENE 295-645 Shoot JEO5:243
SABINENE <0.1 Shoot JEO5:243
TERPINEN-4-OL 75-150 Shoot JEO5:243
TERPINOLENE 11-35 Shoot JEO5:243
VERBENONE <0.1 Shoot JEO5:243

Rosmarinus officinalis L.
"Rosemary"

(+)-LIMONENE 16-76 Plant JAD NAP
(E)-BETA-OCIMENE <0.1-380 Shoot JEO5:243
(Z)-BETA-OCIMENE <0.1-75 Shoot JEO5:243
1,8-CINEOLE 0-8,125 Plant BML JAD JEO3:11 FFJ9:29
1-OCTEN-3-OL Plant NAP
19-ALPHA-HYDROXYURSOLIC-ACID Plant HHB
2-BETA-HYDROXYOLEANOLIC-ACID Plant HHB
3,5,5-TRIMETHYLHEXAN-1-OL 28-133 Plant BML
3-BETA-HYDROXYUREA-12,20(30)-DIEN-17-ON-ACID Plant HHB
3-HEXANONE 74-351 Plant BML
3-METHYL-BUTAN-1-OL Resin, Exudate, Sap NAP
3-O-ACETYLOLEANOLIC-ACID 11 Shoot PM59:276
3-O-ACETYLURSOLIC-ACID 11 Shoot PM59:276
3-OCTANOL 4-19 Plant BML
3-OCTANONE 20-40 Plant NAP
4-HYDROXYBENZOYL-GLUCOSIDE Leaf NAP
4-TERPINENYL-ACETATE 12-110 Leaf FFJ9:29
5-HYDROXY-4',7-DIMETHOXYFLAVONE Plant MPC
6-METHOXY-GENKWANIN Plant CRC
6-METHOXYLUTEOLIN Plant CRC
6-METHOXYLUTEOLIN-7-GLUCOSIDE Plant CRC
6-METHOXYLUTEOLIN-7-METHYL-ETHER Plant CRC
7-ETHOXY-ROSMANOL Leaf NAP
7-METHOXY-ROSMANOL 18 Shoot NAP
ACETIC-ACID Resin, Exudate, Sap NAP
ALLO-AROMADENDRENE 4-5 Leaf FFJ9:29
ALPHA-AMORPHENE 70-665 Plant BML
ALPHA-AMYRENONE 30 Shoot PM59:276
ALPHA-AMYRIN Plant FNF
ALPHA-AMYRIN 13 Shoot PM59:276
ALPHA-COPAENE 2-4 Leaf FFJ9:29
ALPHA-COPAENE Resin, Exudate, Sap FNF
ALPHA-FENCHENE Plant NAP
ALPHA-FENCHOL Leaf tr NAP
ALPHA-FENCHYL-ACETATE 20-95 Plant BML
ALPHA FENCHYL-ALCOHOL 28-133 Plant BML
ALPHA-HUMULENE <0.1-725 Shoot JEO3:11 JEO5:243
ALPHA-HYDROXYHYDROCAFFEIC-ACID Plant MPC
ALPHA-MUUROLENE 2-12 Leaf FFJ9:29
ALPHA-MUUROLENE Plant FNF
ALPHA-PHELLANDRENE <0.1-133 Plant BML JAD NAP JEO5:243
ALPHA-PINENE 235-4,750 Plant BML JAD NAP JEO3:11 JE39:167 FFJ9:29

ALPHA-SELINENE Plant CRC
ALPHA-TERPINENE 4-555 Plant BML JAD JEO3:11JEO5:243 FFJ9:29
ALPHA-TERPINENYL-ACETATE 5-6 Leaf FFJ9:29
ALPHA-TERPINEOL 24-1,555 Plant BML JEO3:11 JE39:167
ALPHA-THUJENE 1-475 Plant JAD JEO3:11
ALPHA-THUJONE 84-399 Plant FNF
APIGENIN Plant HHB
APIGENIN-7-GLUCOSIDE Plant HHB
AR-CURCUMENE 8-38 Plant FNF
ASCORBIC-ACID 612-673 Plant USA
ASH 61,900-75,570 Plant USA
BENZYL-ALCOHOL 7-32 Plant BML
BETA-AMYRENONE 30 Shoot PM59:276
BETA-AMYRIN Plant FNF
BETA-AMYRIN 13 Shoot PM59:276
BETA-CAROTENE 19-21 Plant FNF
BETA-CARYOPHYLLENE 12-2,075 Plant JEO3:11 JEO5:243
BETA-CARYOPHYLLENE 70-2,075 Shoot FNF
BETA-ELEMENE 3-200 Shoot JEO3:11
BETA-FENCHENE Plant FNF
BETA-GURJUNENE 0-0.5 Leaf FFJ9:29
BETA-OCIMENE Resin, Exudate, Sap FNF
BETA-PHELLANDRENE Plant tr BML JAD NAP
BETA-PINENE 17-1,425 Plant BML JAD NAP JEO3:11 FFJ9:29
BETA-THUJONE 11-209 Plant FFJ9:29
BETULIN Plant HHB
BETULIN 12.1 Shoot NAP
BETULINIC-ACID Leaf NAP
BORNEOL 12-4,237 Plant BML JAD JBH NAP FFJ9:29
BORNYL-ACETATE 0-5,054 Plant BML JAD JBH NAP FFJ9:29
BORNYLENE Plant FNF
BORON 22-39 Plant BOB
BUTAN-2-OL Resin, Exudate, Sap NAP
CADALENE Plant BML
CAFFEIC-ACID Plant AYL
CALACORENE Plant BML
CALAMENENE 1-5 Plant BML
CALCIUM 10,919-16,150 Plant USA
CAMPHENE 23-2,350 Plant BML JAD JEO3:11 PT59:465 FFJ9:29
CAMPHOR 60-5,800 Plant JAD JEO3:11 PT59:465
CARBOHYDRATES 640,600-704,660 Plant USA
CARNOSIC-ACID Plant AYL
CARNOSOL 530-9,803 Leaf JBH NAP
CARVACROL 5-6 Leaf FFJ9:29

CARVACROL Plant FNF

CARVONE 16-760 Plant FFJ9:29

CARYOPHYLLENE 16-340 Plant BML

CARYOPHYLLENE-OXIDE <0.1-75 Plant CRC JEO5:243

CHLOROGENIC-ACID Plant FNF

CINEOLE 168-9,728 Plant JAD NAP

CIRSILION Plant FNF

CIRSIMARITRIN 16 Shoot NAP

CIS-ALPHA-BISABOLENE 4-19 Plant BML

CIS-BETA-TERPINEOL 20-95 Plant BML

CIS-MYRTENOL 11-17 Leaf FFJ9:29

CIS-PINAN-3-ONE 17-110 Leaf FFJ9:29

CIS-SABINENE-HYDRATE 0-0.4 Leaf FFJ9:29

CIS-SABINENE-HYDRATE Plant FNF

COPPER 5-6 Plant USA

CUBENENE Plant BML

DECA-TRANS-2,TRANS-4-DIEN-1-AL Resin, Exudate, Sap NAP

DELTA-3-CARENE 0-330 Plant BML JEO5:243

DELTA-4-CARENE Leaf NAP

DELTA-CADINENE <0.1-75 Plant JEO5:243

DELTA-TERPINEOL 7-418 Plant BML JAD NAP

DIMETHYL-STYRENE 1-19 Plant BML FFJ9:29

DIOSMETIN Plant FNF

DIOSMIN Leaf NAP

DIPENTENE Essential Oil NAP

ELEMOL 7-32 Plant BML

EO 3,300-25,000 Plant BML HHB JEO3:11

EPI-ALPHA-AMYRIN Plant MPC

EPI-ALPHA-BISABOLOL 0-3 Leaf FFJ9:29

EPIROSMANOL 26 Leaf NAP

ERIODICTIOL Plant HHB

EUGENOL-METHYL-ETHER 5-7 Leaf FFJ9:29

FAT 134,020-187,418 Plant USA

FENCHONE 250 Plant FNF

FIBER 165,420-206,338 Plant USA

GAMMA-CADINENE 1-5 Leaf FFJ9:29

GAMMA-CADINENE Plant FNF

GAMMA-EUDESMOL 7-32 Plant BML

GAMMA-MUUROLENE 70-665 Plant BML

GAMMA-MUUROLOL 0-1 Leaf FFJ9:29

GAMMA-TERPINENE 4-400 Plant BML JAD JEO3:11

GENKWANIN Leaf NAP

GENKWANIN-4'-METHYL-ETHER Plant CRC

GERANIOL 50-370 Plant JE39:167

GLYCOLIC-ACID Plant FNF

HEPT-TRANS-2-EN-1-AL Resin, Exudate, Sap NAP
HEPTAN-1-AL Resin, Exudate, Sap NAP
HEPTAN-2-OL Resin, Exudate, Sap NAP
HEPTANOIC-ACID Resin, Exudate, Sap NAP
HESPERIDIN Leaf PC37:1463
HEXAN-1-AL Resin, Exudate, Sap NAP
HEXAN-1-OL Resin, Exudate, Sap NAP
HISPIDULIN Plant 411/
HISPIDULOSIDE Plant NAP
HUMULENE-EPOXIDE-I Plant CRC
HUMULENE-EPOXIDE-II Plant CRC
HYDROXYBENZOIC-ACID-4-BETA-D-GLUCOSIDE Leaf NAP
IRON 220-400 Plant FNF
ISOBORNEOL 7-95 Plant BML
ISOBORNYL-ACETATE Plant JFM
ISOBUTYL-ACETATE Plant 411/
ISOPINOCARVEOL 4-19 Plant BML
ISOPULEGOL 4-19 Plant BML
ISOROSMANOL 17 Flower NAP
ISOROSMARICINE Plant AYL
LABIATIC-ACID Plant AYL
LAVANDULOL 7-34 Plant BML
LEDENE Plant CRC
LIMONENE 0-1,950 Plant BML JAD JEO3:11 FFJ9:29
LINALOL <0.1-585 Plant JAD JEO3:11 JE39:167 JEO5:243
LINALOL-ACETATE 32-152 Plant JAD NAP
LONGIFOLENE 20-95 Plant BML
LUTEOLIN Leaf PC37:1463
LUTEOLIN Plant FNF
LUTEOLIN-3'-O-(3"-O-ACETYL)-BETA-D-GLUCURONIDE Leaf PC37:1463
LUTEOLIN-3'-O-(4"-O-ACETYL)-BETA-D-GLUCURONIDE Leaf PC37:1463 LUTEOLIN-3'-O-BETA-D-GLUCURONIDE Leaf PC37:1463
LUTEOLIN-7-GLUCOSIDE Plant FNF
MAGNESIUM 2,142-2,483 Plant USA
MANGANESE 18-19 Plant USA
MESITY-ALCOHOL 40-190 Plant BML
METHYL-ETHER Plant CRC
METHYL-EUGENOL Plant CRC
METHYL-HEPTENONE 8-38 Plant BML
MONOMETHYL-ALKANE Leaf HHB
MYRCENE 25-5,605 Plant BML JAD NAP JEO3:11 JEO5:243
MYRTENOL 8-38 Plant BML
N-METHYL-ROSMARICINE Plant JSG

NEOCHLOROGENIC-ACID Plant AYL
NEOTHUJOL 1.5-5 Leaf FFJ9:29
NEPETIN Plant NAP
NEPETRIN Plant PAS
NIACIN 10-11 Plant USA
NONAN-2-ONE Resin, Exudate, Sap NAP
NOPOL 1-5 Plant BML
O-O-N-TRIMETHYLROSMARICINE Plant FNF
OCTAN-1-OL Resin, Exudate, Sap NAP
OCTANE-2,3-DIONE Resin, Exudate, Sap NAP
OCTANOIC-ACID Resin, Exudate, Sap NAP
OLEANOLIC-ACID 10,500 Plant HHB
OLEANOLIC-ACID 20 Shoot PM59:276
P-CYMEN-8-OL 1-5 Plant BML
P-CYMENE 25-950 Plant BML JAD JEO3:11
P-MENTH-3-EN-1-OL 28-133 Plant BML
P-MENTH-CIS-2-EN-1-OL Leaf tr NAP
P-MENTH-TRANS-2-EN-1-OL Leaf tr NAP
PENTAN-1-AL Resin, Exudate, Sap NAP
PENTAN-1-OL Resin, Exudate, Sap NAP
PENTAN-2-OL Resin, Exudate, Sap NAP
PHOSPHORUS 490-1,000 Plant USA
PHYTOSTEROLS 580-640 Plant USA
PICROSALVIN Plant MPC
PIPERITENONE 4-8 Leaf FFJ9:29
POTASSIUM 8,842-11,284 Plant USA
PRISTANE Leaf HHB
PROTEIN 40,700-62,568 Plant USA
PROTOCATECHUIC-ACID-4-BETA-D-GLUCOSIDE Leaf NAP
ROFFICERONE 20 Shoot PM59:276
ROSMADIAL 30 Leaf NAP
ROSMADIOL Plant JSG
ROSMANOL 92 Leaf NAP
ROSMARIC-ACID 3,000-3,500 Plant ABS
ROSMARICINE Plant AYL
ROSMARIDIPHENOL Plant PAS
ROSMARINIC-ACID 3,500 Leaf NAP
ROSMARINIC-ACID 25,000 Plant JBH FT62:166
ROSMARINIC-ACID 13,500 Shoot NAP
ROSMARINIC-ACID 38,957 Tissue Culture NAP
ROSMARINOL Plant JSG
ROSMARIQUINONE Plant PAS
SABINENE <0.1-190 Plant BML JAD NAP
SABINYL-ACETATE 0-1.5 Leaf FFJ9:29
SAFROLE 32-95 Plant FNF

SALICYLATES 70-680 Leaf JAD85:950
SALICYLIC-ACID-2-BETA-D-GLUCOSIDE Leaf NAP
SALVIGENIN Plant FNF
SANTENE Plant FNF
SINENSETIN Plant 411/
SODIUM 462-592 Plant FNF
SQUALENE Plant FNF
STYRENE Resin, Exudate, Sap NAP
SYRINGIC-ACID-4-BETA-D-GLUCOSIDE Leaf NAP
TERPINEN-4-OL 4-521 Plant JAD NAP
TERPINOLENE 12-350 Plant BML JAD JEO3:11
THIAMIN 5-6 Plant USA
THYMOL Plant FNF
TOLUENE 436-2,071 Plant BML
TRANS-ANETHOLE Plant CRC
TRANS-BETA-TERPINEOL 7-34 Plant BML
TRANS-CARVEOL 1-5 Plant BML
TRANS-MYRTENOL 32 Leaf FFJ9:29
TRANS-OCIMENE 4-130 Plant BML
TRANS-PINOCARVEOL 32-42 Leaf FFJ9:29
TRANS-SABINENE-HYDRATE 0-19 Plant BML JAD NAP
TRICYCLENE Plant tr JAD NAP
TRIMETHYLALKANE Leaf HHB
URSOLIC-ACID 28,000-41,000 Plant HHB
URSOLIC-ACID 20 Shoot PM59:276
VANILLIC-ACID-4-BETA-D-GLUCOSIDE Leaf NAP
VERBENOL Plant FNF
VERBENONE 10-375 Plant BML JAD JEO3:11 JEO5:243
WATER 77,900-108,300 Plant USA
ZINC 30-38 Plant USA
ZINGIBERENE Resin, Exudate, Sap NAP

Rosmarinus officinalis L.
"Rosemary"
Aggregated A

(E)-BETA-OCIMENE 12-25 Shoot JEO5:243
(Z)-BETA-OCIMENE <0.1 Shoot JEO5:243
1,8-CINEOLE 1,400-2,850 Shoot JEO5:243
ALPHA-HUMULENE 19-38 Shoot JEO5:243
ALPHA-PHELLANDRENE <0.1 Shoot JEO5:243
ALPHA-PINENE 1,250-2,545 Shoot JEO5:243
ALPHA-TERPINENE 25-50 Shoot JEO5:243
ALPHA-TERPINEOL 145-290 Shoot JEO5:243
ALPHA-THUJENE 19-38 Shoot JEO5:243
BETA-CARYOPHYLLENE 50-100 Shoot JEO5:243
BETA-PINENE 225-455 Shoot JEO5:243
BORNEOL 225-455 Shoot JEO5:243
BORNYL ACETATE 25-50 Shoot JEO5:243
CAMPHENE 620-1,260 Shoot JEO5:243
CAMPHOR 780-1,590 Shoot JEO5:243
CARYOPHYLLENE-OXIDE 30-60 Shoot JEO5:243
DELTA-3-CARENE 160-330 Shoot JEO5:243
DELTA-CADINENE 25-50 Shoot JEO5:243
EO 6,200-12,600 Shoot JEO5:243
GAMMA-TERPINENE 25-50 Shoot JEO5:243
LIMONENE 225-455 Shoot JEO5:243
LINALOL 30-65 Shoot JEO5:243
MYRCENE 6-13 Shoot JEO5:243
P-CYMENE 75-150 Shoot JEO5:243
SABINENE 25-50 Shoot JEO5:243
TERPINEN-4-OL 55-110 Shoot JEO5:243
TERPINOLENE 25-50 Shoot JEO5:243
VERBENONE 340-695 Shoot JEO5:243

Rosmarinus officinalis L.
"Rosemary"
Aggregated B

(E)-BETA-OCIMENE <0.1-380 Shoot JEO5:243
(Z)-BETA-OCIMENE <0.1-13 Shoot JEO5:243
1,8-CINEOLE 795-2,960 Shoot JEO5:243
ALPHA-HUMULENE <0.1-13 Shoot JEO5:243
ALPHA-PHELLANDRENE 50-125 Shoot JEO5:243
ALPHA-PINENE 1,080-3,000 Shoot JEO5:243
ALPHA-TERPINENE 185-555 Shoot JEO5:243
ALPHA-TERPINEOL 25-290 Shoot JEO5:243
ALPHA-THUJENE 12-50 Shoot JEO5:243
BETA-CARYOPHYLLENE 12-90 Shoot JEO5:243
BETA-PINENE 135-530 Shoot JEO5:243
BORNEOL 100-1,020 Shoot JEO5:243
BORNYL-ACETATE <0.1-90 Shoot JEO5:243
CAMPHENE 355-1,435 Shoot JEO5:243
CAMPHOR 320-1,410 Shoot JEO5:243
CARYOPHYLLENE-OXIDE 12-75 Shoot JEO5:243
DELTA-3-CARENE <0.1 Shoot JEO5:243
DELTA-CADINENE <0.1-38 Shoot JEO5:243
EO 6,200-12,600 Shoot JEO5:243
GAMMA-TERPINENE 37-225 Shoot JEO5:243
LIMONENE 185-530 Shoot JEO5:243
LINALOL <0.1-65 Shoot JEO5:243
MYRCENE <0.1 Shoot JEO5:243
P-CYMENE 100-465 Shoot JEO5:243
SABINENE 19-75 Shoot JEO5:243
TERPINEN-4-OL 25-240 Shoot JEO5:243
TERPINOLENE 19-65 Shoot JEO5:243
VERBENONE 155-905 Shoot JEO5:243

Rosmarinus officinalis L.
"Rosemary"
Aggregated C

(E)-BETA-OCIMENE <0.1-13 Shoot JEO5:243
(Z)-BETA-OCIMENE 6-75 Shoot JEO5:243
1,8-CINEOLE 630-1,360 Shoot JEO5:243
ALPHA-HUMULENE 12-50 Shoot JEO5:243
ALPHA-PHELLANDRENE 19-50 Shoot JEO5:243
ALPHA-PINENE 990-2,180 Shoot JEO5:243
ALPHA-TERPINENE 200-540 Shoot JEO5:243
ALPHA-TERPINEOL 130-1,160 Shoot JEO5:243
ALPHA-THUJENE 37-90 Shoot JEO5:243
BETA-CARYOPHYLLENE 31-150 Shoot JEO5:243
BETA-PINENE 75-200 Shoot JEO5:243
BORNEOL 145-895 Shoot JEO5:243
BORNYL-ACETATE 70-175 Shoot JEO5:243
CAMPHENE 1,035-2,280 Shoot JEO5:243
CAMPHOR 1,885-3,920 Shoot JEO5:243
CARYOPHYLLENE-OXIDE <0.1-38 Shoot JEO5:243
DELTA-3-CARENE <0.1-13 Shoot JEO5:243
DELTA-CADINENE <0.1-75 Shoot JEO5:243
EO 6,200-12,600 Shoot JEO5:243
GAMMA-TERPINENE 105-300 Shoot JEO5:243
LIMONENE 175-365 Shoot JEO5:243
LINALOL 6-13 Shoot JEO5:243
MYRCENE 50-100 Shoot JEO5:243
P-CYMENE 190-480 Shoot JEO5:243
SABINENE <0.1 Shoot JEO5:243
TERPINEN-4-OL 60-190 Shoot JEO5:243
TERPINOLENE 25-65 Shoot JEO5:243
VERBENONE <0.1 Shoot JEO5:243

Rosmarinus officinalis L.
"Rosemary"
Unaggregated A

1,8-CINEOLE 360 Leaf FFJ9:29
1-S-CIS-CALAMENENE 2 Leaf FFJ9:29
2,5-DIMETHYL-STYRENE 18 Leaf FFJ9:29
4-TERPINENYL-ACETATE 110 Leaf FFJ9:29
ALLO-AROMADENDRENE 4 Leaf FFJ9:29
ALPHA-COPAENE 4 Leaf FFJ9:29
ALPHA-HUMULENE 13 Leaf FFJ9:29
ALPHA-MUUROLENE Leaf FFJ9:29
ALPHA-PHELLANDRENE 90 Leaf FFJ9:29
ALPHA-PINENE 375 Leaf FFJ9:29
ALPHA-TERPINENE 15 Leaf FFJ9:29
ALPHA-TERPINENYL-ACETATE 6 Leaf FFJ9:29
ALPHA-TERPINEOL 130 Leaf FFJ9:29
BETA-CARYOPHYLLENE 60 Leaf FFJ9:29
BETA-GURJUNENE Leaf FFJ9:29
BETA-PINENE 70 Leaf FFJ9:29
BETA-THUJONE 11 Leaf FFJ9:29
BORNEOL 12 Leaf FFJ9:29
BORNYL-ACETATE 80 Leaf FFJ9:29
CAMPHENE 145 Leaf FFJ9:29
CAMPHOR 595 Leaf FFJ9:29
CARVACROL 6 Leaf FFJ9:29
CARVONE 16 Leaf FFJ9:29
CARYOPHYLLENE-OXIDE 14 Leaf FFJ9:29
CIS-MYRTENOL 11 Leaf FFJ9:29
CIS-PINAN-3-ONE 110 Leaf FFJ9:29
CIS-SABINENE-HYDRATE Leaf FFJ9:29
DELTA-CADINENE 9 Leaf FFJ9:29
EO 4,000 Leaf FFJ9:29
EPI-ALPHA-BISABOLOL Leaf FFJ9:29
EUGENOL-METHYL-ETHER 7 Leaf FFJ9:29
GAMMA-CADINENE 1 Leaf FFJ9:29
GAMMA-MUUROLOL Leaf FFJ9:29
GAMMA-TERPINENE 4 Leaf FFJ9:29
LIMONENE Leaf FFJ9:29
LINALOL 220 Leaf FFJ9:29
MYRCENE 215 Leaf FFJ9:29
NEO-THUJOL 5 Leaf FFJ9:29
P-CYMENE 250 Leaf FFJ9:29
PIPERITENONE 8 Leaf FFJ9:29
SABINYL-ACETATE Leaf FFJ9:29

TERPINEN-4-OL 85 Leaf FFJ9:29
TRANS-MYRTENOL 32 Leaf FFJ9:29
TRANS-PINOCARVEOL 42 Leaf FFJ9:29
VERBENONE 305 Leaf FFJ9:29

Rosmarinus officinalis L.
"Rosemary"
Unaggregated B

1,8-CINEOLE Leaf FFJ9:29
1-S-CIS-CALAMENENE 2 Leaf FFJ9:29
2,5-DIMETHYL-STYRENE 0.7 Leaf FFJ9:29
4-TERPINENYL-ACETATE 12 Leaf FFJ9:29
ALLO-AROMADENDRENE 5 Leaf FFJ9:29
ALPHA-COPAENE 2 Leaf FFJ9:29
ALPHA-HUMULENE 12 Leaf FFJ9:29
ALPHA-MUUROLENE 2 Leaf FFJ9:29
ALPHA-PHELLANDRENE 6 Leaf FFJ9:29
ALPHA-PINENE 34 Leaf FFJ9:29
ALPHA-TERPINENE 4 Leaf FFJ9:29
ALPHA-TERPINENYL-ACETATE 5 Leaf FFJ9:29
ALPHA-TERPINEOL 70 Leaf FFJ9:29
BETA-CARYOPHYLLENE 38 Leaf FFJ9:29
BETA-GURJUNENE 0.5 Leaf FFJ9:29
BETA-PINENE 17 Leaf FFJ9:29
BETA-THUJONE 13 Leaf FFJ9:29
BORNEOL 24 Leaf FFJ9:29
BORNYL-ACETATE 105 Leaf FFJ9:29
CAMPHENE 23 Leaf FFJ9:29
CAMPHOR 160 Leaf FFJ9:29
CARVACROL 5 Leaf FFJ9:29
CARVONE 23 Leaf FFJ9:29
CARYOPHYLLENE-OXIDE 18 Leaf FFJ9:29
CIS-MYRTENOL 17 Leaf FFJ9:29
CIS-PINAN-3-ONE 17 Leaf FFJ9:29
CIS-SABINENE-HYDRATE 0.4 Leaf FFJ9:29
DELTA-CADINENE 9 Leaf FFJ9:29
EO 1,400 Leaf FFJ9:29
EPI-ALPHA-BISABOLOL 3 Leaf FFJ9:29
EUGENOL-METHYL-ETHER 5 Leaf FFJ9:29
GAMMA-CADINENE 5 Leaf FFJ9:29
GAMMA-MUUROLOL 1 Leaf FFJ9:29
GAMMA-TERPINENE 4 Leaf FFJ9:29
LIMONENE 100 Leaf FFJ9:29
LINALOL 90 Leaf FFJ9:29
MYRCENE 7 Leaf FFJ9:29
NEO-THUJOL 1.5 Leaf FFJ9:29
P-CYMENE 25 Leaf FFJ9:29
PIPERITENONE 4 Leaf FFJ9:29
SABINYL-ACETATE 1.5 Leaf FFJ9:29

TERPINEN-4-OL 40 Leaf FFJ9:29
TRANS-MYRTENOL 32 Leaf FFJ9:29
TRANS-PINOCARVEOL 32 Leaf FFJ9:29
VERBENONE 175 Leaf FFJ9:29

Rosmarinus tomentosus HUBER-MORATH & MAIRE "Hairy Rosemary"

(E)-BETA-OCIMENE <0.1-20 Shoot JEO5:243
(Z)-BETA-OCIMENE <0.1 Shoot JEO5:243
1,8-CINEOLE 530-1,465 Shoot JEO5:243
ALPHA-HUMULENE <0.1 Shoot JEO5:243
ALPHA-PHELLANDRENE <0.1-10 Shoot JEO5:243
ALPHA-PINENE 665-2,830 Shoot JEO5:243
ALPHA-TERPINENE 16-40 Shoot JEO5:243
ALPHA-TERPINEOL 8-90 Shoot JEO5:243
ALPHA-THUJENE 16-30 Shoot JEO5:243
BETA-CARYOPHYLLENE 55-355 Shoot JEO5:243
BETA-PINENE 90-295 Shoot JEO5:243
BORNEOL 110-555 Shoot JEO5:243
BORNYL-ACETATE 120-375 Shoot JEO5:243
CAMPHENE 590-960 Shoot JEO5:243
CAMPHOR 2,745-4,455 Shoot JEO5:243
CARYOPHYLLENE-OXIDE 24-35 Shoot JEO5:243
DELTA-3-CARENE <0.1 Shoot JEO5:243
DELTA-CADINENE <0.1 Shoot JEO5:243
EO 8,000-9,900 Shoot JEO5:243
GAMMA-TERPINENE 24-80 Shoot JEO5:243
LIMONENE 160-295 Shoot JEO5:243
LINALOL 65-100 Shoot JEO5:243
MYRCENE 55-80 Shoot JEO5:243
P-CYMENE 55-210 Shoot JEO5:243
SABINENE <0.1-20 Shoot JEO5:243
TERPINEN-4-OL 48-170 Shoot JEO5:243
TERPINOLENE 16-40 Shoot JEO5:243
VERBENONE <0.1-40 Shoot JEO5:243

Rosmarinus x *lavandulaceus* DE NOE "Lavender Rosemary"

(E)-BETA-OCIMENE <0.1-30 Shoot JEO5:243
(Z)-BETA-OCIMENE 26-110 Shoot JEO5:243
1,8-CINEOLE 470-2,255 Shoot JEO5:243
ALPHA-HUMULENE 3-40 Shoot JEO5:243
ALPHA-PHELLANDRENE 13-220 Shoot JEO5:243
ALPHA-PINENE 405-1,740 Shoot JEO5:243
ALPHA-TERPINENE 13-140 Shoot JEO5:243
ALPHA-TERPINEOL 90-930 Shoot JEO5:243
ALPHA-THUJENE <0.1-30 Shoot JEO5:243
BETA-CARYOPHYLLENE 29-240 Shoot JEO5:243
BETA-PINENE <0.1-210 Shoot JEO5:243
BORNEOL 75-345 Shoot JEO5:243
BORNYL-ACETATE 13-200 Shoot JEO5:243
CAMPHENE 210-1,245 Shoot JEO5:243
CAMPHOR 330-2,335 Shoot JEO5:243
CARYOPHYLLENE-OXIDE 13-60 Shoot JEO5:243
DELTA-3-CARENE 38-455 Shoot JEO5:243
DELTA-CADINENE 6-60 Shoot JEO5:243
EO 3,200-9,900 Shoot JEO5:243
GAMMA-TERPINENE 10-90 Shoot JEO5:243
LIMONENE 110-405 Shoot JEO5:243
LINALOL <0.1-30 Shoot JEO5:243
MYRCENE 19-110 Shoot JEO5:243
P-CYMENE 60-275 Shoot JEO5:243
SABINENE <0.1-295 Shoot JEO5:243
TERPINEN-4-OL 26-130 Shoot JEO5:243
TERPINOLENE 19-80 Shoot JEO5:243
VERBENONE <0.1-385 Shoot JEO5:243

Rosmarinus x *mendizabalii* SAGREDO EX ROSUA "Mendizabali's Rosemary"

(E)-BETA-OCIMENE <0.1-33 Shoot JEO5:243
(Z)-BETA-OCIMENE <0.1 Shoot JEO5:243
1,8-CINEOLE 115-805 Shoot JEO5:243
ALPHA-HUMULENE 4-38 Shoot JEO5:243
ALPHA-PHELLANDRENE 4-22 Shoot JEO5:243
ALPHA-PINENE 680-1,500 Shoot JEO5:243
ALPHA-TERPINENE 43-100 Shoot JEO5:243
ALPHA-TERPINEOL 50-240 Shoot JEO5:243
ALPHA-THUJENE <0.1-22 Shoot JEO5:243
BETA-CARYOPHYLLENE 8-65 Shoot JEO5:243
BETA-PINENE 130-485 Shoot JEO5:243
BORNEOL 23-165 Shoot JEO5:243
BORNYL-ACETATE 4-11 Shoot JEO5:243
CAMPHENE 430-605 Shoot JEO5:243
CAMPHOR 850-1,475 Shoot JEO5:243
CARYOPHYLLENE-OXIDE 13-44 Shoot JEO5:243
DELTA-3-CARENE <0.1 Shoot JEO5:243
DELTA-CADINENE <0.1-16 Shoot JEO5:243
EO 3,900-5,500 Shoot JEO5:243
GAMMA-TERPINENE 23-50 Shoot JEO5:243
LIMONENE 100-370 Shoot JEO5:243
LINALOL 16-33 Shoot JEO5:243
MYRCENE 4-55 Shoot JEO5:243
P-CYMENE 43-60 Shoot JEO5:243
SABINENE <0.1 Shoot JEO5:243
TERPINEN-4-OL 27-55 Shoot JEO5:243
TERPINOLENE 16-33 Shoot JEO5:243
VERBENONE 4-185 Shoot JEO5:243

Salvia canariensis L.
"Canary Island Sage"
Unaggregated A

6,9-GUAIADIENE 230 Leaf FFJ9:201
6-METHYL-HEPT-5-N-2-ONE 35 Leaf FFJ9:201
ALLOAROMADENDRENE 28 Leaf FFJ9:201
ALPHA-ELEMOL <0.1 Leaf FFJ9:201
ALPHA-HUMULENE 110 Leaf FFJ9:201
ALPHA-PHELLANDRENE <0.1 Leaf FFJ9:201
ALPHA-PINENE 670 Leaf FFJ9:201
ALPHA-TERPINEOL <0.1 Leaf FFJ9:201
BETA-CARYOPHYLLENE 2,105 Leaf FFJ9:201
BETA-CARYOPHYLLENE-ALCOHOL <0.1 Leaf FFJ9:201
BETA-EUDESMOL <0.1 Leaf FFJ9:201
BETA-PINENE 28 Leaf FFJ9:201
BORNEOL 160 Leaf FFJ9:201
BORNYL-ACETATE 1,205 Leaf FFJ9:201
CAMPHENE 575 Leaf FFJ9:201
CAMPHOR 28 Leaf FFJ9:201
CARYOPHYLLENE-OXIDE 28 Leaf FFJ9:201
DECANOL <0.1 Leaf FFJ9:201
EO 7,000 Leaf FFJ9:201
FENCHYL-ALCOHOL <0.1 Leaf FFJ9:201
GAMMA-TERPINENE <0.1 Leaf FFJ9:201
GLOBULOL 1,185 Leaf FFJ9:201
MYRCENE <0.1 Leaf FFJ9:201
N-DECYL-ACETATE <0.1 Leaf FFJ9:201
P-CYMENE 42 Leaf FFJ9:201
SABINENE <0.1 Leaf FFJ9:201
SELINENE 21 Leaf FFJ9:201
TERPINEN-4-OL <0.1 Leaf FFJ9:201
TERPINOLENE <0.1 Leaf FFJ9:201
THYMOL <0.1 Leaf FFJ9:201
TRICYCLENE 21 Leaf FFJ9:201
ZINGIBERENE 14 Leaf FFJ9:201

Salvia dorisiana STANDL.
"'Honduran' Sage"

1,8-CINEOLE 19-20 Shoot JEO6:79
2,6-DIMETHYL-10-(P-TOLYL)-UNDECA-2,6-DIENE 6.1-8.5 Shoot JEO6:79
2-OCTANOL Shoot JEO6:79
3-OCTANOL 2.5-2.7 Shoot JEO6:79
ALLOAROMADENDRENE 4-5.2 Shoot JEO6:79
ALPHA-AMORPHENE 0.8-1 Shoot JEO6:79
ALPHA-COPAENE 1.4-1.6 Shoot JEO6:79
ALPHA-CUBEBENE 2.1-3.1 Shoot JEO6:79
ALPHA-GURJUNENE Shoot JEO6:79
ALPHA-HUMULENE 4.3-5.7 Shoot JEO6:79
ALPHA-MUUROLENE 2.8-3.8 Shoot JEO6:79
ALPHA-PHELLANDRENE 0.5-2.1 Shoot JEO6:79
ALPHA-PINENE Shoot JEO6:79
ALPHA-SELINENE 0.4-0.6 Shoot JEO6:79
ALPHA-TERPINENE 2.7-3.7 Shoot JEO6:79
ALPHA-TERPINEOL 25-26 Shoot JEO6:79
ALPHA-THUJENE 2.3-2.9 Shoot JEO6:79
BETA-CARYOPHYLLENE 90-115 Shoot JEO6:79
BETA-CYCLOCITRAL 3.5-3.7 Shoot JEO6:79
BETA-PHENETHYL-BUTYRATE 0.3 Shoot JEO6:79
BETA-PINENE 2.2-2.4 Shoot JEO6:79
BURYRIC-ACID Shoot JEO6:79
CADINA-1,4-DIENE 0.3-0.5 Shoot JEO6:79
CAMPHENE 0.8-1 Shoot JEO6:79
CARYOPHYLLENE-OXIDE 6.9-7.3 Shoot JEO6:79
CIS-ALPHA-BISABOLENE 2.9-3.7 Shoot JEO6:79
CITRONELLOL Shoot JEO6:79
DELTA-CADINENE 11.4-14.2 Shoot JEO6:79
E-NEROLIDOL 2.3-2.6 Shoot JEO6:79
EO 1,000 Shoot JEO6:79
GAMMA-CADINENE 5-5.6 Shoot JEO6:79
GAMMA-MUUROLENE 25-26 Shoot JEO6:79
GAMMA-TERPINENE 2.7-3.7 Shoot JEO6:79
GERANIOL Shoot JEO6:79
GERANYL-ACETATE 2.6-2.8 Shoot JEO6:79
ISOAMYL-ACETATE 1.2 Shoot JEO6:79
ISOCARYOPHYLLENE Shoot JEO6:79
ISOPINOCAMPHONE 2.9-3.1 Shoot JEO6:79
LIMONENE 64-72 Shoot JEO6:79
LINALOL 7-7.4 Shoot JEO6:79
METHYL-PERILLATE 165-180 Shoot JEO6:79

MYRCENE 10.5-12.1 Shoot JEO6:79
MYRTENOL 3.5-3.9 Shoot JEO6:79
MYRTENYL-ACETATE 25-26 Shoot JEO6:79
P-CYMENE 9.6-10.6 Shoot JEO6:79
PERILLALDEHYDE 3.1-3.3 Shoot JEO6:79
PERILLYL-ACETATE 255-280 Shoot JEO6:79
PERILLYL-ALCOHOL 6.7-9.1 Shoot JEO6:79
PERILLYL-BUTYRATE 7.5-8.3 Shoot JEO6:79
SABINENE 1.8-2.4 Shoot JEO6:79
T-CADINOL 1.2-1.8 Shoot JEO6:79
TERPINOLENE 8.6-10 Shoot JEO6:79
TRANS-ALPHA-BERGAMOTENE 3.1-4.1 Shoot JEO6:79
TRANS-CALAMENENE 0.9-1.1 Shoot JEO6:79
VIRIDIFLORENE 4-4.6 Shoot JEO6:79

Salvia gilliesii BENTH.
"'Cordoba' Sage"
Unaggregated A

ALLO-AROMADENDRENE 70 Shoot JEO5:319
ALPHA-COPAENE 41 Shoot JEO5:319
ALPHA-CUBEBENE 5 Shoot JEO5:319
ALPHA-PINENE 55 Shoot JEO5:319
ALPHA-TERPINEOL 7.5 Shoot JEO5:319
ALPHA-THUJENE 2.5 Shoot JEO5:319
ALPHA-THUJONE 28 Shoot JEO5:319
AROMADENDRENE 35 Shoot JEO5:319
BETA-BOURBONENE 44 Shoot JEO5:319
BETA-CARYOPHYLLENE 720 Shoot JEO5:319
BETA-ELEMENE 5.5 Shoot JEO5:319
BETA-PINENE 4.5 Shoot JEO5:319
BICYCLOGERMACENE 1,420 Shoot JEO5:319
CAMPHENE 3.5 Shoot JEO5:319
CAMPHOR 60 Shoot JEO5:319
CARVONE 34 Shoot JEO5:319
CARYOPHYLLENE-OXIDE 205 Shoot JEO5:319
CIS-CALAMENENE <0.1 Shoot JEO5:319
CIS-CARVEOL 20 Shoot JEO5:319
CIS-VERBENOL 2.5 Shoot JEO5:319
DELTA-CADINENE 40 Shoot JEO5:319
DELTA-ELEMENE 185 Shoot JEO5:319
EO 5,000 Shoot JEO5:319
GAMMA-CADINENE 36 Shoot JEO5:319
GAMMA-ELEMENE 27 Shoot JEO5:319
GERMACRENE-D 575 Shoot JEO5:319
ISODIHYDROCARVEOL 4 Shoot JEO5:319
LIMONENE 180 Shoot JEO5:319
MYRCENE 29 Shoot JEO5:319
MYRTENYL-ACETATE 41 Shoot JEO5:319
P-CYMENE 52 Shoot JEO5:319
P-MENTHA-10,3,8-TRIENE 3 Shoot JEO5:319
SPATHULENOL 225 Shoot JEO5:319
TERPINEN-4-OL 7.5 Shoot JEO5:319
TRANS-ALPHA-BERGAMOTENE 10 Shoot JEO5:319
TRANS-CARVYL-ACETATE 10 Shoot JEO5:319
VERBENONE 65 Shoot JEO5:319

Salvia miltiorrhiza BUNGE
"Dan-Shen, Red Sage, Tan-Shen"

ARSENIC 0.34 Root CAL
CALCIUM 2,780 Root CAL
COPPER 8 Root CAL
CRYPTOTANSHINONE Root CCO
DANSHENSPIROKETALLACTONE Plant JSG
HYDROXYTANSHINONE Root CCO
IRON 230 Root CAL
ISOCRYPTOTANSHINONE Root CCO
ISOTANSHINONE-I Root CCO
ISOTANSHINONE-II Root CCO
ISOTANSHINONE-III Plant JSG
MAGNESIUM 3,230 Root CAL
MANGANESE 23 Root CAL
METHYL-TANSHINONATE Root CCO
METHYL-TANSHINONE Root CCO
MILTIRONE Root CCO
POTASSIUM 11,700 Root CAL
SALVIANOLIC-ACID-A Root JBH
SALVIOL Root CCO
SODIUM 70 Root CAL
TANSHINOLACTONE Plant JSG
TANSHINONE-I Root CCO
TANSHINONE-II Root CCO
TANSHINONE-IIA Root CCO
TANSHINONE-IIIB Plant JSG
TANSHINONIC-ACID Plant JSG
ZINC 12 Root CAL

Salvia officinalis L.
"Sage"

(-)-THUJONE 2,500-13,000 Plant HHB
1,8-CINEOLE 390-6,288 Plant BML JAD NAP
1,8-CINEOLE-SYNTHETASE Leaf NAP
1-OCTEN-3-OL Plant FNF
2-ALPHA,3-ALPHA-DIHYDROXY-OLEAN-12-EN-28-OIC-ACID Leaf NAP
2-ALPHA-HYDROXY-3-OXO-OLEAN-12-EN-28-OIC-ACID 1.2-17 Leaf NAP
2-AMINO-ADIPIC-ACID 1.6 Shoot NAP
2-METHYL-3-METHYLENE-5-HEPTANE 50-280 Leaf BML
2-METHYL-3-METHYLENE-HEPT-CIS-5-ENE 40-224 Leaf JAD NAP
2-METHYL-3-METHYLENE-HEPT-TRANS-5-ENE 5-28 Leaf JAD NAP
3-CARENE Plant FNF
3-EPIOLEANOLIC-ACID 1.4 Leaf NAP
3-ISOTHUJONE Leaf NAP
5-METHOXYSALVIGENIN Leaf NAP
6-METHOXYGENKWANIN Plant FNF
6-METHOXYGENKWANIN-7-METHYL-ETHER Plant FNF
6-METHOXYLUTEOLIN Plant NAP
6-METHOXYLUTEOLIN-7-METHYL-ETHER Plant FNF
ACETO-HYDROXY-ROYLEANONE Plant NAP
ALLOAROMADENDRENE 15-84 Leaf BML
ALPHA-AMYRIN 1,800 Leaf NAP
ALPHA-AMYRIN Plant FNF
ALPHA-CEDRENE Plant PAS
ALPHA-COROCALENE Plant FNF
ALPHA-GURJUNENE Plant FNF
ALPHA-HUMULENE 110-616 Leaf BML
ALPHA-OLEANOLIC-ACID Plant FNF
ALPHA-PHELLANDRENE Leaf tr NAP
ALPHA-PINENE 7-1,540 Plant BML JAD NAP
ALPHA-TERPINENE 10-56 Leaf BML JAD NAP
ALPHA-TERPINEOL 5-910 Leaf BML NAP
ALPHA-THUJENE 0-386 Leaf JAD NAP
ALPHA-THUJONE 200-10,172 Leaf BML JAD NAP
ALPHA-URSOLIC-ACID Plant FNF
ALUMINUM 115 Leaf PED
APIGENIN Plant PAS
APIGENIN-7-O-BETA-D-GLUCURONYLPYRANOSIDE Plant NAP
AROMADENDRENE Plant FNF
ASCORBIC-ACID 320-350 Leaf FNF PED

ASH 87,000 Leaf PED
ASPARAGINE Plant FNF
AVENASTEROL Fruit NAP
BETA-AMYRIN 1,000 Leaf NAP
BETA-AMYRIN Plant FNF
BETA-CAROTENE 35-38 Leaf FNF PED
BETA-CARYOPHYLLENE 500-760 Leaf BML
BETA-COPAENE Plant HHB
BETA-OLEANOLIC-ACID Plant FNF
BETA-PHELLANDRENE 5-28 Leaf JAD NAP
BETA-PINENE 20-1,540 Plant BML JAD NAP
BETA-SITOSTEROL 5-2,450 Leaf NAP
BETA-SITOSTEROL 1,214 Stem NAP
BETA-SITOSTEROL-D-GLUCOSIDE Seed FNF
BETA-THUJONE 200-9,968 Leaf BML JAD NAP
BETA-URSOLIC-ACID Plant FNF
BETULIN 15 Leaf NAP
BORNEOL 7,000 Shoot BML JAD NAP
BORNYL-ACETATE 5-1,780 Shoot BML JAD NAP
BORON 25-41 Leaf BOB
CAFFEIC-ACID Shoot NAP
CALAMENE 5-28 Plant FNF
CALCIUM 10,800-17,957 Leaf FNF PED
CAMPESTEROL Fruit NAP
CAMPHENE 20-18,592 Leaf BML JAD NAP
CAMPHOR 0-9,324 Leaf BML JAD NAP
CARBOHYDRATES 660,000 Leaf PED
CARNOSIC-ACID Leaf NAP
CARNOSOL Leaf PED
CARNOSOL Plant JBH PAS
CARNOSOLIC-ACID 2,100 Leaf NAP
CARYOPHYLLENE 1-1,430 Leaf BML
CARYOPHYLLENE-OXIDE 55-308 Plant BML
CATECHIN Plant FNF
CHLOROGENIC-ACID Shoot NAP
CHROMIUM 3 Leaf PED
CHRYSOERIOL Plant PAS
CINAROSIDE Shoot NAP
CINEOLE 500-6,075 Shoot JAD NAP
CIRSILINEOL Plant PAS
CIRSILIOL Plant JBH
CIRSILION Plant BML
CIRSIMARITRIN Plant PAS
CIS-BETA-TERPINEOL 10-56 Leaf NAP
CIS-SABINENE-HYDRATE Leaf tr JAD NAP

CITRAL Plant PAS
COBALT Leaf tr PED
COPPER 7-8 Leaf USA
DELTA-CADINENE 2-14 Plant FNF
DELTA-TERPINEOL Plant tr NAP
DIOSMETIN Plant PAS
DIPENTENE-O-DIPHENOLLACTONE 3,500 Leaf HHB
EO 5,000-28,000 Plant BML GEO NAP
EPIOLEANOLIC-ACID 20 Leaf NAP
FARNESOL Plant FNF
FAT 138,000 Leaf PED
FAT 250,000 Seed FNF
FERULIC-ACID Plant PAS
FIBER 87,000 Leaf PED
FUMARIC-ACID Plant FNF
GALLIC-ACID Plant PAS
GAMMA-CADINENE 2-14 Plant FNF
GAMMA-MAALIENE Plant FNF
GAMMA-TERPINENE 15-140 Leaf BML JAD NAP
GENKWANIN Leaf PED
GERANIOL Plant PAS
GERMANICOL Plant FNF
HISPIDULIN Leaf BML PED
IRON 150-305 Leaf CRC PED
ISOBORNEOL 0-784 Shoot JAD NAP
ISOBORNEOL-ACETATE 168-9,436 Shoot JAD NAP
ISOCARYOPHYLLENE Plant FNF
KILOCALORIES 3,420/kg Leaf PED
LABIATIC-ACID Leaf PED
LEDENE Plant FNF
LIMONENE 39-2,380 Plant BML JAD NAP
LINALOL 0-3,500 Plant BML JAD NAP
LINALYL-ACETATE 0-6,048 Plant BML JAD NAP
LINOLEIC-ACID 73,000 Seed HHB WOI
LINOLENIC-ACID 42,000-86,750 Seed HHB WOI
LUTEOLIN Shoot NAP
LUTEOLIN-7-GLUCOSIDE Plant NAP
LUTEOLIN-7-GLYCOSIDE Plant NAP
LUTEOLIN-7-METHYL-ETHER Plant FNF
LUTEOLIN-7-O-BETA-D-GLUCOSIDE Plant PAS
LUTEOLIN-7-O-GLUCURONIDE Plant JSG
MAGNESIUM 2,830 Leaf PED
MALIC-ACID Plant FNF
MANGANESE 30-31 Leaf PED USA
MASLINIC-ACID 46 Leaf NAP

METHYL-ISOVALERATE 7-42 Plant FNF
MYRCENE 0-336 Plant BML JAD NAP
N-TRIACONTANE Plant FNF
NEPETIN Plant FNF
NIACIN 57-62 Leaf FNF PED
NICOTINIC-ACID-AMIDE 5,000 Plant FNF
OLEANOLIC-ACID 140-786 Leaf NAP
OLEANOLIC-ACID 400 Stem NAP
OLEIC-ACID 35,500 Seed HHB WOI
OXALIC-ACID Plant FNF
P-COUMARIC-ACID Plant PAS
P-CYMEN-8-OL Plant FNF
P-CYMENE 15-495 Shoot BML JAD NAP
PECTOLINAMARIGENIN Plant PAS
PHELLANDRENE 100-560 Leaf JAD NAP
PHOSPHORUS 910-1,280 Leaf FNF PED
PICROSALVIN Leaf PED
PINENE 420-2,352 Leaf JAD NAP
PLANTEOSE Seed NAP
POMOLIC-ACID 2.4-3 Leaf NAP
POTASSIUM 24,700 Leaf PED
POTASSIUM 10,700-11,630 Plant CRC
POTASSIUM-NITRATE Plant FNF
PRISTANE 56 Leaf NAP
PROTEIN 115,000 Leaf PED
PROTEIN 180,000 Seed FNF
RESIN 50,000-60,000 Plant FNF
RIBOFLAVIN 3-4 Leaf FNF
ROSMANOL Plant PAS
ROSMARINIC-ACID 30,000 Plant FT62:166
ROSMARINIC-ACID 2,000-5,800 Shoot JAD JBH NAP
ROYLEANONE 20,300 Root NAP
SABINENE 10-56 Leaf JAD NAP
SABINOL 85-476 Leaf BML
SABINYL-ACETATE 110-616 Plant FNF
SACCHAROPINE 1.6 Shoot NAP
SALICYLIC-ACID Plant PAS
SALVIATANNIN Leaf PED
SALVIGENIN Leaf PED
SALVIGENIN-7-O-GLUCURONIDE Plant JSG
SALVIN Plant JSG
SALVIN-MONOMETHYL-ETHER Leaf NAP
SALVIOL Essential Oil NAP
SAPONIN Plant FNF
SELENIUM Leaf tr PED

SELINA-5,11-DIENE Plant FNF
SILICON 31 Leaf PED
SODIUM 1,080 Leaf PED
SODIUM 11-12 Plant FNF
STIGMASTEROL 5 Leaf NAP
TANNIN 20,000-80,000 Plant FNF
TERPINEN-4-OL 10-1,120 Leaf BML JAD NAP
TERPINEOL Leaf NAP
TERPINOLENE 0-112 Plant BML JAD NAP
THIAMIN 7-8 Leaf FNF
THUJONE 1,453-12,636 Leaf JAD NAP
THYMOL Essential Oil NAP
TIN 8 Leaf PED
TRANS-3-HEXENAL Plant FNF
TRANS-ALLO-OCIMENE Plant FNF
TRANS-OCIMENE Plant FNF
TRANS-SABINENE-HYDRATE 10-56 Leaf JAD NAP
TRANS-SABINOL Leaf tr NAP
TRICYCLENE 15-84 Leaf JAD NAP
URS-12-EN-28-OIC-ACID 14 Leaf NAP
URSOLIC-ACID 1,255-1,300 Leaf NAP
URSOLIC-ACID 200 Stem NAP
UVAOL Plant FNF
VANILLIC-ACID Plant PAS
VIRIDIFLOROL Leaf NAP
WATER 843,000 Leaf PED 1/
ZINC 59 Leaf PED

Salvia sclarea L.
"Clary Sage"

(1R,5R)-EPOXYSALVIAL-4(14)-EN-1-ONE Plant JSG
(2R,5E)-EPOXYCARYOPHYLL-5-EN-12-AL Plant JSG
(2R,5E)-EPOXYCARYOPHYLL-5-ENE Plant JSG
(2S,5E)-EPOXYCARYOPHYLL-5-EN-12-AL Plant JSG
1,8-CINEOLE Plant RIZ
3-OCTANOL Plant CRC
ACETIC-ACID-ESTER Plant CRC
ALPHA-BISABOLOL Plant RIZ
ALPHA-PINENE 0-3 Plant FFJ6:154
ALPHA-TERPINEOL 1-155 Plant FFJ6:154
ALPHA-TERPINYL-ACETATE 1-155 Plant FFJ6:154
BENZALDEHYDE Plant AYL
BETA-CARYOPHYLLENE 1-30 Plant FFJ6:153
BETA-GURJUENE Plant BML
BETA-MYRCENE Plant AYL
BETA-PINENE Plant BML
CADINENE Plant CRC
CAMPHENE Plant BML
CAMPHOR Plant RIZ
CAPRYLIC-ACID 2,926 Plant CRC
CARYOPHYLLENE-OXIDE Plant BML
CEROTINIC-ACID Plant CRC
CIS-3-HEXEN-1-OL Plant FNF
CIS-ALLOOCIMENE Plant AYL
CIS-BETA-OCIMENE 2.4-24 Plant FFJ6:154
CIS-LINALOL-OXIDE Plant FFJ6:153
CITRAL 75-920 Plant FFJ6:154
CITRONELLOL Plant BML
CUMINALDEHYDE Plant AYL
DELTA-3-CARENE Plant CRC
EO 400 5,000 Plant AYL FFJ6:154
FAT 318,000 Seed CRC
FURFUROL Plant BML
GAMMA-TERPINENE 1-21 Plant FFJ6:154
GERANIAL 0.8-580 Plant FFJ6:154
GERANIOL 22-735 Plant FFJ6:154
GERANYL-ACETATE 20-1,150 Plant FFJ6:154
GERMACRENE-D 3-312 Plant FFJ6:154
ISOSPATHULENOL Plant JSG
LIMONENE 0.4-12 Plant FFJ6:154
LINALOL 7-930 Plant FFJ6:153
LINALYL-ACETATE 1-3,750 Plant AYL FFJ6:154

LINOLEIC-ACID 53,488 Seed CRC
LINOLENIC-ACID 161,100 Seed CRC
MANOOL Plant 411/
MYRCENE 1-42 Plant FFJ6:154
N-NONANOL Plant HHB
N-PENTANOL Plant CRC
NERAL 0.4-340 Plant FFJ6:154
NEROL 6-280 Plant FFJ6:154
NEROLIDOL Plant AYL
NERYL-ACETATE 6-144 Plant FFJ6:154
OLEANOLIC-ACID Plant CRC
OLEIC-ACID 69,006 Seed CRC
P-CYMENE 1-27 Plant FFJ6:154
PALMITIC-ACID 22,419 Seed CRC
PHELLANDRENE Plant AYL
PROPIONIC-ACID-ESTER Plant CRC
PROTEIN 207,000 Seed CRC
ROSMARINIC-ACID 34,000 Plant FT62:166
SALVIA-4(14)-EN-1-ONE Plant JSG
SCLAREOL Plant CRC
STEARIC-ACID 8,568 Seed CRC
TERPINEN-4-OL 1-6 Plant FFJ6:154
TERPINOLENE 1-12 Plant FFJ6:154
TRANS-ALLOOCIMENE Plant AYL
TRANS-BETA-OCIMENE 4.4-42 Plant FFJ6:154
TRANS-BETA-TERPINEOL Plant BML
TRANS-LINALOL-OXIDE Plant FFJ6:153
URSOLIC-ACID Plant CRC
VALERIC-ACID-ESTER Plant AYL

Salvia sp.
"Chia"

CAFFEIC-ACID 306 Seed PCF:67
FAT 228,000-360,000 Seed HHB PCF:68
FERULIC-ACID 2,136 Seed PCF:67
GENTISIC-ACID 22 Seed PCF:67
LINOLEIC-ACID 34,885-169,200 Seed HHB PCF:68
LINOLENIC-ACID 14,365-252,000 Seed HHB PCF:68
MUCILAGE Seed HHB
MYRISTIC-ACID 230-360 Seed HHB PCF:68
OLEIC-ACID 15,600-76,680 Seed HHB PCF:68
P-COUMARIC-ACID 19 Seed PCF:67
PALMITIC-ACID 11,855-35,640 Seed HHB PCF:68
PALMITOLIC-ACID Seed tr PCF:68
PROTEIN 175,000 Seed CRC
STEARIC-ACID 6,610-58,320 Seed HHB PCF:68
SYRINGIC-ACID 25 Seed PCF:67
VANILLIC-ACID 69 Seed PCF:67

Salvia triloba L.
"Greek Sage"

1,8-CINEOLE 2,760-18,600 Plant BML HHB
5-HYDROXY-6,7,4'-TRIMETHOXYFLAVONE Plant HHB
6,8-DI-C-GLUCOSYLAPIGENIN Plant JSG
6-HYDROXYLUTEIN-6,3'-DIMETHYL-ETHER Plant JSG
6-METHOXY-APIGENIN-7-GLUCOSIDE Plant JSG
6-METHOXY-APIGENIN-7-GLUCURONIDE Plant JSG
6-METHOXY-LUTEOLIN-7-GLUCOSIDE Plant JSG
6-METHOXY-LUTEOLIN-7-GLUCURONIDE Plant JSG
ALPHA-PINENE 270-1,800 Plant BML HHB
ALPHA-TERPINEOL 140-750 Plant BML HHB
APIGENIN-7-GLUCOSIDE Plant JSG
APIGENIN-7-GLUCURONIDE Plant JSG
BETA-CARYOPHYLLENE 500-2,280 Plant BML HHB
BETA-PINENE 230-1,590 Plant BML HHB
BETA-THUJONE 160-1,500 Plant BML HHB
BETULINIC-ACID Plant HHB
CAMPHENE 170-1,200 Plant BML HHB
CAMPHOR 160-5,040 Plant BML HHB
CARNESOL Plant HHB
CHRYSOERIOL-7-GLUCURONIDE Plant JSG
EO 10,000-30,000 Plant BML HHB
LIMONENE 100-690 Plant BML HHB
LINALOL 700-5,040 Plant BML HHB
LUTEOLIN-7-GLUCOSIDE Plant JSG
LUTEOLIN-7-GLUCURONIDE Plant JSG
LUTEOLIN-7-GLUCURONIDE-3'-GLUCOSIDE Plant JSG
LUTEOLINE-DIGLUCOSIDE Plant JSG
MYRCENE 150-900 Plant BML HHB
OLEANOLIC-ACID 5,200 Plant HHB
PICROSALVIN Plant HHB
SALVIGENIN Plant HHB
THUJONE 100-1,620 Plant BML HHB
URSOLIC-ACID 74,500 Plant HHB

Satureja cilicica P.H. DAVIS
"Turkish Savory"
Unaggregated A

(E)-2-HEXENAL 5 Shoot JEO5:547
(E)-BETA-OCIMENE 55 Shoot JEO5:547
(Z)-BETA-OCIMENE 210 Shoot JEO5:547
(Z)-GERANYL-ACETONE 3 Shoot JEO5:547
1,8-CINEOLE 80 Shoot JEO5:547
1-NONEN-3-OL 6 Shoot JEO5:547
1-OCTYL-3-OL 29 Shoot JEO5:547
6-METHYL-3-HEPTANOL 4 Shoot JEO5:547
ALPHA-COPAENE 3 Shoot JEO5:547
ALPHA-PHELLANDRENE 1 Shoot JEO5:547
ALPHA-PINENE 150 Shoot JEO5:547
ALPHA-TERPINENE 130 Shoot JEO5:547
ALPHA-TERPINEOL 12 Shoot JEO5:547
BETA-BISABOLENE 60 Shoot JEO5:547
BETA-BOURBONENE 2 Shoot JEO5:547
BETA-CARYOPHYLLENE 55 Shoot JEO5:547
BETA-PINENE 22 Shoot JEO5:547
CAMPHENE 60 Shoot JEO5:547
CARVACROL 3,005 Shoot JEO5:547
CIS-DIHYDROCARVONE 28 Shoot JEO5:547
CIS-SABINENE-HYDRATE 21 Shoot JEO5:547
DELTA-CADINENE 18 Shoot JEO5:547
EO 8,000 Shoot JEO5:547
GAMMA-MUUROLENE 5 Shoot JEO5:547
GAMMA-TERPINENE 1,070 Shoot JEO5:547
ISOBORNEOL 275 Shoot JEO5:547
LIMONENE 115 Shoot JEO5:547
LINALOL 13 Shoot JEO5:547
MYRCENE 150 Shoot JEO5:547
P-CYMEN-8-OL 12 Shoot JEO5:547
P-CYMENE 1,150 Shoot JEO5:547
PATCHOULANE 41 Shoot JEO5:547
PULEGONE 7 Shoot JEO5:547
SABINENE 2 Shoot JEO5:547
SPATHULENOL 55 Shoot JEO5:547
TERPINEN-4-OL 55 Shoot JEO5:547
THYMOL 595 Shoot JEO5:547
TRANS-DIHYDROCARVONE 85 Shoot JEO5:547
TRANS-SABINENE-HYDRATE 70 Shoot JEO5:547

Satureja cuneifolia TEN.
"Cuneate Turkish Savory"
Unaggregated A

1,8-CINEOLE 80 Shoot JEO3:365
6-METHYL-3-HEPTANOL 4 Shoot JEO3:365
ACETONE 20 Shoot JEO3:365
ALPHA-PINENE 405 Shoot JEO3:365
ALPHA-TERPINENE 315 Shoot JEO3:365
ALPHA-TERPINEOL 30 Shoot JEO3:365
ALPHA-THUJENE 10 Shoot JEO3:365
BETA-CARYOPHYLLENE 5 Shoot JEO3:365
BETA-PINENE 60 Shoot JEO3:365
BORNEOL 430 Shoot JEO3:365
CAMPHENE 145 Shoot JEO3:365
CARVACROL 8,760 Shoot JEO3:365
CIS-BETA-OCIMENE 525 Shoot JEO3:365
CIS-ISOEUGENOL 15 Shoot JEO3:365
CITRONELLOL 40 Shoot JEO3:365
CITRONELLYL-ACETATE 5 Shoot JEO3:365
DELTA-3-CARENE 15 Shoot JEO3:365
DIHYDROCARVYL-ACETATE 35 Shoot JEO3:365
EO 18,900 Shoot JEO3:365
ETHANOL 255 Shoot JEO3:365
GAMMA-TERPINENE 2,465 Shoot JEO3:365
GERANIOL 115 Shoot JEO3:365
LIMONENE 125 Shoot JEO3:365
LINALOL 60 Shoot JEO3:365
MYRCENE 50-420 Shoot JEO3:365
NONANOL 35 Shoot JEO3:365
P-CYMEN-7-OL 30 Shoot JEO3:365
P-CYMEN-8-OL 60 Shoot JEO3:365
P-CYMENE 2,980 Shoot JEO3:365
SABINENE 5 Shoot JEO3:365
TERPINEN-4-OL 135 Shoot JEO3:365
TERPINOLENE 25 Shoot JEO3:365
THYMOL 45 Shoot JEO3:365
TRANS-BETA-FARNESENE 490 Shoot JEO3:365
TRANS-BETA-OCIMENE 525 Shoot JEO3:365

Satureja douglasii (BENTH.) BRIQ. "Douglas' Savory"

1,8-CINEOLE 39-195 Plant BML
ALPHA-PINENE 299-884 Plant BML
ALPHA-TERPINENE 13 Plant BML
ALPHA-TERPINEOL 1-39 Plant BML
BETA-BOURBONENE 533 Plant BML
BETA-PINENE 130-572 Plant BML
BICYCLOGERMACRENE Plant tr BML
BORNEOL 1-910 Plant BML
CAMPHENE 1,183-2,795 Plant BML
CAMPHOR 2,392-5,941 Plant BML
CARVONE 1,040-2,340 Plant BML
CARYOPHYLLENE 1-39 Plant BML
CIS-OCIMENE 234 Plant BML
DELTA-CADINENE 13 Plant BML
EO 13,000 Plant BML
GAMMA-TERPINENE Plant tr BML
GERMACRENE-D 26 Plant BML
ISOMENTHONE 13-4,277 Plant BML
LIMONENE 234-1,508 Plant BML
LINALOL 26-182 Plant BML
MENTHONE 26-1,950 Plant BML
METHYL-CARVACROL 65 Plant BML
MYRCENE 104 Plant BML
P-CYMENE 26 Plant BML
PIPERITENONE 1-377 Plant BML
PIPERITONE 195-1,443 Plant BML
PULEGONE <1-6,695 Plant BML
SABINENE 26 Plant BML
TERPINEN-4-OL 104-143 Plant BML
TERPINOLENE 39 Plant BML
THYMOL 39 Plant BML
TRANS-OCIMENE 143 Plant BML

Satureja glabella (MICHX.) BRIQ.
"Smooth Savory"

3-OCTANOL 70 Plant BML 1/
ALPHA-PINENE 40 Plant BML 1/
BETA-PINENE 50 Plant BML 1/
EO 10,000-10,600 Plant BML 1/
GERMACRENE-D 30 Plant BML 1/
ISOMENTHOL 410 Plant BML 1/
ISOMENTHONE 7,040 Plant BML 1/
LIMONENE 50 Plant BML 1/
MENTHONE 280 Plant BML 1/
NEOISOMENTHOL 260 Plant BML 1/
NEOMENTHOL 30 Plant BML 1/
PULEGONE 1,550 Plant BML 1/

Satureja grandiflora (L.) SCHEELE
"French Savory"
Unaggregated A

3-OCTANOL 4 Shoot JEO3:361
ALPHA-PINENE 4 Shoot JEO3:361
BETA-PINENE 10 Shoot JEO3:361
BICYCLOGERMACRENE 90 Shoot JEO3:361
EO 3,900 Shoot JEO3:361
GERMACRENE-D 135 Shoot JEO3:361
ISOMENTHONE 965 Shoot JEO3:361
LIMONENE 50 Shoot JEO3:361
MENTHOL 420 Shoot JEO3:361
MENTHONE 35 Shoot JEO3:361
MYRCENE 4 Shoot JEO3:361
NEOISOMENTHOL 925 Shoot JEO3:361
PULEGONE 1,075 Shoot JEO3:361
SABINENE 15 Shoot JEO3:361
TERPINEN-4-OL 30 Shoot JEO3:361

Satureja hortensis L.
"Summer Savory"

ALPHA-PINENE Plant HHB
ALPHA-TERPINENE Plant HHB
ALPHA-TERPINEOL Plant BML
ASH 21,100-105,000 Plant CRC WOI
BETA-CAROTENE 34 Plant CRC
BETA-CARYOPHYLLENE Plant HHB
BETA-PHELLANDRENE Plant HHB
BETA-PINENE Plant HHB
BETA-TERPINENE Plant HHB
BORNEOL Plant BML
BORON 25-37 Plant BOB
CALCIUM 23,429 Plant CRC
CAMPHENE Plant HHB
CAMPHOR Plant HHB
CARBOHYDRATES 755,000 Plant CRC
CARVACROL 450-6,000 Plant WOI
COPPER 8-9 Leaf USA
DIPENTENE Plant HHB
EO 4,000 Leaf WOI
EO 1,500-20,000 Plant HHB WOI
FAT 16,500-65,000 Plant WOI
FAT 420,000 Seed WOI
FIBER 86,000-306,160 Plant WOI
FLOURINE 12 Plant WOI
GERANIOL Plant HHB
IRON 416 Plant CRC
LABIATIC-ACID Plant WOI
LIMONENE Plant 411/
LINOLEIC-ACID 75,600 Seed WOI
LINOLENIC-ACID 260,400 Seed WOI
MANGANESE 61 Leaf USA
MUCILAGE Plant HHB
MYRCENE Plant HHB
NEROL Plant HHB
NIACIN 45 Plant CRC
OLEANOLIC-ACID 600 Plant HHB
OLEIC-ACID 50,400 Seed WOI
P-CYMENE Plant BML
P-CYMOL 300-6,000 Plant HHB
PALMITIC-ACID 8,400 Seed WOI
PENTOSANS 119,500 Leaf WOI
PHOSPHORUS 1,538 Plant CRC

POTASSIUM 11,549 Plant CRC
PROTEIN 41,500-147,740 Plant WOI
PROTEIN 230,000-240,000 Seed CRC WOI
ROSMARINIC-ACID 26,000 Plant FT62:166
SABINENE Plant HHB
SODIUM 264 Plant CRC
STEARIC-ACID 8,400 Seed WOI
SYLVESTRINE Plant WOI
TANNIN 40,000-85,000 Plant HHB
TERPINEN-4-OL Plant BML
THIAMIN 4 Plant CRC
THUJONE Plant HHB
THYMOL Plant BML
URSOLIC-ACID 1,700 Leaf WOI
URSOLIC-ACID 1,000 Stem WOI

***Satureja montana* L.**
"Winter Savory"

1,8-CINEOLE 4-150 Plant BML JEO3:147
1-OCTEN-3-OL 7-360 Plant BML JEO3:147 JE39:167
ALPHA-COPAENE 3-85 Plant BML JEO3:147
ALPHA-HUMULENE 8-325 Plant BML JEO3:147
ALPHA-MUUROLENE 1-15 Plant BML JEO3:147
ALPHA-PHELLANDRENE 1-325 Plant JE39:167 JEO3:147
ALPHA-PINENE 5-245 Plant BML JE39:167 JEO3:147
ALPHA-TERPINENE 15-410 Plant JE39:167 JEO3:147
ALPHA-TERPINEOL 3-2,185 Plant BML CRC JEO3:147
ALPHA-TERPINYL-ACETATE Plant CRC
ALPHA-THUJENE 5-300 Plant GEO HHB JEO3:147
ALPHA-THUJONE 10-385 Plant BML CRC JEO3:147
AROMADENDRENE 2-30 Plant BML JEO3:147
BETA-BISABOLENE 20-400 Plant BML CRC JEO3:147
BETA-BOURBONENE 4-85 Plant BML JEO3:147
BETA-CARYOPHYLLENE 20-1,050 Plant JE39:167
BETA-GURJUNENE 2-45 Plant BML JEO3:147
BETA-PINENE 1-55 Plant BML JE39:167 JEO3:147
BETA-THUJONE 2-65 Plant BML CRC JEO3:147
BORNEOL 15-545 Plant BML JE39:167 JEO3:147
BORNYL-ACETATE 1-30 Plant BML JE39:167 JEO3:147
BORON 25-37 Plant BOB
CAMPHENE 2-85 Plant GEO HHB JEO3:147
CAMPHOR 5-75 Plant BML CRC JEO3:147
CARVACROL 75-17,250 Plant BML CRC HHB JEO3:147
CARVONE <1-95 Plant BML JE39:167 JEO3:147
CARYOPHYLLENE-OXIDE 10-285 Plant BML CRC JEO3:147
CIS-OCIMENE 1-35 Plant BML JEO3:147
CITRONELLOL <1-325 Plant JEO3:153
COPPER 8-9 Leaf USA
DELTA-3-CARENE <1-15 Plant JE39:167 JEO3:147
DELTA-CADINENE 6-135 Plant BML CRC JEO3:147
DIPENTENE 210-3,220 Plant BML
EO 1,500-23,000 Plant GEO HHB JEO3:147
GAMMA-CADINENE 3-70 Plant BML CRC JEO3:147
GAMMA-MUUROLENE 4-90 Plant BML JEO3:147
GAMMA-TERPINENE 15-7,130 Plant BML CRC JEO3:147
GERANIOL Plant BML
GERANYL-ACETATE Plant CRC
GERMACRENE-D 5-95 Plant BML JEO3:147
LIMONENE 4-85 Plant BML JEO3:147
LINALOL 115-14,260 Plant BML CRC JEO3:147

LINALYL-ACETATE 3-240 Plant BML JEO3:147
MANGANESE 61 Plant USA
MENTHADIEN-1,8(9)-OL-(7) Plant HHB
METHYL-CARVACROL 7-115 Plant BML JEO3:147
METHYL-EUGENOL 25-415 Plant JEO3:153
MYRCENE 7-295 Plant BML JE39:167 JEO3:147
NEROL <1-270 Plant JEO3:153
NONANOL 30-415 Plant JEO3:153
OCTANOL <1-55 Plant JEO3:153
OLEANOLIC-ACID Plant BML
P-CYMENE 45-6,210 Plant BML JEO3:147
PERILLYL-ALCOHOL Plant BML
PHENOL Plant BML
POLYISOPRENE Plant BML
ROSMARINIC-ACID 3,000-12,000 Plant ABS FT62:166
SABINENE <1 Plant GEO HHB JEO3:147
TERPENE Plant BML
TERPINEN-4-OL 1-1,610 Plant BML CRC
TERPINEN-4-YL-ACETATE Plant CRC
TERPINOLENE 0.6-25 Plant BML JEO3:147
THUJANOL-4 Plant BML
THYMOL 5-14,030 Plant HHB JE39:167 JEO3:153
THYMOQUINONE <1-75 Plant BML JEO3:147
URSOLIC-ACID 16,000 Plant WOI
VIRIDIFLORENE 4-75 Plant BML JEO3:147
YLANGENE <1-270 Plant JEO3:153

Satureja obovata LAG.
"Iberian Savory, Savory"

ALPHA-PHELLANDRENE <1-30 Shoot FT60:277
ALPHA-PINENE 300-510 Shoot FT60:277
ALPHA-TERPINENE 60-170 Shoot FT60:277
ALPHA-TERPINEOL 775-1,425 Shoot FT60:277
BETA-PINENE 95-145 Shoot FT60:277
CAMPHENE 850-1,305 Shoot FT60:277
CAMPHOR 2,510-4,025 Shoot FT60:277
CARVACROL <1-115 Shoot FT60:277
CARYOPHYLLENE <1-60 Shoot FT60:277
CITRAL 60-110 Shoot FT60:277
CITRONELLOL <1 Shoot FT60:277
DELTA-3-CARENE <1-30 Shoot FT60:277
EO 10,000-11,000 Shoot FT60:277
GAMMA-TERPINENE 60-300 Shoot FT60:277
GERANIOL <1-315 Shoot FT60:277
LIMONENE <1-255 Shoot FT60:277
LINALOL 865-3,370 Shoot FT60:277
MYRCENE <1-30 Shoot FT60:277
P-CYMENE 50-1,235 Shoot FT60:277
SABINENE <1-190 Shoot FT60:277
TERPINEN-4-OL 270-575 Shoot FT60:277
TERPINYL-ACETATE <1-60 Shoot FT60:277
THYMOL 50-850 Shoot FT60:277

Satureja obovata LAG.
"Iberian Savory, Savory"
Unaggregated A

ALPHA-BISABOLOL <0.01 Leaf PC35:83
ALPHA-CUBEBENE <0.01 Leaf PC35:83
ALPHA-PHELLANDRENE 90 Leaf PC35:83
ALPHA-PINENE 425 Leaf PC35:83
ALPHA-TERPINENE <0.01 Leaf PC35:83
ALPHA-TERPINEOL 2,010 Leaf PC35:83
ALPHA-THUJENE <0.01 Leaf PC35:83
ARTEMISIATRIENE <0.01 Leaf PC35:83
BETA-CARYOPHYLLENE 125 Leaf PC35:83
BETA-PHELLANDRENE <0.01 Leaf PC35:83
BETA-PINENE 95 Leaf PC35:83
BETA-TERPINEOL 205 Leaf PC35:83
BORNYL-ACETATE <0.01 Leaf PC35:83
CAMPHENE 880 Leaf PC35:83
CAMPHOR 2,510 Leaf PC35:83
CARVACROL 220 Leaf PC35:83
CARYOPHYLLENE-OXIDE 190 Leaf PC35:83
CIS-CITRAL <0.01 Leaf PC35:83
CIS-LINALOL-OXIDE 100 Leaf PC35:83
CIS-PIPERITOL 285 Leaf PC35:83
CIS-SABINENE-HYDRATE 105 Leaf PC35:83
CITRONELLOL <0.01 Leaf PC35:83
DELTA-3-CARENE <0.01 Leaf PC35:83
EO 16,900 Leaf PC35:83
GAMMA-TERPINENE 325 Leaf PC35:83
GERANIOL <0.01 Leaf PC35:83
GERMACRENE-B 290 Leaf PC35:83
ISOCARYOPHYLLENE 425 Leaf PC35:83
LEDOL <0.01 Leaf PC35:83
LINALOL 750 Leaf PC35:83
LINALYL-ACETATE <0.01 Leaf PC35:83
MYRCENE 305 Leaf PC35:83
P-CYMEN-8-OL 145 Leaf PC35:83
P-CYMENE 550 Leaf PC35:83
PHELLANDROL <0.01 Leaf PC35:83
SABINENE <0.01 Leaf PC35:83
T-CADINOL <0.01 Leaf PC35:83
TERPINEN-4-OL 390 Leaf PC35:83
TERPINOLENE 170 Leaf PC35:83
THYMOL 955 Leaf PC35:83

Satureja obovata LAG.
"Iberian Savory, Savory"
Unaggregated B

ALPHA-BISABOLOL <0.01 Leaf PC35:83
ALPHA-CUBEBENE 785 Leaf PC35:83
ALPHA-PHELLANDRENE <0.01 Leaf PC35:83
ALPHA-PINENE 330 Leaf PC35:83
ALPHA-TERPINENE 235 Leaf PC35:83
ALPHA-TERPINEOL 2,135 Leaf PC35:83
ALPHA-THUJENE <0.01 Leaf PC35:83
ARTEMISIATRIENE <0.01 Leaf PC35:83
BETA-CARYOPHYLLENE <0.01 Leaf PC35:83
BETA-PHELLANDRENE 360 Leaf PC35:83
BETA-PINENE <0.01 Leaf PC35:83
BETA-TERPINEOL 315 Leaf PC35:83
BORNYL-ACETATE <0.01 Leaf PC35:83
CAMPHENE 520 Leaf PC35:83
CAMPHOR 2,660 Leaf PC35:83
CARVACROL 260 Leaf PC35:83
CARYOPHYLLENE-OXIDE 175 Leaf PC35:83
CIS-CITRAL <0.01 Leaf PC35:83
CIS-LINALOL-OXIDE <0.01 Leaf PC35:83
CIS-PIPERITOL 470 Leaf PC35:83
CIS-SABINENE-HYDRATE <0.01 Leaf PC35:83
CITRONELLOL <0.01 Leaf PC35:83
DELTA-3-CARENE <0.01 Leaf PC35:83
EO 22,600 Leaf PC35:83
GAMMA-TERPINENE 1,715 Leaf PC35:83
GERANIOL <0.01 Leaf PC35:83
GERMACRENE-B 525 Leaf PC35:83
ISOCARYOPHYLLENE 520 Leaf PC35:83
LEDOL <0.01 Leaf PC35:83
LINALOL 1,940 Leaf PC35:83
LINALYL-ACETATE <0.01 Leaf PC35:83
MYRCENE <0.01 Leaf PC35:83
P-CYMEN-8-OL <0.01 Leaf PC35:83
P-CYMENE 990 Leaf PC35:83
PHELLANDROL <0.01 Leaf PC35:83
SABINENE <0.01 Leaf PC35:83
T-CADINOL <0.01 Leaf PC35:83
TERPINEN-4-OL 825 Leaf PC35:83
TERPINOLENE 190 Leaf PC35:83
THYMOL 2,910 Leaf PC35:83

Satureja obovata LAG.
"Iberian Savory, Savory"
Unaggregated C

ALPHA-BISABOLOL <0.01 Leaf PC35:83
ALPHA-CUBEBENE 200 Leaf PC35:83
ALPHA-PHELLANDRENE 220 Leaf PC35:83
ALPHA-PINENE 130 Leaf PC35:83
ALPHA-TERPINENE 120 Leaf PC35:83
ALPHA-TERPINEOL 1,365 Leaf PC35:83
ALPHA-THUJENE <0.01 Leaf PC35:83
ARTEMISIATRIENE <0.01 Leaf PC35:83
BETA-CARYOPHYLLENE 985 Leaf PC35:83
BETA-PHELLANDRENE 300 Leaf PC35:83
BETA-PINENE <0.01 Leaf PC35:83
BETA-TERPINEOL 845 Leaf PC35:83
BORNYL-ACETATE 135 Leaf PC35:83
CAMPHENE 130 Leaf PC35:83
CAMPHOR 1,225 Leaf PC35:83
CARVACROL 680 Leaf PC35:83
CARYOPHYLLENE-OXIDE <0.01 Leaf PC35:83
CIS-CITRAL <0.01 Leaf PC35:83
CIS-LINALOL-OXIDE 165 Leaf PC35:83
CIS-PIPERITOL 135 Leaf PC35:83
CIS-SABINENE-HYDRATE <0.01 Leaf PC35:83
CITRONELLOL <0.01 Leaf PC35:83
DELTA-3-CARENE <0.01 Leaf PC35:83
EO 24,100 Leaf PC35:83
GAMMA-TERPINENE 1,590 Leaf PC35:83
GERANIOL <0.01 Leaf PC35:83
GERMACRENE-B 210 Leaf PC35:83
ISOCARYOPHYLLENE 1,465 Leaf PC35:83
LEDOL <0.01 Leaf PC35:83
LINALOL 945 Leaf PC35:83
LINALYL-ACETATE 230 Leaf PC35:83
MYRCENE 215 Leaf PC35:83
P-CYMEN-8-OL <0.01 Leaf PC35:83
P-CYMENE 1,230 Leaf PC35:83
PHELLANDROL <0.01 Leaf PC35:83
SABINENE <0.01 Leaf PC35:83
T-CADINOL <0.01 Leaf PC35:83
TERPINEN-4-OL 970 Leaf PC35:83
TERPINOLENE <0.01 Leaf PC35:83
THYMOL 7,765 Leaf PC35:83

Satureja obovata LAG.
"Iberian Savory, Savory"
Unaggregated D

ALPHA-BISABOLOL <0.01 Leaf PC35:83
ALPHA-CUBEBENE 285 Leaf PC35:83
ALPHA-PHELLANDRENE 265 Leaf PC35:83
ALPHA-PINENE 225 Leaf PC35:83
ALPHA-TERPINENE 55 Leaf PC35:83
ALPHA-TERPINEOL 1,700 Leaf PC35:83
ALPHA-THUJENE 65 Leaf PC35:83
ARTEMISIATRIENE 70 Leaf PC35:83
BETA-CARYOPHYLLENE 95 Leaf PC35:83
BETA-PHELLANDRENE 110 Leaf PC35:83
BETA-PINENE <0.01 Leaf PC35:83
BETA-TERPINEOL 220 Leaf PC35:83
BORNYL-ACETATE 60 Leaf PC35:83
CAMPHENE 170 Leaf PC35:83
CAMPHOR 2,040 Leaf PC35:83
CARVACROL 160 Leaf PC35:83
CARYOPHYLLENE-OXIDE 60 Leaf PC35:83
CIS-CITRAL <0.01 Leaf PC35:83
CIS-LINALOL-OXIDE 60 Leaf PC35:83
CIS-PIPERITOL <0.01 Leaf PC35:83
CIS-SABINENE-HYDRATE <0.01 Leaf PC35:83
CITRONELLOL <0.01 Leaf PC35:83
DELTA-3-CARENE <0.01 Leaf PC35:83
EO 9,800 Leaf PC35:83
GAMMA-TERPINENE 250 Leaf PC35:83
GERANIOL <0.01 Leaf PC35:83
GERMACRENE-B 70 Leaf PC35:83
ISOCARYOPHYLLENE 70 Leaf PC35:83
LEDOL <0.01 Leaf PC35:83
LINALOL 1,580 Leaf PC35:83
LINALYL-ACETATE 55 Leaf PC35:83
MYRCENE 70 Leaf PC35:83
P-CYMEN-8-OL <0.01 Leaf PC35:83
P-CYMENE 100 Leaf PC35:83
PHELLANDROL <0.01 Leaf PC35:83
SABINENE 65 Leaf PC35:83
T-CADINOL <0.01 Leaf PC35:83
TERPINEN-4-OL 855 Leaf PC35:83
TERPINOLENE 215 Leaf PC35:83
THYMOL 95 Leaf PC35:83

Satureja odora (GRIS.) EPL.
"Scented Pampa Savory"
Unaggregated A

1,8-CINEOLE 160 Shoot JEO5:549
1-NONENE Shoot JEO5:549
ALPHA-PINENE 22 Shoot JEO5:549
ALPHA-TERPINEOL Shoot JEO5:549
BETA-PINENE 18 Shoot JEO5:549
CAMPHENE 18 Shoot JEO5:549
CEDROL 525 Shoot JEO5:549
CITRONELLAL 10 Shoot JEO5:549
CITRONELLOL Shoot JEO5:549
DECANOIC-ACID 25 Shoot JEO5:549
DODECANAL Shoot JEO5:549
EO 5,000 Shoot JEO5:549
ISOBORNEOL 250 Shoot JEO5:549
ISOEUGENOL 525 Shoot JEO5:549
LIMONENE 175 Shoot JEO5:549
MENTHOL Shoot JEO5:549
METHYL-ESTER-DECANOATE 25 Shoot JEO5:549
MYRCENE Shoot JEO5:549
NEMTHONE 145 Shoot JEO5:549
P-CYMENE 50 Shoot JEO5:549
PATCHOULI-ALCOHOL 100 Shoot JEO5:549
PIPERITENONE-OXIDE 105 Shoot JEO5:549
PIPERITONE 1,380 Shoot JEO5:549
PIPERITONE-OXIDE 105 Shoot JEO5:549
PULEGONE 1,050 Shoot JEO5:549
SPATHULENOL Shoot JEO5:549

Satureja parvifolia (PHIL.) EPL.
"Small-Leaf Pampa Savory"
Unaggregated A

1,8-CINEOLE 55 Shoot JEO5:549
1-NONENE 36 Shoot JEO5:549
ALPHA-PINENE 20 Shoot JEO5:549
ALPHA-TERPINEOL 100 Shoot JEO5:549
BETA-PINENE 22 Shoot JEO5:549
CAMPHENE 19 Shoot JEO5:549
CEDROL Shoot JEO5:549
CITRONELLAL 15 Shoot JEO5:549
CITRONELLOL 150 Shoot JEO5:549
DECANOIC-ACID Shoot JEO5:549
DODECANAL 190 Shoot JEO5:549
EO 5,000 Shoot JEO5:549
ISOBORNEOL Shoot JEO5:549
ISOEUGENOL Shoot JEO5:549
LIMONENE 255 Shoot JEO5:549
MENTHOL 1,010 Shoot JEO5:549
MENTHONE 305 Shoot JEO5:549
METHYL-ESTER-DECANOATE Shoot JEO5:549
MYRCENE 22 Shoot JEO5:549
P-CYMENE Shoot JEO5:549
PATCHOULI-ALCOHOL Shoot JEO5:549
PIPERITENONE-OXIDE 750 Shoot JEO5:549
PIPERITONE 95 Shoot JEO5:549
PIPERITONE-OXIDE 1,500 Shoot JEO5:549
PULEGONE 180 Shoot JEO5:549
SPATHULENOL 150 Shoot JEO5:549

Satureja subspicata ssp. *liburnica*
"Yugoslav Savory"
Unaggregated A

1,8-CINEOLE 5 Plant JEO3:153
ALPHA-COPAENE 15 Plant JEO3:153
ALPHA-PINENE 30 Plant JEO3:153
ALPHA-TERPINENE 4 Plant JEO3:153
ALPHA-TERPINEOL 25 Plant JEO3:153
BETA-CARYOPHYLLENE 65 Plant JEO3:153
BORNEOL 25 Plant JEO3:153
BORNYL-ACETATE 135 Plant JEO3:153
CADINOL 20 Plant JEO3:153
CAMPHENE 4 Plant JEO3:153
CARVACROL 220 Plant JEO3:153
CARVONE 10 Plant JEO3:153
CITRONELLOL 40 Plant JEO3:153
EO 1,000 Plant JEO3:153
GAMMA-MUUROLENE 30 Plant JEO3:153
GAMMA-TERPINENE 60 Plant JEO3:153
LIMONENE 25 Plant JEO3:153
LINALOL 35 Plant JEO3:153
LINALYL-ACETATE 15 Plant JEO3:153
METHYL-EUGENOL 60 Plant JEO3:153
NONANOL 15 Plant JEO3:153
OCTANOL 6 Plant JEO3:153
P-CYMENE 50 Plant JEO3:153
THYMOL 75 Plant JEO3:153
YLANGENE 15 Plant JEO3:153

Satureja thymbra L.
"Goat Oregano"
Unaggregated A

4-TERPINEOL 185 Shoot ZLU197:20
ALPHA-CARYOPHYLLENE 37 Shoot ZLU197:20
ALPHA-HUMULENE 75 Shoot ZLU197:20
ALPHA-MUUROLENE <0.1 Shoot ZLU197:20
ALPHA-PHELLANDRENE 75 Shoot ZLU197:20
ALPHA-PINENE 480 Shoot ZLU197:20
ALPHA-TERPINENE 555 Shoot ZLU197:20
ALPHA-TERPINEOL <0.1 Shoot ZLU197:20
ALPHA-THUJENE 295 Shoot ZLU197:20
AROMADENDRENE 37 Shoot ZLU197:20
BETA-BISABOLENE Shoot ZLU197:20
BETA-CARYOPHYLLENE 1,590 Shoot ZLU197:20
BETA-PINENE 185 Shoot ZLU197:20
BORNEOL 185 Shoot ZLU197:20
BORNYL-ACETATE 37 Shoot ZLU197:20
CAMPHENE 150 Shoot ZLU197:20
CARVACROL 19,830 Shoot ZLU197:20
CARVACROL-METHYL-ETHER Shoot ZLU197:20
CARVONE 150 Shoot ZLU197:20
CIS-OCIMENE 37 Shoot ZLU197:20
CIS-PINANE Shoot ZLU197:20
CIS-THUJANOL 110 Shoot ZLU197:20
DELTA-3-CARENE <0.1 Shoot ZLU197:20
EO 37,000 Shoot ZLU197:20
GAMMA-CADINENE 37 Shoot ZLU197:20
GAMMA-GURJUNENE 37 Shoot ZLU197:20
GAMMA-TERPINENE 6,660 Shoot ZLU197:20
LIMONENE Shoot ZLU197:20
MYRCENE 590 Shoot ZLU197:20
P-CYMENE 5,070 Shoot ZLU197:20
TERPINOLENE 37 Shoot ZLU197:20
THYMOL 37 Shoot ZLU197:20
TRANS-OCIMENE 37 Shoot ZLU197:20
TRANS-THUJANOL 370 Shoot ZLU197:20

Schizonepeta tenuifolia BRIQ. "Ching-Chieh, Jing-Jie"

APIGENIN-7-O-BETA-D-GLUCOSIDE Plant FNF
CALCIUM 10,900 Plant FNF
COPPER 23 Plant FNF
D-LIMONENE Plant FNF
D-MENTHONE Plant FNF
DL-MENTHONE Plant FNF
EO 18,000 Plant FNF
IRON 1,700 Plant FNF
L-PULEGONE Plant FNF
LUTEOLIN-7-O-BETA-D-GLUCOSIDE Plant FNF
MAGNESIUM 3,390 Plant FNF
MANGANESE 68 Plant FNF
POTASSIUM 10,600 Plant FNF
SCHIZONEPETOSIDE-A 400 Flower FNF
SCHIZONEPETOSIDE-B 40 Flower FNF
SODIUM 118 Plant FNF
ZINC 28 Plant FNF

Scutellaria baicalensis GEORGI
"Chinese Skullcap, Huang Qin"

(2R,3R)-2',3,5,6',7-PENTAHYDROXYFLAVONONE Root CCO
2S-2',5,6',7-TETRAHYDROXYFLAVONONE Root ABS
5,2',5'-TRIHYDROXY-6,7,8-TRIMETHOXYFLAVONE Root ABS
5,7,2',5'-TETRAHYDROXY-8,6'-DIMETHOXYFLAVONE Root ABS
5,7,2'-TRIHYDROXY-8-METHOXYFLAVONE Root ABS
5,7,2'-TRIHYDROXYFLAVONE Root ABS
ARSENIC 0.18 Root CAL
BAICALEIN Root CCO
BAICALEIN-7-O-BETA-O-D-GLUCOPYRANOSIDE Root ABS
BAICALEIN-7-O-GLUCOSIDE Root CCO
BAICALIN 43,000 Root CCO JBH
BAICALIN-7-GLUCURONIDE Root CCO
BAICALIN-A Plant CMB39:119
CALCIUM 4,630-10,300 Root CAL SMO
CAMPESTEROL Root CCO
CARTHAMIDIN Plant JSG
CHRYSIN Root CCO
COPPER 13-18 Root CAL SMO
D-GLUCOSE Root CCO
DIHYDROBAICALIN Root ABS
DIHYDROOROXYLIN Plant FNF
DIHYDROOROXYLIN-A Plant FNF
HUANGQIN Root CCO
IRON 170-220 Root CAL SMO
ISOCARTHAMIDIN Plant JSG
ISOWOGONIN Plant JSG
KOGANEBANANIN Root CCO
MAGNESIUM 7,170-7,220 Root CAL SMO
MANGANESE 18-29 Root CAL SMO
MERCURY 0.08 Root CAL
NEOBAICALEIN Plant HHB
NORWOGONIN Root ABS
OROXYLIN-A Root CCO
OROXYLIN-A-7-O-GLUCURONIDE Root CCO
POTASSIUM 6,200-9,120 Root CAL SMO
SKULLCAPFLAVONE-I Root CCO
SKULLCAPFLAVONE-II Root CCO JBH
SODIUM 740-991 Root CAL SMO
WOGONIN 5,000 Root CCO JBH
WOGONIN-7-O-GLUCURONIDE Root CCO
ZINC 17-18 Root CAL SMO

Scutellaria churchilliana FERNALD "Hybrid Skullcap"

1-OCTEN-3-OL 58 Plant BML
ALPHA-HUMULENE 11 Plant BML
CARYOPHYLLENE 68 Plant BML
CARYOPHYLLENE-OXIDE 14 Plant BML
DELTA-CADINENE 9 Plant BML
EO 300 Plant BML
GERMACRENE-D 19 Plant BML
LIGNIN Plant FNF
LINALOL 26 Plant BML
SCUTELLAREIN Plant FNF
SCUTELLARIN Plant FNF

Scutellaria discolor COLEBR.
"Discolor Skullcap"

(.+-.)-5,2'-DIHYDROXY-6,7,6'-TRIMETHOXYFLAVONONE Plant FNF
(.+-.)-5,2'-DIHYDROXY-7,8,6'-TRIMETHOXYFLAVANONE Plant FNF
2',4'-DIHYDROXY-2,3',6'-TRIMETHOXYCHALCONE Plant FNF
5,7,2'-TRIHYDROXY-8-METHOXYFLAVONE Plant FNF
NORWOGONIN Plant FNF
WOGONIN Plant FNF

Scutellaria galericulata L.
"Marsh Skullcap"

6-HYDROXYLUTEIN Plant ABS
ALPHA-HUMULENE 16 Plant BML
APIGENIN Plant ABS
BAICALEIN Plant ABS
BAICALEIN-7-BETA-L-RHAMNOSIDE Plant HHB JBH
BAICALIN 10,000 Leaf HHB
BETA-CUBEBENE 20 Plant BML
CARYOPHYLLENE 176 Plant BML
CHRYSIN Plant ABS JBH
CHRYSIN-7-GLUCURONIDE 27,000 Plant HHB
DIHYDROBAICALEIN Plant ABS
DIHYDRONORWOGONIN Plant ABS
EO 600 Plant BML
GERMACRENE-D 13 Plant BML
LIMONENE 13 Plant BML
LUTEOLIN Plant ABS
MENTHONE 62 Plant BML
OROXYLIN-A Plant ABS
SCUTELLAREIN Plant ABS
TANNIN 28,600-35,000 Plant BML
TRANS-BETA-FARNESENE 102 Plant BML
WOGONIN Plant ABS

***Scutellaria lateriflora* L.**
"Maddog Skullcap"

2-METHYL-ALKANES Plant JNP
3,9-DIMETHYL-ALKANES Plant JNP
3-METHYL-ALKANES Plant JNP
ALKANES Plant FNF
ALUMINUM 258 Plant PED
ASCORBIC-ACID 1,025 Plant PED
ASH 96,000 Plant PED
BETA-CAROTENE 18-19 Plant PED
CALCIUM 4,550 Plant PED
CARBOHYDRATES 780,000 Plant PED
CHROMIUM 7 Plant PED
COBALT 11 Plant PED
FAT 15,000 Plant PED
FIBER 82,000 Plant PED
IRON 250 Plant PED
KILOCALORIES 3,140/kg Plant PED
LIGNIN Plant JAD
MAGNESIUM 1,130 Plant PED
MANGANESE 47 Plant PED
N-HENTRIACONTANE Plant JNP
N-NONACOSANE Plant JNP
N-PENTATRIACONTANE Plant JNP
N-TRITRIACONTANE Plant JNP
NIACIN 35 Plant PED
PHOSPHORUS 1,810 Plant PED
POTASSIUM 21,800 Plant PED
PROTEIN 109,000 Plant PED
RIBOFLAVIN 2.2 Plant PED
SCUTELLARIN Plant HHB
SELENIUM 8.3 Plant PED
SILICON 48 Plant PED
SODIUM 160 Plant PED
THIAMIN 2.8 Plant PED
TIN 1.2 Plant PED
WATER 863,000 Plant PED
WAX 12,000 Plant JNP
ZINC 86 Plant PED

Scutellaria parvula MICHX.
"Small Skullcap"
Unaggregated A

1-OCTEN-3-OL 2 Plant BML
10-ALPHA-CADINOL 27 Plant BML
ACETOPHENONE 14 Plant BML
ALPHA-BISABOLOL 185 Plant BML
ALPHA-HUMULENE 19 Plant BML
ALPHA-TERPINEOL 36 Plant BML
BETA-BISABOLENE 54 Plant BML
BETA-SESQUIPHELLANDRENE 13 Plant BML
CARYOPHYLLENE 47 Plant BML
CARYOPHYLLENE-OXIDE 20 Plant BML
CIS-ALPHA-BISABOLENE 32 Plant BML
DELTA-CADINENE 36 Plant BML
EO 900 Plant BML
GAMMA-CADINENE 9 Plant BML
GERANIOL 1 Plant BML
LINALOL 10 Plant BML
TRANS-ALPHA-BERGAMOTENE 121 Plant BML
TRANS-BETA-FARNESENE 23 Plant BML

Sideritis athoa PAPANIKOLAOU ET KOKKINI
"Kedi Kuyrugu Cayi"
Unaggregated A

(E)-2-HEXENAL 4 Shoot JEO5:669
1,8-CINEOLE 8 Shoot JEO5:669
1-OCTEN-3-OL 2.5 Shoot JEO5:669
3-HEXENOL 0.2 Shoot JEO5:669
5-METHYL-3-HEPTANONE 0.5 Shoot JEO5:669
6-METHYL-3-HEPTANOL 0.5 Shoot JEO5:669
6-METHYL-5-HEPTEN-2-ONE 0.2 Shoot JEO5:669
ACORADIENE 105 Shoot JEO5:669
ALPHA-BERGAMOTENE 5 Shoot JEO5:669
ALPHA-COPAENE 7 Shoot JEO5:669
ALPHA-PINENE 160 Shoot JEO5:669
ALPHA-TERPINEOL 100 Shoot JEO5:669
AR-CURCUMENE 165 Shoot JEO5:669
BETA-BOURBONENE 4 Shoot JEO5:669
BETA-CARYOPHYLLENE 7 Shoot JEO5:669
BETA-HIMACHALENE 55 Shoot JEO5:669
BETA-PINENE 310 Shoot JEO5:669
BORNYL-ACETATE 2.5 Shoot JEO5:669
CALAMENENE 11 Shoot JEO5:669
CAMPHENE 2 Shoot JEO5:669
CAMPHOR 0.5 Shoot JEO5:669
CEDR-8-ENE 26 Shoot JEO5:669
DELTA-3-CARENE 8 Shoot JEO5:669
EO 2,500 Shoot JEO5:669
GERMACRENE-D 45 Shoot JEO5:669
ISOBORNEOL 55 Shoot JEO5:669
ISOPULEGOL 0.8 Shoot JEO5:669
LIMONENE 26 Shoot JEO5:669
LINALOL 10 Shoot JEO5:669
MYRCENE 975 Shoot JEO5:669
NEROLIDOL 1 Shoot JEO5:669
NONANAL 6 Shoot JEO5:669
NONANOIC-ACID 3.5 Shoot JEO5:669
OCTANOL 0.8 Shoot JEO5:669
P-CYMEN-8-OL 0.8 Shoot JEO5:669
P-CYMENE 2 Shoot JEO5:669
PERILLENE 1 Shoot JEO5:669
SABINENE 7 Shoot JEO5:669
SPATHULENOL 26 Shoot JEO5:669
TERPINEN-4-OL 0.2 Shoot JEO5:669
THYMOL 3.5 Shoot JEO5:669

Sideritis germanicolpitana BORNM

1,8-CINEOLE 4-15 Plant JEO4:533
1-OCTEN-3-OL 2-3 Plant JEO4:533
2-HEXANAL 3-4 Plant JEO4:533
6-METHYL-3-HEPTANOL 5-6 Plant JEO4:533
ALPHA-COPAENE 4-15 Plant JEO4:533
ALPHA-CUBEBENE 1-6 Plant JEO4:533
ALPHA-PINENE 88-104 Plant JEO4:533
ALPHA-TERPINENE 2-12 Plant JEO4:533
ALPHA-TERPINEOL 6-14 Plant JEO4:533
BETA-BISABOLENE 17-41 Plant JEO4:533
BETA-BOURBONENE 6-9 Plant JEO4:533
BETA-CARYOPHYLLENE 23-47 Plant JEO4:533
BETA-CUBEBENE 1-3 Plant JEO4:533
BETA-EUDESMOL 75-105 Plant JEO4:533
BETA-FARNESENE 8-19 Plant JEO4:533
BETA-PINENE 122-123 Plant JEO4:533
CALAMENENE 20-21 Plant JEO4:533
CIS-SABINENE-HYDRATE 1-2 Plant JEO4:533
DELTA-CADINENE 2-18 Plant JEO4:533
ELEMOL 80-213 Plant JEO4:533
EO 3,300 Plant JEO4:533
GAMMA-TERPINENE 4-20 Plant JEO4:533
GERANYL-ACETONE 2-3 Plant JEO4:533
GERMACRENE-D 22-42 Plant JEO4:533
HEXANOL 1 Plant JEO4:533
KAUR-15-ENE 5-24 Plant JEO4:533
KAUR-16-ENE 14-15 Plant JEO4:533
LIMONENE 35-45 Plant JEO4:533
LINALOL 3-4 Plant JEO4:533
MYRCENE 1,285-1,625 Plant JEO4:533
P-CYMENE 2 Plant JEO4:533
PATCHOULANE 6-11 Plant JEO4:533
PERILLENE 1-2 Plant JEO4:533
SABINENE 87-693 Plant JEO4:533
TERPINEN-4-OL 7-36 Plant JEO4:533
TERPINOLENE 10-11 Plant JEO4:533
TRANS-SABINENE-HYDRATE 3-4 Plant JEO4:533

Sideritis mugronensis

1,8-CINEOLE 355-1,255 Flower JEO3:395
1,8-CINEOLE 125-275 Leaf JEO3:395
ALLO-AROMADENDRENE 1-2 Flower JEO3:395
ALLO-AROMADENDRENE 10-20 Leaf JEO3:395
ALPHA-BISABOLOL 105-710 Flower JEO3:395
ALPHA-BISABOLOL 60-410 Leaf JEO3:395
ALPHA-COPAENE 10-35 Flower JEO3:395
ALPHA-COPAENE 5-15 Leaf JEO3:395
ALPHA-CUBEBENE 30-45 Flower JEO3:395
ALPHA-CUBEBENE 15-25 Leaf JEO3:395
ALPHA-CURCUMENE 20-30 Flower JEO3:395
ALPHA-CURCUMENE 5-10 Leaf JEO3:395
ALPHA-HUMULENE 10-15 Flower JEO3:395
ALPHA-HUMULENE 10-15 Leaf JEO3:395
ALPHA-MUUROLENE <1 Flower JEO3:395
ALPHA-MUUROLENE 35-50 Leaf JEO3:395
ALPHA-PHELLANDRENE 35-270 Flower JEO3:395
ALPHA-PHELLANDRENE 5-10 Leaf JEO3:395
ALPHA-PINENE 15-410 Flower JEO3:395
ALPHA-PINENE 75-110 Leaf JEO3:395
ALPHA-TERPINENE 1 Flower JEO3:395
ALPHA-TERPINENE 10-15 Leaf JEO3:395
ALPHA-TERPINEOL 20-150 Flower JEO3:395
ALPHA-TERPINEOL 5-20 Leaf JEO3:395
ALPHA-THUJENE 35-45 Flower JEO3:395
ALPHA-THUJENE 4-6 Leaf JEO3:395
BETA-BOURBONENE 10-15 Flower JEO3:395
BETA-BOURBONENE 15-25 Leaf JEO3:395
BETA-CARYOPHYLLENE 15-20 Flower JEO3:395
BETA-CARYOPHYLLENE 20-65 Leaf JEO3:395
BETA-GURJUNENE 30-45 Flower JEO3:395
BETA-GURJUNENE 10-15 Leaf JEO3:395
BORNYL-ACETATE 30-45 Flower JEO3:395
BORNYL-ACETATE 10-20 Leaf JEO3:395
CADINOL-I 15-225 Flower JEO3:395
CADINOL-I 15-75 Leaf JEO3:395
CADINOL-II 15-160 Flower JEO3:395
CADINOL-II 15-60 Leaf JEO3:395
CADINOL-III 1-390 Flower JEO3:395
CADINOL-III 40-115 Leaf JEO3:395
CALACORENE 5-280 Flower JEO3:395
CALACORENE 20-50 Leaf JEO3:395
CALACORENE-ISOMER 125-170 Flower JEO3:395

CALACORENE-ISOMER 20-35 Leaf JEO3:395
CALAMENENE <1-90 Flower JEO3:395
CALAMENENE <1-10 Leaf JEO3:395
CAMPHENE 2-3 Flower JEO3:395
CAMPHENE 0.5-1 Leaf JEO3:395
DELTA-CADINENE 70-390 Flower JEO3:395
DELTA-CADINENE 30-85 Leaf JEO3:395
EO 3,200-4,400 Flower JEO3:395
EO 1,100-1,600 Leaf JEO3:395
FARNESOL 90-325 Flower JEO3:395
FARNESOL 10-125 Leaf JEO3:395
FENCHONE 15-25 Flower JEO3:395
GAMMA-TERPINENE 30-75 Flower JEO3:395
GAMMA-TERPINENE 15-20 Leaf JEO3:395
GERMACRENE-B 45-250 Flower JEO3:395
GERMACRENE-B 15-30 Leaf JEO3:395
LFNCHONE 1-5 Leaf JEO3:395
LIMONENE <1-155 Flower JEO3:395
LIMONENE <1 Leaf JEO3:395
LINALOL 10-25 Flower JEO3:395
LINALOL 5-10 Leaf JEO3:395
MYRCENE 15-100 Flower JEO3:395
MYRCENE 15-20 Leaf JEO3:395
NEROL 160-220 Flower JEO3:395
NEROL 55-80 Leaf JEO3:395
NEROLIDOL 10-205 Flower JEO3:395
NEROLIDOL 15-90 Leaf JEO3:395
P-CYMENE 35-75 Flower JEO3:395
P-CYMENE 3-5 Leaf JEO3:395
SABINENE 85-670 Flower JEO3:395
SABINENE 15-170 Leaf JEO3:395
TERPINEN-4-OL <1-95 Flower JEO3:395
TERPINEN-4-OL 20-50 Leaf JEO3:395
TERPINOLENE 5-10 Flower JEO3:395
TERPINOLENE 5-90 Leaf JEO3:395

Sideritis pauli PAU
"El Molinillo Sideritis"
Unaggregated A

1,8-CINEOLE 65 Shoot JJF7:47
3-BETA-HYDROXY-DELTA-CADINENE 29 Shoot JJF7:47
ALPHA-CADINOL 32 Shoot JJF7:47
ALPHA-COPAENE 95 Shoot JJF7:47
ALPHA-CUBEBENE 6 Shoot JJF7:47
ALPHA-GURJUNENE 6 Shoot JJF7:47
ALPHA-HUMULENE 3 Shoot JJF7:47
ALPHA-MUUROLENE 13 Shoot JJF7:47
ALPHA-PHELLANDRENE 26 Shoot JJF7:47
ALPHA-PINENE 1,540 Shoot JJF7:47
ALPHA-TERPINENE 3 Shoot JJF7:47
ALPHA-TERPINEOL 16 Shoot JJF7:47
ALPHA-THUJENE 10 Shoot JJF7:47
AROMADENDRENE 6 Shoot JJF7:47
BETA-BOURBONENE 10 Shoot JJF7:47
BETA-CARYOPHYLLENE 13 Shoot JJF7:47
BETA-PINENE 110 Shoot JJF7:47
CALACORENE 6 Shoot JJF7:47
CAMPHENE 6 Shoot JJF7:47
CAMPHOR 3 Shoot JJF7:47
CARYOPHYLLENE-EPOXIDE 26 Shoot JJF7:47
DELTA-CADINENE 220 Shoot JJF7:47
ENT-13-EPI-MANOYL-OXIDE 16 Shoot JJF7:47
ENT-8-HYDROXYLABDA-13(16),14-DIENE 225 Shoot JJF7:47
EO 3,200 Shoot JJF7:47
GAMMA-TERPINENE 19 Shoot JJF7:47
GAMMA-TERPINEOL 3 Shoot JJF7:47
GAMMA-TERPINYL-ACETATE 19 Shoot JJF7:47
GERANYL-ACETATE 3 Shoot JJF7:47
GERMACRENE-D 95 Shoot JJF7:47
LIMONENE 55 Shoot JJF7:47
LINALOL 3 Shoot JJF7:47
MYRCENE 22 Shoot JJF7:47
P-CYMENE 3 Shoot JJF7:47
SABINENE 95 Shoot JJF7:47
T-MUUROLOL 205 Shoot JJF7:47
TERPINEN-4-OL 35 Shoot JJF7:47
TERPINOLENE 10 Shoot JJF7:47

Sideritis scardica GRISEB.
"Balkan Sideritis"
Unaggregated A

1-NONEN-3-OL 110 Shoot PM57:A137
9-EICOSINE 275 Shoot PM57:A137
ALPHA-PINENE-3-OL 65 Shoot PM57:A137
ALPHA-TERPINEOL 70 Shoot PM57:A137
ANETHOLE 80 Shoot PM57:A137
BENZYL-BENZOATE 36 Shoot PM57:A137
BETA-CARYOPHYLLENE 105 Shoot PM57:A137
BETA-FARNESENE 49 Shoot PM57:A137
BETA-IONONE 41 Shoot PM57:A137
CAMPHOR 75 Shoot PM57:A137
CARVONE 20 Shoot PM57:A137
CINEOLE 27 Shoot PM57:A137
COPAENE 34 Shoot PM57:A137
EO 4,350 Shoot PM57:A137
FARNESOL 80 Shoot PM57:A137
GERANIOL 245 Shoot PM57:A137
HUMULENE 22 Shoot PM57:A137
MENTHOL 370 Shoot PM57:A137
MENTHYL-ACETATE 60 Shoot PM57:A137
NEROL 65 Shoot PM57:A137
NEROLIDOL 145 Shoot PM57:A137
PATCHOULENE 16 Shoot PM57:A137
PATCHOULI ALCOHOL 57 Shoot PM57:A137
PENTYLCYCLOPROPANE 135 Shoot PM57:A137
THUJONE 20 Shoot PM57:A137
THUJOPSENE 47 Shoot PM57:A137
THYMOL 21 Shoot PM57:A137
TRANS-MENTHONE 105 Shoot PM57:A137
YLANGENE 18 Shoot PM57:A137

Stachys germanica L.
"Downy Woundwort"
Unaggregated A

1,8-CINEOLE 48 Plant BML
ALPHA-PINENE 67 Plant BML
ALPHA-TERPINEOL 6 Plant BML
CIS-OCIMENE 65 Plant BML
EO 800 Plant BML
GERMACRENE-D 43 Plant BML
LIMONENE 18 Plant BML
MYRCENE 7 Plant BML
P-CYMENE 13 Plant BML
SABINENE 14 Plant BML
TRANS-BETA-FARNESENE 56 Plant BML

Stachys officinalis (L.) TREVISAN "Betony"

(+)-OXYSTACHYDRINE Plant CRC
(-)-OXYSTACHYDRINE Plant CRC
4-CAFFEOYLQUINIC-ACID Plant HHB
ACHILLEIN Plant CRC
BETAINE 5,000 Plant CRC
BETONICIN Plant CRC
CAFFEIC-ACID 5,000 Plant CRC
CHLOROGENIC-ACID Plant CRC
HARPAGIDE Plant CRC
NEOCHLOROGENIC-ACID Plant CRC
ROSMARINIC-ACID Plant CRC
STACHYDRINE Plant CRC
TANNIN 150,000 Plant CRC
TURICINE Plant CRC

Teucrium arduini L.
"'Croatian Alpine' Germander"
Unaggregated A

1,8-CINEOLE 3 Shoot JEO4:223
ALLOAROMADENDRENE 43 Shoot JEO4:223
ALPHA-CADINOL 2 Shoot JEO4:223
ALPHA-CEDRENE <0.1 Shoot JEO4:223
ALPHA-FENCHYL-ALCOHOL 2 Shoot JEO4:223
ALPHA-HUMULENE 37 Shoot JEO4:223
BETA-CARYOPHYLLENE 245 Shoot JEO4:223
BETA-CUBEBENE 4 Shoot JEO4:223
BISABOLOL <0.1 Shoot JEO4:223
CARYOPHYLLENE-OXIDE 3 Shoot JEO4:223
CEDRAN-5-ONE 8 Shoot JEO4:223
CYPERENE 8 Shoot JEO4:223
DELTA-CADINENE 8 Shoot JEO4:223
EO 1,800 Shoot JEO4:223
EPI-ALPHA-CADINOL 16 Shoot JEO4:223
GAMMA-CADINENE 23 Shoot JEO4:223
GAMMA-TERPINENE 27 Shoot JEO4:223
GERMACRENE-B 180 Shoot JEO4:223
GERMACRENE-D 1,040 Shoot JEO4:223
LEDOL 3 Shoot JEO4:223
LIMONENE 3 Shoot JEO4:223
LINALOL 4 Shoot JEO4:223
NEROLIDOL 8 Shoot JEO4:223
OCIMENE 4 Shoot JEO4:223
P-CYMENE 16 Shoot JEO4:223
SPATHULENOL 12 Shoot JEO4:223
TERPINEN-4-OL 6 Shoot JEO4:223

Teucrium arduini L.
"'Croatian Alpine' Germander"
Unaggregated B

1,8-CINEOLE 3 Shoot JEO4:223
ALLOAROMADENDRENE 20 Shoot JEO4:223
ALPHA-CADINOL 14 Shoot JEO4:223
ALPHA-CEDRENE 4 Shoot JEO4:223
ALPHA-FENCHYL-ALCOHOL <0.1 Shoot JEO4:223
ALPHA-HUMULENE 20 Shoot JEO4:223
BETA-CARYOPHYLLENE 120 Shoot JEO4:223
BETA-CUBEBENE 8 Shoot JEO4:223
BISABOLOL 13 Shoot JEO4:223
CARYOPHYLLENE-OXIDE 13 Shoot JEO4:223
CEDRAN-5-ONE 6 Shoot JEO4:223
CYPERENE 58 Shoot JEO4:223
DELTA-CADINENE 8 Shoot JEO4:223
EO 700 Shoot JEO4:223
EPI-ALPHA-CADINOL 17 Shoot JEO4:223
GAMMA-CADINENE 68 Shoot JEO4:223
GAMMA-TERPINENE 13 Shoot JEO4:223
GERMACRENE-B 20 Shoot JEO4:223
GERMACRENE-D 165 Shoot JEO4:223
LEDOL <0.1 Shoot JEO4:223
LIMONENE 3 Shoot JEO4:223
LINALOL 2 Shoot JEO4:223
NEROLIDOL 7 Shoot JEO4:223
OCIMENE <0.1 Shoot JEO4:223
P-CYMENE 4 Shoot JEO4:223
SPATHULENOL 18 Shoot JEO4:223
TERPINEN-4-OL 3 Shoot JEO4:223

Teucrium asiaticum L.
Unaggregated A

1,8-CINEOLE 0.62 Shoot PC29:1165
1-OCTEN-3-OL <0.1 Shoot PC29:1165
3 BETA-HYDROXY-DELTA-CADINENE <0.1 Shoot PC29:1165
3-BETA-HYDROXY-MUUROLENE <0.1 Shoot PC29:1165
3-OCTANOL <0.1 Shoot PC29:1165
ALLOAROMADENDRENE 1.92 Shoot PC29:1165
ALPHA-BISABOLOL <0.1 Shoot PC29:1165
ALPHA-CADINOL 0.34 Shoot PC29:1165
ALPHA-CALACORENE 8.65 Shoot PC29:1165
ALPHA-CEDRENE <0.1 Shoot PC29:1165
ALPHA-COPAENE 1.75 Shoot PC29:1165
ALPHA-CUBEBENE 0.3 Shoot PC29:1165
ALPHA-HUMULENE 1.92 Shoot PC29:1165
ALPHA-MUUROLENE 9.94 Shoot PC29:1165
ALPHA-NEROLIDOL 0.29 Shoot PC29:1165
ALPHA-PHELLANDRENE 0.17 Shoot PC29:1165
ALPHA-PINENE 0.18 Shoot PC29:1165
ALPHA-TERPINEOL 3.21 Shoot PC29:1165
ALPHA-TERPINYL-ACETATE <0.1 Shoot PC29:1165
ALPHA-THUJENE <0.1 Shoot PC29:1165
AR-CURCUMENE <0.1 Shoot PC29:1165
ARISTOLENE 0.15 Shoot PC29:1165
AROMADENDRENE 1.16 Shoot PC29:1165
BETA-BISABOLOL 0.62 Shoot PC29:1165
BETA-BOURBONENE 0.87 Shoot PC29:1165
BETA-CARYOPHYLLENE 0.15 Shoot PC29:1165
BETA-PINENE 1.03 Shoot PC29:1165
BORNEOL <0.1 Shoot PC29:1165
CADALENE 1.24 Shoot PC29:1165
CADIN-3-EN-9ALPHA-OL-2-ONE <0.1 Shoot PC29:1165
CADIN-3-EN-9BETA-OL-2-ONE <0.1 Shoot PC29:1165
CADINENOL 0.48 Shoot PC29:1165
CALAMENENE 23.11 Shoot PC29:1165
CAMPHENE Shoot PC29:1165
CAMPHOR <0.1 Shoot PC29:1165
CARVONE <0.1 Shoot PC29:1165
CARYOPHYLLENE-EPOXIDE 0.35 Shoot PC29:1165
CEDROL 0.14 Shoot PC29:1165
CIS-LINALYL-OXIDE 0.63 Shoot PC29:1165
CUBENOL 0.15 Shoot PC29:1165
DELTA-CADINENE 0.27 Shoot PC29:1165
EIPCUBENOL 0.58 Shoot PC29:1165

EO 100 Shoot PC29:1165
FARNESYL-ACETATE 0.34 Shoot PC29:1165
FENCHYL-ALCOHOL <0.1 Shoot PC29:1165
GAMMA-CADINENE <0.1 Shoot PC29:1165
GAMMA-TERPINENE 0.75 Shoot PC29:1165
GAMMA-TERPINEOL <0.1 Shoot PC29:1165
GERMACRENE-D 5.17 Shoot PC29:1165
HUMULENE-EPOXIDE 0.14 Shoot PC29:1165
LIMONENE 0.62 Shoot PC29:1165
LINALOL 15.38 Shoot PC29:1165
LINALYL-ACETATE 6.25 Shoot PC29:1165
LONGIPINENE 1.39 Shoot PC29:1165
MONOTERPENES 33.16 Shoot PC29:1165
MUUROL-3-EN-9BETA-OL-2-ONE <0.1 Shoot PC29:1165
MYRCENE 1.24 Shoot PC29:1165
NOOTKATONE 0.37 Shoot PC29:1165
NORCAMPHOR <0.1 Shoot PC29:1165
P-CYMENE 0.61 Shoot PC29:1165
SABINENE <0.1 Shoot PC29:1165
SESQUITERPENES 63.29 Shoot PC29:1165
SPATHULENOL 0.35 Shoot PC29:1165
T-CADINOL 1.16 Shoot PC29:1165
TERPINEN-4-OL 1.73 Shoot PC29:1165
THYMOL <0.1 Shoot PC29:1165
THYMYL-ACETATE <0.1 Shoot PC29:1165
TRANS-ALPHA-OCIMENE 0.74 Shoot PC29:1165
TRANS-LINALYL-OXIDE <0.1 Shoot PC29:1165

Teucrium chamaedrys L. "Wall Germander"

6-EPITEUCRIN-A Plant JSG
ACETYL-HARPAGIDE Plant 411/
AJUGOSE Plant HHB
BETA-AMYRIN Plant ABS
BETA-CARYOPHYLLENE 420 Plant HHB
BORNYL-ACETATE 55 Plant RIZ
CAFFEIC-ACID Plant ABS
CHLOROGENIC-ACID 200 Plant ABS RIZ
CLERODANE Plant 411/
DIOSMIN Plant HHB
EO 700 Plant WOI
GLUCOSE Plant HHB
HARPAGIDE Plant 411/
ISOQUERCITRIN Plant HHB
ISOTEUFLIDIN Plant 411/
MARRUBIIN Plant HHB
NEOCLERODANE Plant 411/
PARAFFIN 1,000 Plant HHB
POLYPHENOLS 81,000-96,000 Plant ABS
SACCHAROSE Plant HHB
SAPONIN Plant HHB
SCUTELLARIN Plant HHB
STACHYOSE Plant HHB
TANNIN 50,000 Plant HHB
TEUCHAMAEDRYINS Plant JSG
TEUCRIN-A Plant HHB
TEUCRIN-B Plant HHB
TEUCRIN-C Plant HHB
TEUCRIN-D Plant HHB
TEUCRIN-F Plant HHB
TEUCRIN-G Plant HHB
TEUCROSIDE Plant 411/
TEUCVIDIN Plant 411/
TEUCVIN Plant 411/
TEUFLIDIN Plant 411/
TEUFLIN Plant 411/
TEUGIN Plant JSG
URSOLIC-ACID Plant ABS
VERBASCOSE Plant HHB

Teucrium cyprium BOISS.
"'Cyprus' Germander"
Unaggregated A

1,8-CINEOLE Leaf JE35:105
1-OCTEN-3-OL Leaf JE35:105
ALLOAROMADENDRENE Leaf JE35:105
ALPHA-CADINENE 1,090 Leaf JE35:105
ALPHA-CARYOPHYLLENE 375 Leaf JE35:105
ALPHA-CEDRENE 195 Leaf JE35:105
ALPHA-COPAENE 65 Leaf JE35:105
ALPHA-HIMACHALENE Leaf JE35:105
ALPHA-HUMULENE 50 Leaf JE35:105
ALPHA-PINENE 430 Leaf JE35:105
ALPHA-TERPINENE 130 Leaf JE35:105
ALPHA-TERPINEOL 145 Leaf JE35:105
ALPHA-TERPINOLENE Leaf JE35:105
ANETHOLE 90 Leaf JE35:105
BETA-BISABOLENE Leaf JE35:105
BETA-BOURBONENE 105 Leaf JE35:105
BETA-CARYOPHYLLENE 600 Leaf JE35:105
BETA-CUBEBENE Leaf JE35:105
BETA-GUAIENE 610 Leaf JE35:105
BETA-PHELLANDRENE 50 Leaf JE35:105
BETA-PINENE 115 Leaf JE35:105
BORNEOL 65 Leaf JE35:105
CAMPHENE Leaf JE35:105
CARVACROL 155 Leaf JE35:105
CARYOPHYLLENE-OXIDE 275 Leaf JE35:105
CITRONELLAL 80 Leaf JE35:105
DELTA-CADINENE 1,250 Leaf JE35:105
EO 13,000 Leaf JE35:105
GAMMA-CADINENE Leaf JE35:105
GAMMA-ELEMENE Leaf JE35:105
GAMMA-TERPINENE 210 Leaf JE35:105
ISOPULEGOL Leaf JE35:105
LIMONENE 39 Leaf JE35:105
LINALOL 80 Leaf JE35:105
MYRCENE 39 Leaf JE35:105
MYRTENAL 65 Leaf JE35:105
MYRTENOL 180 Leaf JE35:105
NEROL 50 Leaf JE35:105
P-CYMENE 170 Leaf JE35:105
PINOCARVONE 39 Leaf JE35:105
SABINENE 2,770 Leaf JE35:105

SPATHULENOL Leaf JE35:105
TERPINEN-4-OL Leaf JE35:105
THYMOL Leaf JE35:105
TRANS-BETA-BERGAMOTENE 105 Leaf JE35:105
TRANS-BETA-FARNESENE 155 Leaf JE35:105
TRANS-NEROLIDOL Leaf JE35:105
TRANS-PINOCARVEOL 220 Leaf JE35:105

Teucrium divaricatum var. *canescens* (CELAK.) HOLMBOE
"Hoary Divaricate Germander"
Unaggregated A

1,8-CINEOLE 6 Leaf JE35:105
1-OCTEN-3-OL 18 Leaf JE35:105
ALLOAROMADENDRENE 15 Leaf JE35:105
ALPHA-CADINENE Leaf JE35:105
ALPHA-CARYOPHYLLENE 210 Leaf JE35:105
ALPHA-CEDRENE Leaf JE35:105
ALPHA-COPAENE 21 Leaf JE35:105
ALPHA-HIMACHALENE 120 Leaf JE35:105
ALPHA-HUMULENE Leaf JE35:105
ALPHA-PINENE 365 Leaf JE35:105
ALPHA-TERPINENE Leaf JE35:105
ALPHA-TERPINEOL 27 Leaf JE35:105
ALPHA-TERPINOLENE Leaf JE35:105
ANETHOLE Leaf JE35:105
BETA-BISABOLENE Leaf JE35:105
BETA-BOURBONENE 90 Leaf JE35:105
BETA-CARYOPHYLLENE 530 Leaf JE35:105
BETA-CUBEBENE 800 Leaf JE35:105
BETA-GUAIENE Leaf JE35:105
BETA-PHELLANDRENE 12 Leaf JE35:105
BETA-PINENE 185 Leaf JE35:105
BORNEOL Leaf JE35:105
CAMPHENE 3 Leaf JE35:105
CARVACROL 9 Leaf JE35:105
CARYOPHYLLENE-OXIDE 80 Leaf JE35:105
CITRONELLAL Leaf JE35:105
DELTA-CADINENE 60 Leaf JE35:105
EO 3,000 Leaf JE35:105
GAMMA-CADINENE Leaf JE35:105
GAMMA-ELEMENE 120 Leaf JE35:105
GAMMA-TERPINENE 15 Leaf JE35:105
ISOPULEGOL Leaf JE35:105
LIMONENE 125 Leaf JE35:105
LINALOL 36 Leaf JE35:105
MYRCENE 15 Leaf JE35:105
MYRTENAL 9 Leaf JE35:105
MYRTENOL Leaf JE35:105
NEROL Leaf JE35:105
P-CYMENE Leaf JE35:105

PINOCARVONE Leaf JE35:105
SABINENE 3 Leaf JE35:105
SPATHULENOL Leaf JE35:105
TERPINEN-4-OL Leaf JE35:105
THYMOL Leaf JE35:105
TRANS-BETA-BERGAMOTENE Leaf JE35:105
TRANS-BETA-FARNESENE 33 Leaf JE35:105
TRANS-NEROLIDOL Leaf JE35:105
TRANS-PINOCARVEOL 33 Leaf JE35:105

Teucrium gnaphalodes L'HER.
"Iberian Germander"
Unaggregated A

(E,Z)-FARNESOL Shoot JEO5:397
(Z)-BETA-FARNESENE 0.5 Shoot JEO5:397
(Z)-NEROLIDOL 0.5 Shoot JEO5:397
(Z,E)-FARNESOL Shoot JEO5:397
1,8-CINEOLE Shoot JEO5:397
3,3-DIMETHYL-2-ISOPROPYL-CYCLOPENTANE Shoot JEO5:397
3,6,6-TRIMETHYLBICYCLO[3.1.0]HEXAN-3-CARBOXALDEHYDE 6 Shoot JEO5:397
3-OCTANOL 1 Shoot JEO5:397
3-OCTYL-ACETATE Shoot JEO5:397
ALLOAROMADENDRENE 17 Shoot JEO5:397
ALPHA-BERGAMOTENE Shoot JEO5:397
ALPHA-BISABOLENE Shoot JEO5:397
ALPHA-CADINOL 2.5 Shoot JEO5:397
ALPHA-CAMPHOLENAL 2 Shoot JEO5:397
ALPHA-CEDRENE Shoot JEO5:397
ALPHA-CEDROL 4 Shoot JEO5:397
ALPHA-COPAENE 2.5 Shoot JEO5:397
ALPHA-PINENE 8 Shoot JEO5:397
ALPHA-TERPINENE 0.5 Shoot JEO5:397
ALPHA-TERPINEOL 5 Shoot JEO5:397
ALPHA-THUJENE 5 Shoot JEO5:397
ALPHA-YLANGENE Shoot JEO5:397
AR-CURCUMENE 0.5 Shoot JEO5:397
AROMADENDRENE-EPOXIDE Shoot JEO5:397
ARTEMISIA-KETONE 3 Shoot JEO5:397
BETA-BISABOLENE Shoot JEO5:397
BETA-CARYOPHYLLENE 60 Shoot JEO5:397
BETA-CUBEBENE Shoot JEO5:397
BETA-GUAIENE Shoot JEO5:397
BETA-PINENE 35 Shoot JEO5:397
BETA-THUJONE Shoot JEO5:397
BICYCLOGERMACRENE 1.5 Shoot JEO5:397
CADIN-4-EN-1-OL 2.5 Shoot JEO5:397
CADINENOL Shoot JEO5:397
CALACORENE 0.5 Shoot JEO5:397
CALAMENENE Shoot JEO5:397
CALARENE Shoot JEO5:397
CAMPHOR 6 Shoot JEO5:397
CARVACROL 0.5 Shoot JEO5:397
CARVENONE 9 Shoot JEO5:397

CARVONE 4 Shoot JEO5:397
CARYOPHYLLENE-EPOXIDE 17 Shoot JEO5:397
CHRYSANTHENONE Shoot JEO5:397
CIS-VERBENOL 13 Shoot JEO5:397
CUMINYL-ALDEHYDE 1 Shoot JEO5:397
DEHYDROSABINAKETONE 20 Shoot JEO5:397
DEHYDROSABINENE 2.5 Shoot JEO5:397
DELTA-CADINENE 2 Shoot JEO5:397
ENDOFENCHOL Shoot JEO5:397
EO 500 Shoot JEO5:397
EPI-ALPHA-BISABOLOL Shoot JEO5:397
EUCARVONE 0.5 Shoot JEO5:397
GAMMA-CADINENE 1.5 Shoot JEO5:397
GAMMA-MUUROLENE 0.5 Shoot JEO5:397
GAMMA-TERPINENE Shoot JEO5:397
GAMMA-TERPINEOL Shoot JEO5:397
GAMMA-TERPINYL-ACETATE 1.5 Shoot JEO5:397
GERANYL-ACETONE Shoot JEO5:397
GERMACRENE-D 1.5 Shoot JEO5:397
HUMULENE-EPOXIDE 2 Shoot JEO5:397
ISOVALERALDEHYDE 0.5 Shoot JEO5:397
LEDOL 20 Shoot JEO5:397
LIMONENE 7 Shoot JEO5:397
LINALOL 12 Shoot JEO5:397
LONGIFOLENE Shoot JEO5:397
MUUROLOL 7 Shoot JEO5:397
MYRCENE 8 Shoot JEO5:397
MYRTENAL 28 Shoot JEO5:397
NEODIHYDROCARVEOL Shoot JEO5:397
P-CYMEN-7-OL 0.5 Shoot JEO5:397
P-CYMEN-8-OL 22 Shoot JEO5:397
P-CYMENE Shoot JEO5:397
P-MENTHA-1(7),2-DIEN-1,8-DIOL 5 Shoot JEO5:397
P-MENTHA-1,4-DIEN-7-OL Shoot JEO5:397
PHELLANDRAL 1 Shoot JEO5:397
PINOCARVONE 5 Shoot JEO5:397
SABINAKETONE 10 Shoot JEO5:397
SABINENE 44 Shoot JEO5:397
SABINYL-ACETATE 4 Shoot JEO5:397
SPATHULENOL 0.5 Shoot JEO5:397
T-CADINOL 2.5 Shoot JEO5:397
TERPINEN-4-OL 2.5 Shoot JEO5:397
TERPINOLENE 10 Shoot JEO5:397
THUJYL-ALCOHOL Shoot JEO5:397
THYMOL 0.5 Shoot JEO5:397

TRANS-CARVEOL 4 Shoot JEO5:397
TRANS-CARVYL-ACETATE 1 Shoot JEO5:397
TRANS-LINALOL-OXIDE 5 Shoot JEO5:397
TRANS-PINOCARVEOL 39 Shoot JEO5:397
UMBELLULONE Shoot JEO5:397
VALENCENE 2 Shoot JEO5:397
VERBENENE 5 Shoot JEO5:397
VERBENONE Shoot JEO5:397
VIRIDIFLOROL 4 Shoot JEO5:397

Teucrium kotschyanum POECH "Stavros Kotschy's Germander" Unaggregated A

1,8-CINEOLE Leaf JE35:105
1-OCTEN-3-OL 27 Leaf JE35:105
ALLOAROMADENDRENE 65 Leaf JE35:105
ALPHA-CADINENE 70 Leaf JE35:105
ALPHA-CARYOPHYLLENE Leaf JE35:105
ALPHA-CEDRENE Leaf JE35:105
ALPHA-COPAENE 160 Leaf JE35:105
ALPHA-HIMACHALENE Leaf JE35:105
ALPHA-HUMULENE 550 Leaf JE35:105
ALPHA-PINENE Leaf JE35:105
ALPHA-TERPINENE Leaf JE35:105
ALPHA-TERPINEOL 80 Leaf JE35:105
ALPHA-TERPINOLENE Leaf JE35:105
ANETHOLE 18 Leaf JE35:105
BETA-BISABOLENE Leaf JE35:105
BETA-BOURBONENE 640 Leaf JE35:105
BETA-CARYOPHYLLENE 245 Leaf JE35:105
BETA-CUBEBENE 125 Leaf JE35:105
BETA-GUAIENE 55 Leaf JE35:105
BETA-PHELLANDRENE 18 Leaf JE35:105
BETA-PINENE Leaf JE35:105
BORNEOL Leaf JE35:105
CAMPHENE Leaf JE35:105
CARVACROL 360 Leaf JE35:105
CARYOPHYLLENE-OXIDE 27 Leaf JE35:105
CITRONELLAL Leaf JE35:105
DELTA-CADINENE 290 Leaf JE35:105
EO 9,000 Leaf JE35:105
GAMMA-CADINENE Leaf JE35:105
GAMMA-ELEMENE Leaf JE35:105
GAMMA-TERPINENE Leaf JE35:105
ISOPULEGOL Leaf JE35:105
LIMONENE 45 Leaf JE35:105
LINALOL 135 Leaf JE35:105
MYRCENE Leaf JE35:105
MYRTENAL 55 Leaf JE35:105
MYRTENOL Leaf JE35:105
NEROL Leaf JE35:105
P-CYMENE Leaf JE35:105
PINOCARVONE Leaf JE35:105
SABINENE 36 Leaf JE35:105

SPATHULENOL Leaf JE35:105
TERPINEN-4-OL Leaf JE35:105
THYMOL Leaf JE35:105
TRANS-BETA-BERGAMOTENE Leaf JE35:105
TRANS-BETA-FARNESENE Leaf JE35:105
TRANS-NEROLIDOL Leaf JE35:105
TRANS-PINOCARVEOL 245 Leaf JE35:105

Teucrium marum L.
"Cat Thyme"

ALLODOLICHOLACTONE Plant JSG
DOLICHODIAL Plant JBH
DOLICHOLACTONE Plant JSG GEO
MARRUBIIN Plant HHB
SAPONIN Plant FNF

Teucrium micropodioides ROUY
"Small 'Cyprus' Germander"
Unaggregated A

1,8-CINEOLE 12 Leaf JE35:105
1-OCTEN-3-OL 8 Leaf JE35:105
ALLOAROMADENDRENE Leaf JE35:105
ALPHA-CADINENE Leaf JE35:105
ALPHA-CARYOPHYLLENE Leaf JE35:105
ALPHA-CEDRENE 150 Leaf JE35:105
ALPHA-COPAENE 32 Leaf JE35:105
ALPHA-HIMACHALENE 105 Leaf JE35:105
ALPHA-HUMULENE 44 Leaf JE35:105
ALPHA-PINENE 265 Leaf JE35:105
ALPHA-TERPINENE Leaf JE35:105
ALPHA-TERPINEOL 85 Leaf JE35:105
ALPHA-TERPINOLENE Leaf JE35:105
ANETHOLE 12 Leaf JE35:105
BETA-BISABOLENE 255 Leaf JE35:105
BETA-BOURBONENE 20 Leaf JE35:105
BETA-CARYOPHYLLENE 170 Leaf JE35:105
BETA-CUBEBENE Leaf JE35:105
BETA-GUAIENE Leaf JE35:105
BETA-PHELLANDRENE Leaf JE35:105
BETA-PINENE 705 Leaf JE35:105
BORNEOL Leaf JE35:105
CAMPHENE 8 Leaf JE35:105
CARVACROL 32 Leaf JE35:105
CARYOPHYLLENE-OXIDE 68 Leaf JE35:105
CITRONELLAL Leaf JE35:105
DELTA-CADINENE 65 Leaf JE35:105
EO 4,000 Leaf JE35:105
GAMMA-CADINENE Leaf JE35:105
GAMMA-ELEMENE 140 Leaf JE35:105
GAMMA-TERPINENE 8 Leaf JE35:105
ISOPULEGOL Leaf JE35:105
LIMONENE 170 Leaf JE35:105
LINALOL 20 Leaf JE35:105
MYRCENE 24 Leaf JE35:105
MYRTENAL 48 Leaf JE35:105
MYRTENOL Leaf JE35:105
NEROL 80 Leaf JE35:105
P-CYMENE Leaf JE35:105
PINOCARVONE 44 Leaf JE35:105
SABINENE 165 Leaf JE35:105

SPATHULENOL 24 Leaf JE35:105
TERPINEN-4-OL 12 Leaf JE35:105
THYMOL 32 Leaf JE35:105
TRANS-BETA-BERGAMOTENE 70 Leaf JE35:105
TRANS-BETA-FARNESENE 65 Leaf JE35:105
TRANS-NEROLIDOL 32 Leaf JE35:105
TRANS-PINOCARVEOL 140 Leaf JE35:105

Teucrium montanum L.
"Mountain Germander"
Unaggregated A

DIOSMIN Leaf HHB
EO 1,000 Plant HHB
FRUCTOSE 33,000 Leaf HHB
GALACTOSE 3,000 Leaf HHB
GLUCOSE 260,000 Leaf HHB
ISOQUERCITRIN Plant HHB
MONTANIN-A Plant JBH
QUERCETIN Plant HHB
SAPONIN Plant HHB
SEDOHEPTULOSE <0.1 Leaf HHB
STARCH 22,000 Leaf HHB
SUCROSE 16,000 Leaf HHB
TANNIN 22,000 Plant HHB
VERBASCOSE <0.1 Leaf HHB

Teucrium oxylepis ssp. *marianum* RUIZ DE LA TORRE Unaggregated A

1,8-CINEOLE 0.99 Shoot PC29:1165
1-OCTEN-3-OL Shoot tr PC29:1165
3-BETA-HYDROXY-DELTA-CADINENE 0.59 Shoot PC29:1165
3-BETA-HYDROXY-MUUROLENE Shoot tr PC29:1165
3-OCTANOL Shoot tr PC29:1165
ALLOAROMADENDRENE 0.96 Shoot PC29:1165
ALPHA-BISABOLOL 0.15 Shoot PC29:1165
ALPHA-CADINOL 8.77 Shoot PC29:1165
ALPHA-CALACORENE 1.55 Shoot PC29:1165
ALPHA-CEDRENE 0.5 Shoot PC29:1165
ALPHA-COPAENE 3.31 Shoot PC29:1165
ALPHA-CUBEBENE 0.19 Shoot PC29:1165
ALPHA-HUMULENE 0.96 Shoot PC29:1165
ALPHA-MUUROLENE 2.4 Shoot PC29:1165
ALPHA-NEROLIDOL Shoot tr PC29:1165
ALPHA-PHELLANDRENE 0.22 Shoot PC29:1165
ALPHA-PINENE 0.18 Shoot PC29:1165
ALPHA-TERPINEOL 1.45 Shoot PC29:1165
ALPHA-TERPINYL-ACETATE 0.14 Shoot PC29:1165
ALPHA-THUJENE Shoot tr PC29:1165
AR-CURCUMENE 0.85 Shoot PC29:1165
ARISTOLENE 1.74 Shoot PC29:1165
AROMADENDRENE 1.16 Shoot PC29:1165
BETA-BISABOLOL 0.05 Shoot PC29:1165
BETA-BOURBONENE 0.34 Shoot PC29:1165
BETA-CARYOPHYLLENE 1.74 Shoot PC29:1165
BETA-PINENE 0.15 Shoot PC29:1165
BORNEOL Shoot tr PC29:1165
CADALENE 0.71 Shoot PC29:1165
CADIN-3-EN-9ALPHA-OL-2-ONE 0.12 Shoot PC29:1165
CADIN-3-EN-9BETA-OL-2-ONE Shoot tr PC29:1165
CADINENOL 0.87 Shoot PC29:1165
CALAMENENE 7.86 Shoot PC29:1165
CAMPHENE 0.2 Shoot PC29:1165
CAMPHOR 0.22 Shoot PC29:1165
CARVONE 0.59 Shoot PC29:1165
CARYOPHYLLENE-EPOXIDE 2.45 Shoot PC29:1165
CEDROL 1.65 Shoot PC29:1165
CIS-LINALYL-OXIDE 0.21 Shoot PC29:1165
CUBENOL 1.14 Shoot PC29:1165
DELTA-CADINENE 0.5 Shoot PC29:1165
EO 100 Shoot PC29:1165

EPICUBENOL 1.7 Shoot PC29:1165
FARNESYL-ACETATE 0.56 Shoot PC29:1165
FENCHYL-ALCOHOL Shoot PC29:1165
GAMMA-CADINENE 7.9 Shoot PC29:1165
GAMMA-TERPINENE 1.1 Shoot PC29:1165
GAMMA-TERPINEOL 0.23 Shoot PC29:1165
GERMACRENE-D 5.93 Shoot PC29:1165
HUMULENE-EPOXIDE 1.65 Shoot PC29:1165
LIMONENE 0.99 Shoot PC29:1165
LINALOL 12.87 Shoot PC29:1165
LINALYL-ACETATE 0.12 Shoot PC29:1165
LONGIPINENE 0.34 Shoot PC29:1165
MONOTERPENES 25 Shoot PC29:1165
MUUROL-3-EN-9BETA-OL-2-ONE 0.46 Shoot PC29:1165
MYRCENE 1.48 Shoot PC29:1165
NOOTKATONE 0.45 Shoot PC29:1165
NORCAMPHOR Shoot tr PC29:1165
P-CYMENE 0.78 Shoot PC29:1165
SABINENE 0.9 Shoot PC29:1165
SESQUITERPENES 71.23 Shoot PC29:1165
SPATHULENOL 2.45 Shoot PC29:1165
T-CADINOL 9.24 Shoot PC29:1165
TERPINEN-4-OL 1.1 Shoot PC29:1165
THYMOL 0.27 Shoot PC29:1165
THYMYL-ACETATE 0.15 Shoot PC29:1165
TRANS-ALPHA-OCIMENE 0.25 Shoot PC29:1165
TRANS-LINALYL-OXIDE 0.41 Shoot PC29.1165

Teucrium oxylepis ssp. *oxylepis*
Unaggregated A

1,8-CINEOLE 1.83 Shoot PC29:1165
1-OCTEN-3-OL Shoot tr PC29:1165
3-BETA-HYDROXY-DELTA-CADINENE Shoot tr PC29:1165
3-BETA-HYDROXY-MUUROLENE Shoot tr PC29:1165
3-OCTANOL Shoot tr PC29:1165
ALLOAROMADENDRENE Shoot tr PC29:1165
ALPHA-BISABOLOL Shoot tr PC29:1165
ALPHA-CADINOL 12.84 Shoot PC29:1165
ALPHA-CALACORENE Shoot tr PC29:1165
ALPHA-CEDRENE 0.8 Shoot PC29:1165
ALPHA-COPAENE 4.8 Shoot PC29:1165
ALPHA-CUBEBENE 8.5 Shoot PC29:1165
ALPHA-HUMULENE Shoot tr PC29:1165
ALPHA-MUUROLENE 0.54 Shoot PC29:1165
ALPHA-NEROLIDOL 0.54 Shoot PC29:1165
ALPHA-PHELLANDRENE 0.27 Shoot PC29:1165
ALPHA-PINENE 0.35 Shoot PC29:1165
ALPHA-TERPINEOL 1.88 Shoot PC29:1165
ALPHA-TERPINYL-ACETATE 0.06 Shoot PC29:1165
ALPHA-THUJENE 0.07 Shoot PC29:1165
AR-CURCUMENE 1.03 Shoot PC29:1165
ARISTOLENE 5.2 Shoot PC29:1165
AROMADENDRENE 3.6 Shoot PC29:1165
BETA-BISABOLOL Shoot tr PC29:1165
BETA-BOURBONENE 2.92 Shoot PC29:1165
BETA-CARYOPHYLLENE 5.2 Shoot PC29:1165
BETA-PINENE 0.75 Shoot PC29:1165
BORNEOL Shoot tr PC29:1165
CADALENE Shoot tr PC29:1165
CADIN-3-EN-9ALPHA-OL-2-ONE Shoot tr PC29:1165
CADIN-3-EN-9BETA-OL-2-ONE Shoot tr PC29:1165
CADINENOL 0.99 Shoot PC29:1165
CALAMENENE 3.66 Shoot PC29:1165
CAMPHENE 0.39 Shoot PC29:1165
CAMPHOR 0.06 Shoot PC29:1165
CARVONE 0.08 Shoot PC29:1165
CARYOPHYLLENE-EPOXIDE 1.18 Shoot PC29:1165
CEDROL 0.88 Shoot PC29:1165
CIS-LINALYL-OXIDE 0.27 Shoot PC29:1165
CUBENOL 0.86 Shoot PC29:1165
DELTA-CADINENE 7.18 Shoot PC29:1165
EO 100 Shoot PC29:1165

EPICUBENOL 7.69 Shoot PC29:1165
FARNESYL-ACETATE 0.06 Shoot PC29:1165
FENCHYL-ALCOHOL 0.29 Shoot PC29:1165
GAMMA-CADINENE 0.99 Shoot PC29:1165
GAMMA-TERPINENE 1.36 Shoot PC29:1165
GAMMA-TERPINEOL 0.18 Shoot PC29:1165
GERMACRENE-D 3.3 Shoot PC29:1165
HUMULENE-EPOXIDE 0.88 Shoot PC29:1165
LIMONENE 1.83 Shoot PC29:1165
LINALOL 2.69 Shoot PC29:1165
LINALYL-ACETATE 0.06 Shoot PC29:1165
LONGIPINENE 0.05 Shoot PC29:1165
MONOTERPENES 23.14 Shoot PC29:1165
MUUROL-3-EN-9BETA-OL-2-ONE Shoot tr PC29:1165
MYRCENE 2.02 Shoot PC29:1165
NOOTKATONE Shoot tr PC29:1165
NORCAMPHOR Shoot tr PC29:1165
P-CYMENE 1.36 Shoot PC29:1165
SABINENE 3.47 Shoot PC29:1165
SESQUITERPENES 75.85 Shoot PC29:1165
SPATHULENOL 1.18 Shoot PC29:1165
T-CADINOL 1 Shoot PC29:1165
TERPINEN-4-OL 2.41 Shoot PC29:1165
THYMOL 0.74 Shoot PC29:1165
THYMYL-ACETATE Shoot tr PC29:1165
TRANS-ALPHA-OCIMENE 0.21 Shoot PC29:1165
TRANS-LINALYL OXIDE 0.51 Shoot PC29.1165

Teucrium polium var. *valentinum*
"Iberian Golden Germander"
Unaggregated A

(E,Z)-FARNESOL 3 Shoot JEO5:397
(Z)-BETA-FARNESENE 0.5 Shoot JEO5:397
(Z)-NEROLIDOL 1 Shoot JEO5:397
(Z,E)-FARNESOL 2 Shoot JEO5:397
1,8-CINEOLE 1 Shoot JEO5:397
3,3-DIMETHYL-2-ISOPROPYL-CYCLOPENTANE 6 Shoot JEO5:397
3,6,6-TRIMETHYLBICYCLO[3.1.0]HEXAN-3-CARBOXALDEHYDE 1 Shoot JEO5:397
3-OCTANOL 2 Shoot JEO5:397
3-OCTYL-ACETATE 1 Shoot JEO5:397
ALLOAROMADENDRENE 0.5 Shoot JEO5:397
ALPHA-BERGAMOTENE 1 Shoot JEO5:397
ALPHA-BISABOLENE 1 Shoot JEO5:397
ALPHA-CADINOL 1.5 Shoot JEO5:397
ALPHA-CAMPHOLENAL 5 Shoot JEO5:397
ALPHA-CEDRENE 0.5 Shoot JEO5:397
ALPHA-CEDROL 0.5 Shoot JEO5:397
ALPHA-COPAENE 1.5 Shoot JEO5:397
ALPHA-PINENE 80 Shoot JEO5:397
ALPHA-TERPINENE 2 Shoot JEO5:397
ALPHA-TERPINEOL 4 Shoot JEO5:397
ALPHA-THUJENE 7 Shoot JEO5:397
ALPHA-YLANGENE 0.5 Shoot JEO5:397
AR-CURCUMENE 0.5 Shoot JEO5:397
AROMADENDRENE-EPOXIDE 0.5 Shoot JEO5:397
ARTEMISIA-KETONE 2 Shoot JEO5:397
BETA-BISABOLENE 12 Shoot JEO5:397
BETA-CARYOPHYLLENE 4 Shoot JEO5:397
BETA-CUBEBENE 2 Shoot JEO5:397
BETA-GUAIENE 0.5 Shoot JEO5:397
BETA-PINENE 60 Shoot JEO5:397
BETA-THUJONE 1 Shoot JEO5:397
BICYCLOGERMACRENE 5 Shoot JEO5:397
CADIN-4-EN-1-OL 0.5 Shoot JEO5:397
CADINENOL 2 Shoot JEO5:397
CALACORENE 1 Shoot JEO5:397
CALAMENENE 0.5 Shoot JEO5:397
CALARENE 0.5 Shoot JEO5:397
CAMPHOR 1 Shoot JEO5:397
CARVACROL 2 Shoot JEO5:397
CARVENONE 2 Shoot JEO5:397

CARVONE 4 Shoot JEO5:397
CARYOPHYLLENE-EPOXIDE 1 Shoot JEO5:397
CHRYSANTHENONE 2 Shoot JEO5:397
CIS-VERBENOL 2 Shoot JEO5:397
CUMINYL-ALDEHYDE 2 Shoot JEO5:397
DEHYDROSABINAKETONE 2 Shoot JEO5:397
DEHYDROSABINENE 4 Shoot JEO5:397
DELTA-CADINENE 5 Shoot JEO5:397
ENDOFENCHOL 1 Shoot JEO5:397
EO 500 Shoot JEO5:397
EPI-ALPHA-BISABOLOL 2 Shoot JEO5:397
EUCARVONE 1 Shoot JEO5:397
GAMMA-CADINENE 0.5 Shoot JEO5:397
GAMMA-MUUROLENE 0.5 Shoot JEO5:397
GAMMA-TERPINENE 7 Shoot JEO5:397
GAMMA-TERPINEOL 2 Shoot JEO5:397
GAMMA-TERPINYL-ACETATE 3 Shoot JEO5:397
GERANYL-ACETONE 1 Shoot JEO5:397
GERMACRENE-D 4 Shoot JEO5:397
HUMULENE-EPOXIDE 1.5 Shoot JEO5:397
ISOVALERALDEHYDE 7 Shoot JEO5:397
LEDOL Shoot JEO5:397
LIMONENE 16 Shoot JEO5:397
LINALOL 12 Shoot JEO5:397
LONGIFOLENE 0.5 Shoot JEO5:397
MUUROLOL 5 Shoot JEO5:397
MYRCENE 15 Shoot JEO5:397
MYRTENAL 12 Shoot JEO5:397
NEODIHYDROCARVEOL 0.5 Shoot JEO5:397
P-CYMEN-7-OL 1 Shoot JEO5:397
P-CYMEN-8-OL 10 Shoot JEO5:397
P-CYMENE 3-19 Shoot JEO5:397
P-MENTHA-1(7),2-DIEN-1,8-DIOL 4 Shoot JEO5:397
P-MENTHA-1,4-DIEN-7-OL 2 Shoot JEO5:397
PHELLANDRAL 2 Shoot JEO5:397
PINOCARVONE 3 Shoot JEO5:397
SABINAKETONE 4 Shoot JEO5:397
SABINENE 36 Shoot JEO5:397
SABINYL-ACETATE Shoot JEO5:397
SPATHULENOL 2 Shoot JEO5:397
T-CADINOL 3 Shoot JEO5:397
TERPINEN-4-OL 22 Shoot JEO5:397
TERPINOLENE 5 Shoot JEO5:397
THUJYL-ALCOHOL 1 Shoot JEO5:397
THYMOL 2 Shoot JEO5:397

TRANS-CARVEOL 2 Shoot JEO5:397
TRANS-CARVYL-ACETATE 2 Shoot JEO5:397
TRANS-LINALOL-OXIDE 2 Shoot JEO5:397
TRANS-PINOCARVEOL 22 Shoot JEO5:397
UMBELLULONE 4 Shoot JEO5:397
VALENCENE 0.5 Shoot JEO5:397
VERBENENE 6 Shoot JEO5:397
VERBENONE 2 Shoot JEO5:397
VIRIDIFLOROL 3.5 Shoot JEO5:397

Teucrium polium L.
"Golden Germander"

19-ACETYLGNAPHALIN Plant RIZ
19-ACETYLTEUPOLIN Plant JSG
6-ACETYLPICROPOLIN Plant JBH RIZ
6-METHOXYGENKWANIN Plant RIZ
8-ACETYLHARPAGIDE Plant RIZ
ACACETIN Plant ABS
ALPHA-AMYRIN Plant RIZ
ALPHA-HUMULENE Plant RIZ
ALPHA-PINENE Plant RIZ
APIGENIN Plant ABS
AUROPOLIN Plant RIZ
BETA-AMYRIN Plant RIZ
BETA-CARYOPHYLLENE Plant RIZ
BETA-EUDESMOL Plant RIZ
BETA-PINENE Plant RIZ
BORNEOL Plant RIZ
BORNYL-ACETATE Plant RIZ
CADINENE Plant RIZ
CAMPHOR Plant RIZ
CAPRIC-ACID Plant RIZ
CIRSIDIOL Plant JSG
CIRSILIOL Plant RIZ
CIRSIMARITRIN Plant HHB
COPAENE Plant RIZ
CYASTERON Plant RIZ
CYCLOPENTYLALDEHYDE-DERIVATIVE Plant FNF
EO 400-900 Shoot RIZ
FAT 146,000 Seed CRC
GAMMA-TERPINENE Plant RIZ
GNAPHALIDIN Plant RIZ
HARPAGIDE Plant RIZ
HEDERAGENIN Plant RIZ
ISOPICROPOLIN Plant RIZ
KAEMPFEROL Plant ABS
LIMONENE Plant RIZ
LUTEOLIN Plant ABS
MARRUBIIN Plant HHB
MENTHOFURAN Plant HHB
MONTANIN-B Plant RIZ
MYRCENE Plant RIZ
OCIMENE Plant RIZ
P-CYMENE Plant RIZ

PHELLANDRENE Plant RIZ
PICROPOLIN Plant HHB
PIPERITONE Plant RIZ
PROTEIN 148,000 Seed CRC
PULEGONE Plant RIZ
RAFFINOSE Plant HHB
SACCHAROSE Plant HHB
SALVIGENIN Plant RIZ
STACHYDRINE Plant HHB
TERPINEOL Plant RIZ
TEUCARDOSIDE Plant RIZ
TEUCRIN-H-3 Plant RIZ
TEUPOLINS Plant RIZ
URSOLIC-ACID Plant RIZ

Teucrium pseudoscorodonia DESF. Unaggregated A

1,8-CINEOLE 0.281 Shoot PC29:1165
1-OCTEN-3-OL Shoot tr PC29:1165
3-BETA-HYDROXY-DELTA-CADINENE Shoot tr PC29:1165
3-BETA-HYDROXY-MUUROLENE Shoot tr PC29:1165
3-OCTANOL Shoot tr PC29:1165
ALLOAROMADENDRENE 1.52 Shoot PC29:1165
ALPHA-BISABOLOL Shoot tr PC29:1165
ALPHA-CADINOL 0.66 Shoot PC29:1165
ALPHA-CALACORENE 0.27 Shoot PC29:1165
ALPHA-CEDRENE Shoot tr PC29:1165
ALPHA-COPAENE 1.8 Shoot PC29:1165
ALPHA-CUBEBENE 4.46 Shoot PC29:1165
ALPHA-HUMULENE 1.52 Shoot PC29:1165
ALPHA-MUUROLENE 2.24 Shoot PC29:1165
ALPHA-NEROLIDOL Shoot tr PC29:1165
ALPHA-PHELLANDRENE 0.24 Shoot PC29:1165
ALPHA-PINENE Shoot tr PC29:1165
ALPHA-TERPINEOL 0.15 Shoot PC29:1165
ALPHA-TERPINYL-ACETATE 0.06 Shoot PC29:1165
ALPHA-THUJENE Shoot tr PC29:1165
AR-CURCUMENE Shoot tr PC29:1165
ARISTOLENE 19.84 Shoot PC29:1165
AROMADENDRENE 14.01 Shoot PC29:1165
BETA-BISABOLOL 0.06 Shoot PC29:1165
BETA-BOURBONENE 2.45 Shoot PC29:1165
BETA-CARYOPHYLLENE 19.84 Shoot PC29:1165
BETA-PINENE Shoot tr PC29:1165
BORNEOL Shoot tr PC29:1165
CADALENE 0.13 Shoot PC29:1165
CADIN-3-EN-9ALPHA-OL-2-ONE 0.19 Shoot PC29:1165
CADIN-3-EN-9BETA-OL-2-ONE 0.31 Shoot PC29:1165
CADINENOL Shoot tr PC29:1165
CALAMENENE 1.42 Shoot PC29:1165
CAMPHENE 0.1 Shoot PC29:1165
CAMPHOR 0.1 Shoot PC29:1165
CARVONE 0.12 Shoot PC29:1165
CARYOPHYLLENE-EPOXIDE 2.48 Shoot PC29:1165
CEDROL 0.56 Shoot PC29:1165
CIS-LINALYL-OXIDE Shoot tr PC29:1165
CUBENOL 0.27 Shoot PC29:1165
DELTA-CADINENE 0.56 Shoot PC29:1165
EO 100 Shoot PC29:1165

EPICUBENOL 0.74 Shoot PC29:1165
FARNESYL-ACETATE 0.75 Shoot PC29:1165
FENCHYL-ALCOHOL 0.14 Shoot PC29:1165
GAMMA-CADINENE 0.56 Shoot PC29:1165
GAMMA-TERPINENE Shoot tr PC29:1165
GAMMA-TERPINEOL 0.14 Shoot PC29:1165
GERMACRENE-D 11.33 Shoot PC29:1165
HUMULENE-EPOXIDE 2.25 Shoot PC29:1165
LIMONENE 0.28 Shoot PC29:1165
LINALOL 1.18 Shoot PC29:1165
LINALYL-ACETATE 0.14 Shoot PC29:1165
LONGIPINENE 0.37 Shoot PC29:1165
MONOTERPENES 4.63 Shoot PC29:1165
MUUROL-3-EN-9BETA-OL-2-ONE 0.07 Shoot PC29:1165
MYRCENE 0.75 Shoot PC29:1165
NOOTKATONE 0.1 Shoot PC29:1165
NORCAMPHOR Shoot tr PC29:1165
P-CYMENE 0.08 Shoot PC29:1165
SABINENE Shoot tr PC29:1165
SESQUITERPENES 94.05 Shoot PC29:1165
SPATHULENOL 2.48 Shoot PC29:1165
T-CADINOL 0.82 Shoot PC29:1165
TERPINEN-4-OL Shoot tr PC29:1165
THYMOL 0.34 Shoot PC29:1165
THYMYL-ACETATE 0.14 Shoot PC29:1165
TRANS-ALPHA-OCIMENE 0.4 Shoot PC29:1165
TRANS-LINALYL-OXIDE Shoot tr PC29:1165

Teucrium salviastrum SCHREBER
Unaggregated A

1,8-CINEOLE 0.98 Shoot PC29:1165
1-OCTEN-3-OL Shoot tr PC29:1165
3-BETA-HYDROXY-DELTA-CADINENE Shoot tr PC29:1165
3-BETA-HYDROXY-MUUROLENE Shoot tr PC29:1165
3-OCTANOL Shoot tr PC29:1165
ALLOAROMADENDRENE 9.5 Shoot PC29:1165
ALPHA-BISABOLOL Shoot tr PC29:1165
ALPHA-CADINOL 0.82 Shoot PC29:1165
ALPHA-CALACORENE Shoot tr PC29:1165
ALPHA-CEDRENE 0.93 Shoot PC29:1165
ALPHA-COPAENE 1.27 Shoot PC29:1165
ALPHA-CUBEBENE 1.09 Shoot PC29:1165
ALPHA-HUMULENE 9.5 Shoot PC29:1165
ALPHA-MUUROLENE 0.87 Shoot PC29:1165
ALPHA-NEROLIDOL Shoot tr PC29:1165
ALPHA-PHELLANDRENE 0.11 Shoot PC29:1165
ALPHA-PINENE 0.2 Shoot PC29:1165
ALPHA-TERPINEOL 0.21 Shoot PC29:1165
ALPHA-TERPINYL-ACETATE Shoot tr PC29:1165
ALPHA-THUJENE Shoot tr PC29:1165
AR-CURCUMENE Shoot tr PC29:1165
ARISTOLENE 10.8 Shoot PC29:1165
AROMADENDRENE 2.76 Shoot PC29:1165
BETA-BISABOLOL Shoot tr PC29:1165
BETA-BOURBONENE 1.56 Shoot PC29:1165
BETA-CARYOPHYLLENE 10.8 Shoot PC29:1165
BETA-PINENE 1.71 Shoot PC29:1165
BORNEOL Shoot tr PC29:1165
CADALENE 0.24 Shoot PC29:1165
CADIN-3-EN-9ALPHA-OL-2-ONE Shoot tr PC29:1165
CADIN-3-EN-9BETA-OL-2-ONE Shoot tr PC29:1165
CADINENOL Shoot tr PC29:1165
CALAMENENE 0.84 Shoot PC29:1165
CAMPHENE Shoot tr PC29:1165
CAMPHOR Shoot tr PC29:1165
CARVONE Shoot tr PC29:1165
CARYOPHYLLENE-EPOXIDE 9.3 Shoot PC29:1165
CEDROL 0.85 Shoot PC29:1165
CIS-LINALYL-OXIDE 0.4 Shoot PC29:1165
CUBENOL 1.38 Shoot PC29:1165
DELTA-CADINENE 1.73 Shoot PC29:1165
EO 100 Shoot PC29:1165

EPICUBENOL 0.97 Shoot PC29:1165
FARNESYL-ACETATE 0.78 Shoot PC29:1165
FENCHYL-ALCOHOL Shoot tr PC29:1165
GAMMA-CADINENE 1.9 Shoot PC29:1165
GAMMA-TERPINENE 0.17 Shoot PC29:1165
GAMMA-TERPINEOL Shoot tr PC29:1165
GERMACRENE-D 7.75 Shoot PC29:1165
HUMULENE-EPOXIDE 3.02 Shoot PC29:1165
LIMONENE 0.98 Shoot PC29:1165
LINALOL 0.74 Shoot PC29:1165
LINALYL-ACETATE Shoot tr PC29:1165
LONGIPINENE 0.22 Shoot PC29:1165
MONOTERPENES 6.91 Shoot PC29:1165
MUUROL-3-EN-9BETA-OL-2-ONE Shoot tr PC29:1165
MYRCENE 1.05 Shoot PC29:1165
NOOTKATONE 0.24 Shoot PC29:1165
NORCAMPHOR Shoot tr PC29:1165
P-CYMENE 0.25 Shoot PC29:1165
SABINENE Shoot tr PC29:1165
SESQUITERPENES 89.06 Shoot PC29:1165
SPATHULENOL 9.3 Shoot PC29:1165
T-CADINOL 0.65 Shoot PC29:1165
TERPINEN-4-OL Shoot tr PC29:1165
THYMOL Shoot tr PC29:1165
THYMYL-ACETATE Shoot tr PC29:1165
TRANS-ALPHA-OCIMENE 0.11 Shoot PC29:1165
TRANS-LINALYL-OXIDE Shoot tr PC29:1165

Teucrium scordium
"Water Germander"

BETA-AMYRIN Plant WOI
CHLOROGENIC-ACID Plant WOI
EO 1,500 Plant HHB
ISOQUERCITRIN Plant HHB
QUERCETIN Plant HHB
RUTIN Plant HHB
SAPONIN Plant HHB
SCORDEIN Plant HHB
URSOLIC-ACID Plant WOI

Teucrium scorodonia L.
"Germander, Wood Germander"
Unaggregated A

1,8-CINEOLE 0.51 Shoot PC29:1165
1-OCTEN-3-OL Shoot tr PC29:1165
3-BETA-HYDROXY-DELTA-CADINENE Shoot tr PC29:1165
3-BETA-HYDROXY-MUUROLENE Shoot tr PC29:1165
3-OCTANOL Shoot tr PC29:1165
ALLOAROMADENDRENE 1.11 Shoot PC29:1165
ALPHA-BISABOLOL Shoot tr PC29:1165
ALPHA-CADINOL 0.92 Shoot PC29:1165
ALPHA-CALACORENE Shoot tr PC29:1165
ALPHA-CEDRENE 0.65 Shoot PC29:1165
ALPHA-COPAENE 4.77 Shoot PC29:1165
ALPHA-CUBEBENE 8.5 Shoot PC29:1165
ALPHA-HUMULENE 1.11 Shoot PC29:1165
ALPHA-MUUROLENE 1.93 Shoot PC29:1165
ALPHA-NEROLIDOL Shoot tr PC29:1165
ALPHA-PHELLANDRENE Shoot tr PC29:1165
ALPHA-PINENE 0.08 Shoot PC29:1165
ALPHA-TERPINEOL 0.53 Shoot PC29:1165
ALPHA-TERPINYL-ACETATE Shoot tr PC29:1165
ALPHA-THUJENE 0.13 Shoot PC29:1165
AR-CURCUMENE Shoot tr PC29:1165
ARISTOLENE 10.52 Shoot PC29:1165
AROMADENDRENE 5.1 Shoot PC29:1165
BETA-BISABOLOL Shoot tr PC29:1165
BETA-BOURBONENE 4.88 Shoot PC29:1165
BETA-CARYOPHYLLENE 10.52 Shoot PC29:1165
BETA-PINENE 5.91 Shoot PC29:1165
BORNEOL Shoot tr PC29:1165
CADALENE Shoot tr PC29:1165
CADIN-3-EN-9ALPHA-OL-2-ONE 0.14 Shoot PC29:1165
CADIN-3-EN-9BETA-OL-2-ONE Shoot tr PC29:1165
CADINENOL Shoot tr PC29:1165
CALAMENENE 2.56 Shoot PC29:1165
CAMPHENE 0.05 Shoot PC29:1165
CAMPHOR 0.28 Shoot PC29:1165
CARVONE 0.78 Shoot PC29:1165
CARYOPHYLLENE-EPOXIDE 3.03 Shoot PC29:1165
CEDROL 0.34 Shoot PC29:1165
CIS-LINALYL-OXIDE 0.56 Shoot PC29:1165
CUBENOL 0.65 Shoot PC29:1165
DELTA-CADINENE 2.22 Shoot PC29:1165

EO 100 Shoot PC29:1165
EPICUBENOL 0.69 Shoot PC29:1165
FARNESYL-ACETATE 0.91 Shoot PC29:1165
FENCHYL-ALCOHOL 0.91 Shoot PC29:1165
GAMMA-CADINENE 1.22 Shoot PC29:1165
GAMMA-TERPINENE 0.18 Shoot PC29:1165
GAMMA-TERPINEOL Shoot tr PC29:1165
GERMACRENE-D 12.76 Shoot PC29:1165
HUMULENE-EPOXIDE 2.61 Shoot PC29:1165
LIMONENE 0.51 Shoot PC29:1165
LINALOL 1.28 Shoot PC29:1165
LINALYL-ACETATE 0.35 Shoot PC29:1165
LONGIPINENE 0.4 Shoot PC29:1165
MONOTERPENES 13.62 Shoot PC29:1165
MUUROL-3-EN-9BETA-OL-2-ONE 0.12 Shoot PC29:1165
MYRCENE Shoot tr PC29:1165
NOOTKATONE Shoot tr PC29:1165
NORCAMPHOR Shoot tr PC29:1165
P-CYMENE 0.06 Shoot PC29:1165
SABINENE Shoot tr PC29:1165
SESQUITERPENES 81.71 Shoot PC29:1165
SPATHULENOL 3.03 Shoot PC29:1165
T-CADINOL 1.02 Shoot PC29:1165
TERPINEN-4-OL 0.24 Shoot PC29:1165
THYMOL Shoot tr PC29:1165
THYMYL-ACETATE Shoot tr PC29:1165
TRANS-ALPHA-OCIMENE 0.24 Shoot PC29:1165
TRANS-LINALYL-OXIDE Shoot tr PC29:1165

Teucrium scorodonia L.
"Germander, Wood Germander"
Aggregated

5-KAEMPFEROL-GLYCOSIDE Leaf HHB
ARABINOSE Plant HHB
CAFFEIC-ACID 20,000 Leaf HHB
EO 2,900 Shoot HHB
FLAVONOIDS 17,000 Leaf HHB
GLUCOSE Plant HHB
ISOQUERCITRIN Plant HHB
LABIATIC-ACID 12,000 Shoot HHB
MARRUBIIN 8 Shoot HHB
PHENOL-CARBONIC-ACID Plant HHB
RAFFINOSE Plant HHB
RIBOSE Plant HHB
ROSMARINIC-ACID Plant JBH
RUTIN Plant HHB
SCORDEIN Plant HHB
SEDOHEPTULOSE Plant HHB
STACHYOSE Plant HHB
TANNIN 80,000-90,000 Leaf HHB
TETRALOSE Plant HHB
URSOLIC-ACID 3,000 Leaf HHB

Thymus broussonettii BOISS.
"Moroccan Thyme"
Unaggregated A

3-OCTANOL 250 Shoot JEO5:45
ALPHA-PINENE 860 Shoot JEO5:45
ALPHA-TERPINENE 410 Shoot JEO5:45
ALPHA-TERPINEOL 70 Shoot JEO5:45
BETA-PINENE 40 Shoot JEO5:45
BORNEOL 70 Shoot JEO5:45
BORNYL-ACETATE Shoot JEO5:45
CAMPHENE 20 Shoot JEO5:45
CAMPHOR Shoot JEO5:45
CARVACROL 5,330 Shoot JEO5:45
CIS-LINALOL-OXIDE 130 Shoot JEO5:45
CIS-SABINENE-HYDRATE Shoot JEO5:45
DIHYDROCARVONE <0.1 Shoot JEO5:45
EO (ASSUMED) 10,000 Shoot JEO5:45
GAMMA-TERPINENE 20 Shoot JEO5:45
LIMONENE 30 Shoot JEO5:45
LINALOL 50 Shoot JEO5:45
LINALYL-ACETATE 60 Shoot JEO5:45
METHYL-CARVACROL 10 Shoot JEO5:45
MYRCENE 110 Shoot JEO5:45
P-CYMEN-8-OL 30 Shoot JEO5:45
P-CYMENE 1,350 Shoot JEO5:45
TERPINEN-4-OL <0.1 Shoot JEO5:45
TERPINOLENE 30 Shoot JEO5:45
THYMOL 20 Shoot JEO5:45
TRANS-PINOCARVEOL Shoot JEO5:45

Thymus capitatus (L.) HOFFM.
"'Sicilian' Thyme, Spanish Origanum, Spanish Thyme"

1,8-CINEOLE 10-30 Plant BML
1-METHYL-4-ISOPROPYL-4-(2-METHYL-5-ISOPROPYLPHENOXY)-CYCLO... Plant FNF
1-OCTEN-3-OL 20-40 Plant BML
1-OCTEN-3-YL-ACETATE Plant BML
3-OCTANOL Plant BML
ALLO-AROMADENDRENE Plant BML
ALPHA-HUMULENE 10 Plant BML
ALPHA-PHELLANDRENE 1-80 Plant BML
ALPHA-PINENE 10-150 Plant BML
ALPHA-TERPINENE 10-120 Plant BML
ALPHA-TERPINEOL 10-20 Plant BML
ALPHA-THUJENE 60-190 Plant BML
AROMADENDRENE Plant BML
BETA-BISABOLENE 20-40 Plant BML
BETA-PHELLANDRENE Plant BML
BETA-PINENE 10 Plant BML
BORNEOL 20-330 Plant BML
BORNYL-ACETATE Plant BML
CAMPHENE 10 Plant BML
CAMPHOR Plant BML
CARVACROL 6,700-7,900 Plant BML
CARVACRYL-ACETATE Plant BML
CARVENONE Plant BML
CARVONE Plant BML
CARYOPHYLLENE 300-310 Plant BML
CARYOPHYLLENE-OXIDE Plant BML
CIS-DIHYDROCARVONE 10 Plant BML
DELTA-3-CARENE 1-20 Plant BML
DELTA-CADINENE 10-70 Plant BML
FENCHONE Plant BML
GAMMA-MUUROLENE Plant BML
GAMMA-TERPINENE 490-530 Plant BML
GERANIAL Plant BML
GERANIOL 40 Plant BML
GERMACRENE-D 20-30 Plant BML
ISOCARYOPHYLLENE Plant BML
ISOGERANIOL Plant BML
LIMONENE 10-30 Plant BML
LINALOL 30-180 Plant BML
MYRCENE 150-180 Plant BML
NERAL Plant BML

P-CYMEN-8-OL 1-50 Plant BML
P-CYMENE 10-550 Plant BML
SABINENE 1-10 Plant BML
TERPINEN-4-OL 50-70 Plant BML
TERPINOLENE 1-30 Plant BML
THYMOL 10-980 Plant BML
TRANS-ALPHA-BERGAMOTENE Plant BML
TRANS-DIHYDROCARVONE 10 Plant BML
TRANS-P-MENTH-3-EN-9-AL Plant BML
TRANS-SABINENE-HYDRATE 30-60 Plant BML

Thymus capitatus (L.) HOFFM.
"'Sicilian' Thyme, Spanish Origanum, Spanish Thyme"
Unaggregated A

1,8-CINEOLE Shoot FFJ8:331
3-CARENE Shoot FFJ8:331
4,5-DIMETHYL-2-ETHYLPHENOL 50 Shoot FFJ8:331
ALPHA,ALPHA-DIMETHYLBENZYLALCOHOL Shoot FFJ8:331
ALPHA-BISABOLENE 26 Shoot FFJ8:331
ALPHA-CUBEBENE Shoot FFJ8:331
ALPHA-ELEMENOL Shoot FFJ8:331
ALPHA-ELEMOL 95 Shoot FFJ8:331
ALPHA-HUMULENE 10 Shoot FFJ8:331
ALPHA-HUMULENE-EPOXIDE 6 Shoot FFJ8:331
ALPHA-PHELLANDRENE 7 Shoot FFJ8:331
ALPHA-PINENE 4 Shoot FFJ8:331
ALPHA-TERPINENE 4 Shoot FFJ8:331
ALPHA-TERPINEOL 43 Shoot FFJ8:331
ALPHA-THUJONE Shoot FFJ8:331
BENZALDEHYDE Shoot FFJ8:331
BETA-BERGAMOTENE Shoot FFJ8:331
BETA-BISABOLENE 45 Shoot FFJ8:331
BETA-CUBEBENE Shoot FFJ8:331
BETA-PHELLANDRENE Shoot FFJ8:331
BETA-PINENE Shoot FFJ8:331
BETA-TERPINENE <0.1 Shoot FFJ8:331
BETA-TERPINEOL 26 Shoot FFJ8:331
BETA-THUJONE Shoot FFJ8:331
BORNEOL 60 Shoot FFJ8:331
BORNYL-ACETATE Shoot FFJ8:331
CAMPHENE 7 Shoot FFJ8:331
CAMPHOR 9 Shoot FFJ8:331
CARVACROL 6,130 Shoot FFJ8:331
CARVACROL-METHYLETHER Shoot FFJ8:331
CARVACRYL-ACETATE Shoot FFJ8:331
CARVONE 21 Shoot FFJ8:331
CARYOPHYLLENE 200 Shoot FFJ8:331
CARYOPHYLLENE-OXIDE Shoot FFJ8:331
CEDRANOXIDE Shoot FFJ8:331
CIS-LINALOL-OXIDE Shoot FFJ8:331
CIS-OCIMENE 10 Shoot FFJ8:331
CIS-SABINENE-HYDRATE Shoot FFJ8:331
CUMINYL-ALCOHOL 15 Shoot FFJ8:331
DIHYDROCARVEOL Shoot FFJ8:331
ELEMENAL Shoot FFJ8:331

ELEMENE Shoot FFJ8:331
EO 7,100 Shoot FFJ8:331
EUGENOL 21 Shoot FFJ8:331
EUGENOL-METHYL-ETHER Shoot FFJ8:331
FARNESOL Shoot FFJ8:331
GAMMA-CADINENE <0.1 Shoot FFJ8:331
GAMMA-ELEMENE Shoot FFJ8:331
GAMMA-TERPINENE Shoot FFJ8:331
GERANIOL Shoot FFJ8:331
GERANYL-ACETATE Shoot FFJ8:331
HEPT-1-EN-3-OL <0.1 Shoot FFJ8:331
HEX-2-ENAL 4 Shoot FFJ8:331
HEX-3-EN-1-OL Shoot FFJ8:331
HEXANAL Shoot FFJ8:331
ISOPULEGOL Shoot FFJ8:331
LIMONENE 4 Shoot FFJ8:331
LINALOL 11 Shoot FFJ8:331
LINALYL-ACETATE Shoot FFJ8:331
MANOYL-OXIDE Shoot FFJ8:331
MUUROLENE Shoot FFJ8:331
MYRCENE 21 Shoot FFJ8:331
MYRCENYL-ACETATE Shoot FFJ8:331
NERYL-ACETATE Shoot FFJ8:331
OCT-1-EN-3-OL Shoot FFJ8:331
P-CYMENE 45 Shoot FFJ8:331
PATCHOULENE Shoot FFJ8:331
SABINENE 4 Shoot FFJ8:331
TERPINEN-4-OL 37 Shoot FFJ8:331
TERPINYL-ACETATE Shoot FFJ8:331
THUJENE Shoot FFJ8:331
THYMOL 9 Shoot FFJ8:331
THYMOL-METHYLETHER Shoot FFJ8:331
THYMYL-ACETATE Shoot FFJ8:331
TRANS-OCIMENE Shoot FFJ8:331
TRANS-SABINENE-HYDRATE Shoot FFJ8:331
TRICYCLENE Shoot FFJ8:331
ZINGIBERONE Shoot FFJ8:331

Thymus cilicicus BOISS. & BAL.
"'Anatolian' Thyme"
Unaggregated A

(Z)-E-HEXENOL 6 Shoot JEO6:97
1,8-CINEOLE 1,515 Shoot JEO6:97
1-OCTEN-3-OL 35 Shoot JEO6:97
3,5,5-TRIMETHYL-2-CYCLOHEXANONE 15 Shoot JEO6:97
5-METHYL-3-HEPTANONE 4 Shoot JEO6:97
6-METHYL-3-HEPTANOL 22 Shoot JEO6:97
ALPHA-TERPINENE 39 Shoot JEO6:97
ALPHA-TERPINEOL 535 Shoot JEO6:97
BETA-BOURBONENE 47 Shoot JEO6:97
BETA-CARYOPHYLLENE 200 Shoot JEO6:97
BORNYL-ACETATE 12 Shoot JEO6:97
CAMPHOR 940 Shoot JEO6:97
CARVACROL <0.1 Shoot JEO6:97
CARVONE 7 Shoot JEO6:97
CARYOPHYLLENE-OXIDE 370 Shoot JEO6:97
CIS-CARVEOL 4 Shoot JEO6:97
CIS-LINALOL-OXIDE 13 Shoot JEO6:97
CIS-SABINENE-HYDRATE 64 Shoot JEO6:97
CIS-VERBENOL 1,195 Shoot JEO6:97
DOCOSANE 15 Shoot JEO6:97
EO 14,600 Shoot JEO6:97
EUGENOL <0.1 Shoot JEO6:97
GAMMA-ELEMENE 7 Shoot JEO6:97
GAMMA-TERPINENE 90 Shoot JEO6:97
GERMACRENE-D 160 Shoot JEO6:97
ISOBORNEOL 1,350 Shoot JEO6:97
LIMONENE 570 Shoot JEO6:97
LINALOL 120 Shoot JEO6:97
MYRTENAL 29 Shoot JEO6:97
MYRTENOL 22 Shoot JEO6:97
OCTANOL 1.5 Shoot JEO6:97
P-CYMEN-8-OL 38 Shoot JEO6:97
P-CYMENE 170 Shoot JEO6:97
PERILLENE 13 Shoot JEO6:97
SPATHULENOL 42 Shoot JEO6:97
TERPINEN-4-OL 450 Shoot JEO6:97
TERPINOLENE 26 Shoot JEO6:97
TRANS-CAMPHOLENIC-ALDEHYDE 30 Shoot JEO6:97
TRANS-LINALOL-OXIDE 12 Shoot JEO6:97
TRANS-PINOCARVEOL 90 Shoot JEO6:97
TRANS-SABINENE-HYDRATE 145 Shoot JEO6:97
TRANS-VERBENOL 900 Shoot JEO6:97

Thymus funkii COUSS.
"Funk's Thyme"
Unaggregated A

1,8-CINEOLE 1,960 Shoot FFJ10:379
10-EPI-CADINOL 7 Shoot FFJ10:379
10-EPI-GAMMA-EUDESMOL 14 Shoot FFJ10:379
3-OCTANOL <4 Shoot FFJ10:379
7-EPI-ALPHA-EUDESMOL 7 Shoot FFJ10:379
ALLO-AROMADENDRENE 21 Shoot FFJ10:379
ALPHA-CADINOL 14 Shoot FFJ10:379
ALPHA-ELEMOL 7 Shoot FFJ10:379
ALPHA-EUDESMOL 14 Shoot FFJ10:379
ALPHA-HUMULENE 14 Shoot FFJ10:379
ALPHA-PINENE 380 Shoot FFJ10:379
ALPHA-TERPENYL-ACETATE 7 Shoot FFJ10:379
ALPHA-TERPINEOL 90 Shoot FFJ10:379
ALPHA-THUJENE 70 Shoot FFJ10:379
BETA-BISABOLENE Shoot FFJ10:379
BETA-BOURBONENE Shoot FFJ10:379
BETA-CARYOPHYLLENE 90 Shoot FFJ10:379
BETA-CURCUMENE 7 Shoot FFJ10:379
BETA-ELEMENE Shoot FFJ10:379
BETA-ELEMOL 35 Shoot FFJ10:379
BETA-EUDESMOL 14 Shoot FFJ10:379
BETA-GURJUNENE 7 Shoot FFJ10:379
BETA-PINENE 405 Shoot FFJ10:379
BETA-TERPENYL-ACETATE <4 Shoot FFJ10:379
BICYCLOGERMACRENE 210 Shoot FFJ10:379
BORNEOL 415 Shoot FFJ10:379
BORNYL-ACETATE 35 Shoot FFJ10:379
BORNYL-BUTYRATE <4 Shoot FFJ10:379
BORNYL-FORMATE 7 Shoot FFJ10:379
BORNYL-ISOVALERATE <4 Shoot FFJ10:379
BORNYL-PROPIONATE <4 Shoot FFJ10:379
CAMPHENE 440 Shoot FFJ10:379
CAMPHOLENAL <4 Shoot FFJ10:379
CAMPHOR 1,240 Shoot FFJ10:379
CARVACROL <4 Shoot FFJ10:379
CARVONE <4 Shoot FFJ10:379
CARYOPHYLLENE-OXIDE 50 Shoot FFJ10:379
CIS-SABINENE-HYDRATE 55 Shoot FFJ10:379
CITRONELLOL 7 Shoot FFJ10:379
CUBENOL 7 Shoot FFJ10:379
CUPARENE <4 Shoot FFJ10:379

DELTA-CADINENE 14 Shoot FFJ10:379
DELTA-ELEMENE <4 Shoot FFJ10:379
DELTA-TERPINEOL 14 Shoot FFJ10:379
EO 7,000 Shoot FFJ10:379
EUGENOL <4 Shoot FFJ10:379
GAMMA-CADINENE 14 Shoot FFJ10:379
GAMMA-EUDESMOL 14 Shoot FFJ10:379
GAMMA-TERPINENE 20 Shoot FFJ10:379
GERANIAL 7 Shoot FFJ10:379
GERANIOL 7 Shoot FFJ10:379
GERANYL-ACETATE <4 Shoot FFJ10:379
GERANYL-BUTYRATE 7 Shoot FFJ10:379
GERANYL-CAPROATE <4 Shoot FFJ10:379
GERANYL-ISOBUTYRATE 7 Shoot FFJ10:379
GERANYL-ISOVALERATE <4 Shoot FFJ10:379
GERANYL-PROPIONATE 7 Shoot FFJ10:379
GERANYL-VALERATE <4 Shoot FFJ10:379
GERMACRENE-D 14 Shoot FFJ10:379
GLOBULOL <4 Shoot FFJ10:379
ISOCARYOPHYLLENE Shoot FFJ10:379
LIMONENE <4 Shoot FFJ10:379
LINALOOL 35 Shoot FFJ10:379
MYRCENE 260 Shoot FFJ10:379
MYRTENAL 7 Shoot FFJ10:379
MYRTENOL 7 Shoot FFJ10:379
NERAL 7 Shoot FFJ10:379
NEROL 7 Shoot FFJ10:379
NERYL-ACETATE <4 Shoot FFJ10:379
NERYL-BUTYRATE Shoot FFJ10:379
NERYL-ISOBUTYRATE <4 Shoot FFJ10:379
NERYL-PROPIONATE <4 Shoot FFJ10:379
OCT-1-EN-3-OL 7 Shoot FFJ10:379
OCT-1-EN-3-YL-ACETATE Shoot FFJ10:379
P-CYMEN-8-OL 7 Shoot FFJ10:379
P-CYMENE 50 Shoot FFJ10:379
PINOCARVONE 7 Shoot FFJ10:379
ROSE-OXIDE <4 Shoot FFJ10:379
SABINENE 225 Shoot FFJ10:379
SPATHULENOL 145 Shoot FFJ10:379
T-CADINOL 35 Shoot FFJ10:379
TERPINEN-4-OL 110 Shoot FFJ10:379
TERPINOLENE 7 Shoot FFJ10:379
THYMOL <4 Shoot FFJ10:379
TRANS-BETA-FARNESENE Shoot FFJ10:379
TRANS-BETA-OCIMENE <4 Shoot FFJ10:379

TRANS-CARVEOL 7 Shoot FFJ10:379
TRANS-PINOCARVEOL 7 Shoot FFJ10:379
TRANS-SABINENE-HYDRATE 55 Shoot FFJ10:379
TRICYCLENE 28 Shoot FFJ10:379
VERBENONE <4 Shoot FFJ10:379
VIRIDIFLOROL 7 Shoot FFJ10:379

Thymus funkii COUSS.
"Funk's Thyme"
Unaggregated B

1,8-CINEOLE 2,200 Shoot FFJ10:379
10-EPI-CADINOL 8 Shoot FFJ10:379
10-EPI-GAMMA-EUDESMOL 8 Shoot FFJ10:379
3-OCTANOL <5 Shoot FFJ10:379
7-EPI-ALPHA-EUDESMOL Shoot FFJ10:379
ALLO-AROMADENDRENE 32 Shoot FFJ10:379
ALPHA-CADINOL 32 Shoot FFJ10:379
ALPHA-ELEMOL Shoot FFJ10:379
ALPHA-EUDESMOL Shoot FFJ10:379
ALPHA-HUMULENE 16 Shoot FFJ10:379
ALPHA-PINENE 425 Shoot FFJ10:379
ALPHA-TERPENYL-ACETATE <5 Shoot FFJ10:379
ALPHA-TERPINEOL 175 Shoot FFJ10:379
ALPHA-THUJENE 65 Shoot FFJ10:379
BETA-BISABOLENE <5 Shoot FFJ10:379
BETA-BOURBONENE 8 Shoot FFJ10:379
BETA-CARYOPHYLLENE 210 Shoot FFJ10:379
BETA-CURCUMENE 8 Shoot FFJ10:379
BETA-ELEMENE Shoot FFJ10:379
BETA-ELEMOL 16 Shoot FFJ10:379
BETA-EUDESMOL Shoot FFJ10:379
BETA-GURJUNENE Shoot FFJ10:379
BETA-PINENE 490 Shoot FFJ10:379
BETA-TERPENYL-ACETATE Shoot FFJ10:379
BICYCLOGERMACRENE 170 Shoot FFJ10:379
BORNEOL 550 Shoot FFJ10:379
BORNYL-ACETATE 32 Shoot FFJ10:379
BORNYL-BUTYRATE <5 Shoot FFJ10:379
BORNYL-FORMATE 8 Shoot FFJ10:379
BORNYL-ISOVALERATE <5 Shoot FFJ10:379
BORNYL-PROPIONATE <5 Shoot FFJ10:379
CAMPHENE 470 Shoot FFJ10:379
CAMPHOLENAL 8 Shoot FFJ10:379
CAMPHOR 1,135 Shoot FFJ10:379
CARVACROL <5 Shoot FFJ10:379
CARVONE 8 Shoot FFJ10:379
CARYOPHYLLENE-OXIDE 135 Shoot FFJ10:379
CIS-SABINENE-HYDRATE 65 Shoot FFJ10:379
CITRONELLOL 16 Shoot FFJ10:379
CUBENOL 8 Shoot FFJ10:379
CUPARENE Shoot FFJ10:379

DELTA-CADINENE 16 Shoot FFJ10:379
DELTA-ELEMENE Shoot FFJ10:379
DELTA-TERPINEOL 24 Shoot FFJ10:379
EO 8,000 Shoot FFJ10:379
EUGENOL <5 Shoot FFJ10:379
GAMMA-CADINENE 16 Shoot FFJ10:379
GAMMA-EUDESMOL Shoot FFJ10:379
GAMMA-TERPINENE 16 Shoot FFJ10:379
GERANIAL 8 Shoot FFJ10:379
GERANIOL 8 Shoot FFJ10:379
GERANYL-ACETATE Shoot FFJ10:379
GERANYL-BUTYRATE 8 Shoot FFJ10:379
GERANYL-CAPROATE 8 Shoot FFJ10:379
GERANYL-ISOBUTYRATE 8 Shoot FFJ10:379
GERANYL-ISOVALERATE <5 Shoot FFJ10:379
GERANYL-PROPIONATE <5 Shoot FFJ10:379
GERANYL-VALERATE <5 Shoot FFJ10:379
GERMACRENE-D 16 Shoot FFJ10:379
GLOBULOL 16 Shoot FFJ10:379
ISOCARYOPHYLLENE <5 Shoot FFJ10:379
LIMONENE <5 Shoot FFJ10:379
LINALOOL 32 Shoot FFJ10:379
MYRCENE 280 Shoot FFJ10:379
MYRTENAL 8 Shoot FFJ10:379
MYRTENOL 16 Shoot FFJ10:379
NERAL <5 Shoot FFJ10:379
NEROL 16 Shoot FFJ10:379
NERYL-ACETATE Shoot FFJ10:379
NERYL-BUTYRATE <5 Shoot FFJ10:379
NERYL-ISOBUTYRATE Shoot FFJ10:379
NERYL-PROPIONATE Shoot FFJ10:379
OCT-1-EN-3-OL 8 Shoot FFJ10:379
OCT-1-EN-3-YL-ACETATE <5 Shoot FFJ10:379
P-CYMEN-8-OL 8 Shoot FFJ10:379
P-CYMENE 65 Shoot FFJ10:379
PINOCARVONE 8 Shoot FFJ10:379
ROSE-OXIDE <5 Shoot FFJ10:379
SABINENE 280 Shoot FFJ10:379
SPATHULENOL 190 Shoot FFJ10;379
T-CADINOL 25 Shoot FFJ10:379
TERPINEN-4-OL 105 Shoot FFJ10:379
TERPINOLENE <5 Shoot FFJ10:379
THYMOL <5 Shoot FFJ10:379
TRANS-BETA-FARNESENE <5 Shoot FFJ10:379
TRANS-BETA-OCIMENE <5 Shoot FFJ10:379

TRANS-CARVEOL 8 Shoot FFJ10:379
TRANS-PINOCARVEOL <5 Shoot FFJ10:379
TRANS-SABINENE-HYDRATE 90 Shoot FFJ10:379
TRICYCLENE 24 Shoot FFJ10:379
VERBENONE <5 Shoot FFJ10:379
VIRIDIFLOROL 8 Shoot FFJ10:379

Thymus longicaulis C. Presl
"Kekik, Tas Kekik"
Unaggregated A

(E)-2-HEXENAL 2 Shoot JEO5:291
(Z)-3-HEXENOL 0.6 Shoot JEO5:291
(Z)-3-HEXENYL-ACETATE 0.3 Shoot JEO5:291
(Z)-BETA-FARNESENE Shoot JEO5:291
(Z)-BETA-OCIMENE Shoot JEO5:291
1,8-CINEOLE 5 Shoot JEO5:291
1-DODECANOL Shoot JEO5:291
1-NONEN-3-OL 10 Shoot JEO5:291
1-OCTEN-3-OL 44 Shoot JEO5:291
3-NONANONE 0.3 Shoot JEO5:291
5-METHYL-3-HEPTANONE 6 Shoot JEO5:291
6-METHYL-3-HEPTANOL 23 Shoot JEO5:291
6-METHYL-5-HEPTEN-2-ONE Shoot JEO5:291
ALPHA-BERGAMOTENE 1.5 Shoot JEO5:291
ALPHA-CADINENE 3 Shoot JEO5:291
ALPHA-ELEMENE Shoot JEO5:291
ALPHA-HUMULENE 0.3 Shoot JEO5:291
ALPHA-PINENE 16 Shoot JEO5:291
ALPHA-TERPINENE 22-115 Shoot JEO5:291
ALPHA-TERPINYL-ACETATE Shoot JEO5:291
BETA-BISABOLENE 100 Shoot JEO5:291
BETA-BOURBONENE Shoot JEO5:291
BETA-CARYOPHYLLENE 39 Shoot JEO5:291
BETA-PHELLANDRENE Shoot JEO5:291
BETA-PINENE 3 Shoot JEO5:291
BETA-SESQUIPHELLANDRENE Shoot JEO5:291
BORNYL-ACETATE Shoot JEO5:291
BUTYL-ACETATE Shoot JEO5:291
CAMPHENE 8 Shoot JEO5:291
CAMPHOR 1 Shoot JEO5:291
CARVACROL 70 Shoot JEO5:291
CIS-P-MENTHA-2,8-DIENE-1-OL 1 Shoot JEO5:291
CIS-SABINENE-HYDRATE 4 Shoot JEO5:291
CUMIN-ALCOHOL 1.5 Shoot JEO5:291
CUMINALDEHYDE 3 Shoot JEO5:291
EO 3,000 Shoot JEO5:291
GERANIAL Shoot JEO5:291
GERANIOL 8 Shoot JEO5:291
GERANYL-ACETATE Shoot JEO5:291
GERMACRENE-D 6 Shoot JEO5:291
HEXAHYDROFARNESYL-ACETONE 4 Shoot JEO5:291

HEXANAL 1 Shoot JEO5:291
IONOL 3 Shoot JEO5:291
ISOAMYL-ACETATE 1 Shoot JEO5:291
ISOBORNEOL 70 Shoot JEO5:291
LIMONENE 9 Shoot JEO5:291
LIMONENE-OXIDE-II Shoot JEO5:291
LINALOL 14 Shoot JEO5:291
METHYL-ISOVALERATE 0.9 Shoot JEO5:291
MYRCENE 15 Shoot JEO5:291
MYRTENOL 1 Shoot JEO5:291
NERAL Shoot JEO5:291
NEROL Shoot JEO5:291
NONANAL 2 Shoot JEO5:291
NONANOIC-ACID 7.5 Shoot JEO5:291
OCTANOIC-ACID 2 Shoot JEO5:291
OCTANOL 5 Shoot JEO5:291
OCTENYL-ACETATE Shoot JEO5:291
P-CYMEN-8-OL 4 Shoot JEO5:291
P-CYMENE 550 Shoot JEO5:291
PATCHOULANE 25 Shoot JEO5:291
SABINENE 0.3 Shoot JEO5:291
SPATHULENOL 2.4 Shoot JEO5:291
TERPINEN-4-OL 40 Shoot JEO5:291
TERPINOLENE 3 Shoot JEO5:291
THYMOL 1,590 Shoot JEO5:291
TRANS-DIHYDROCARVONE 1 Shoot JEO5:291
TRANS-SABINENE-HYDRATE 11 Shoot JEO5:291
TRICYCLENE Shoot JEO5:291

Thymus longicaulis C. Presl
"Kekik, Tas Kekik"
Unaggregated B

(E)-2-HEXENAL 2 Shoot JEO5:291
(Z)-3-HEXENOL Shoot JEO5:291
(Z)-3-HEXENYL-ACETATE 3 Shoot JEO5:291
(Z)-BETA-FARNESENE Shoot JEO5:291
(Z)-BETA-OCIMENE Shoot JEO5:291
1,8-CINEOLE 24 Shoot JEO5:291
1-DODECANOL 7 Shoot JEO5:291
1-NONEN-3-OL 2 Shoot JEO5:291
1-OCTEN-3-OL Shoot JEO5:291
3-NONANONE Shoot JEO5:291
5-METHYL-3-HEPTANONE 3 Shoot JEO5:291
6-METHYL-3-HEPTANOL Shoot JEO5:291
6-METHYL-5-HEPTEN-2-ONE Shoot JEO5:291
ALPHA-BERGAMOTENE Shoot tr JEO5:291
ALPHA-CADINENE Shoot JEO5:291
ALPHA-ELEMENE 24 Shoot JEO5:291
ALPHA-HUMULENE Shoot JEO5:291
ALPHA-PINENE 635 Shoot JEO5:291
ALPHA-TERPINENE 7-60 Shoot JEO5:291
ALPHA-TERPINYL-ACETATE 13,950 Shoot JEO5:291
BETA-BISABOLENE 355 Shoot JEO5:291
BETA-BOURBONENE Shoot JEO5:291
BETA-CARYOPHYLLENE 22 Shoot JEO5:291
BETA-PHELLANDRENE 3 Shoot JEO5:291
BETA-PINENE 24 Shoot JEO5:291
BETA-SESQUIPHELLANDRENE Shoot JEO5:291
BORNYL-ACETATE 85 Shoot JEO5:291
BUTYL-ACETATE Shoot JEO5:291
CAMPHENE 40 Shoot JEO5:291
CAMPHOR Shoot JEO5:291
CARVACROL 8 Shoot JEO5:291
CIS-P-MENTHA-2,8-DIENE-1-OL Shoot JEO5:291
CIS-SABINENE-HYDRATE 7 Shoot JEO5:291
CUMIN-ALCOHOL Shoot JEO5:291
CUMINALDEHYDE Shoot JEO5:291
EO 17,000 Shoot JEO5:291
GERANIAL Shoot JEO5:291
GERANIOL Shoot JEO5:291
GERANYL-ACETATE Shoot JEO5:291
GERMACRENE-D 14 Shoot JEO5:291
HEXAHYDROFARNESYL-ACETONE Shoot JEO5:291

HEXANAL Shoot JEO5:291
IONOL 14 Shoot JEO5:291
ISOAMYL-ACETATE Shoot JEO5:291
ISOBORNEOL Shoot JEO5:291
LIMONENE 770 Shoot JEO5:291
LIMONENE-OXIDE-II 10 Shoot JEO5:291
LINALOL 24 Shoot JEO5:291
METHYL-ISOVALERATE Shoot JEO5:291
MYRCENE 115 Shoot JEO5:291
MYRTENOL Shoot JEO5:291
NERAL Shoot JEO5:291
NEROL Shoot JEO5:291
NONANAL Shoot JEO5:291
NONANOIC-ACID Shoot JEO5:291
OCTANOIC-ACID Shoot JEO5:291
OCTANOL Shoot JEO5:291
OCTENYL-ACETATE 165 Shoot JEO5:291
P-CYMEN-8-OL Shoot JEO5:291
P-CYMENE 40 Shoot JEO5:291
PATCHOULANE 7 Shoot JEO5:291
SABINENE 185 Shoot JEO5:291
SPATHULENOL Shoot JEO5:291
TERPINEN-4-OL 30 Shoot JEO5:291
TERPINOLENE 120 Shoot JEO5:291
THYMOL 12 Shoot JEO5:291
TRANS-DIHYDROCARVONE Shoot JEO5:291
TRANS-SABINENE-HYDRATE 5 Shoot JEO5:291
TRICYCLENE 1.7 Shoot JEO5:291

Thymus longicaulis C. Presl
"Kekik, Tas Kekik"
Unaggregated C

(E)-2-HEXENAL 0.8 Shoot JEO5:291
(Z)-3-HEXENOL 1.5 Shoot JEO5:291
(Z)-3-HEXENYL-ACETATE Shoot JEO5:291
(Z)-BETA-FARNESENE 9 Shoot JEO5:291
(Z)-BETA-OCIMENE 0.8 Shoot JEO5:291
1,8-CINEOLE 0.8 Shoot JEO5:291
1-DODECANOL Shoot JEO5:291
1-NONEN-3-OL 5 Shoot JEO5:291
1-OCTEN-3-OL 19 Shoot JEO5:291
3-NONANONE Shoot JEO5:291
5-METHYL-3-HEPTANONE 5 Shoot JEO5:291
6-METHYL-3-HEPTANOL 17 Shoot JEO5:291
6-METHYL-5-HEPTEN-2-ONE 1.5 Shoot JEO5:291
ALPHA-BERGAMOTENE 6 Shoot JEO5:291
ALPHA-CADINENE Shoot JEO5:291
ALPHA-ELEMENE Shoot JEO5:291
ALPHA-HUMULENE Shoot JEO5:291
ALPHA-PINENE 3 Shoot JEO5:291
ALPHA-TERPINENE 0-2.5 Shoot JEO5:291
ALPHA-TERPINYL-ACETATE Shoot JEO5:291
BETA-BISABOLENE 345 Shoot JEO5:291
BETA-BOURBONENE 9 Shoot JEO5:291
BETA-CARYOPHYLLENE 28 Shoot JEO5:291
BETA-PHELLANDRENE Shoot JEO5:291
BETA-PINENE Shoot tr JEO5:291
BETA-SESQUIPHELLANDRENE 60 Shoot JEO5:291
BORNYL-ACETATE Shoot JEO5:291
BUTYL-ACETATE 0.8 Shoot JEO5:291
CAMPHENE 3 Shoot JEO5:291
CAMPHOR 3 Shoot JEO5:291
CARVACROL 8 Shoot JEO5:291
CIS-P-MENTHA-2,8-DIENE-1-OL Shoot JEO5:291
CIS-SABINENE-HYDRATE Shoot JEO5:291
CUMIN-ALCOHOL Shoot JEO5:291
CUMINALDEHYDE Shoot JEO5:291
EO 8,000 Shoot JEO5:291
GERANIAL 52 Shoot JEO5:291
GERANIOL 5,500 Shoot JEO5:291
GERANYL-ACETATE 1,315 Shoot JEO5:291
GERMACRENE-D 33 Shoot JEO5:291
HEXAHYDROFARNESYL-ACETONE 1.5 Shoot JEO5:291

HEXANAL 0.8 Shoot JEO5:291
IONOL 9 Shoot JEO5:291
ISOAMYL-ACETATE Shoot JEO5:291
ISOBORNEOL 36 Shoot JEO5:291
LIMONENE 5 Shoot JEO5:291
LIMONENE-OXIDE-II Shoot JEO5:291
LINALOL 40 Shoot JEO5:291
METHYL-ISOVALERATE Shoot JEO5:291
MYRCENE 4 Shoot JEO5:291
MYRTENOL Shoot JEO5:291
NERAL 42 Shoot JEO5:291
NEROL 140 Shoot JEO5:291
NONANAL 0.8 Shoot JEO5:291
NONANOIC-ACID Shoot JEO5:291
OCTANOIC-ACID Shoot JEO5:291
OCTANOL 3 Shoot JEO5:291
OCTENYL-ACETATE Shoot JEO5:291
P-CYMEN-8-OL Shoot JEO5:291
P-CYMENE 3 Shoot JEO5:291
PATCHOULANE 5 Shoot JEO5:291
SABINENE Shoot JEO5:291
SPATHULENOL Shoot JEO5:291
TERPINEN-4-OL 1.5 Shoot JEO5:291
TERPINOLENE Shoot JEO5:291
THYMOL 16 Shoot JEO5:291
TRANS-DIHYDROCARVONE 5 Shoot JEO5:291
TRANS-SABINENE-HYDRATE Shoot JEO5:291
TRICYCLENE Shoot JEO5:291

Thymus mastichina L.
"Spanish Marjoram"

1,8-CINEOLE 300-7,500 Plant BML
3-TERPINEN-1-OL 20 Plant BML
3-TERPINEN-1-YL-ACETATE 20 Plant BML
ACETIC-ACID Plant BML
ALLO-AROMADENDRENE 90 Plant BML
ALPHA-GURJUNENE 1 Plant BML
ALPHA-PINENE 260 Plant BML
ALPHA-TERPINENE 1 Plant BML
ALPHA-TERPINEOL 210-800 Plant BML
ALPHA-TERPINYL-ACETATE 310 Plant BML
ALPHA-THUJENE 40 Plant BML
AROMADENDRENE 20 Plant BML
BETA-BOURBONENE 10 Plant BML
BETA-GURJUNENE 30 Plant BML
BETA-PINENE 200-270 Plant BML
BORNEOL 1-350 Plant BML
BORNYL-ACETATE 20 Plant BML
CAMPHENE 20-140 Plant BML
CAMPHOR 1-450 Plant BML
CARYOPHYLLENE 130 Plant BML
CARYOPHYLLENE-OXIDE 1 Plant BML
CIS-BETA-TERPINEOL 1 Plant BML
CIS-BETA-TERPINYL-ACETATE 1 Plant BML
CIS-NEROLIDOL 10 Plant BML
COPAENE 1 Plant BML
DELTA-CADINENE 5 Plant BML
DELTA-TERPINEOL 1 Plant BML
ELEMOL 1 Plant BML
GAMMA-CADINENE 5 Plant BML
GAMMA-TERPINENE 1-60 Plant BML
GERANIOL 20 Plant BML
GERANYL-ACETATE 10 Plant BML
ISOVALERIC-ACID Plant GEO
LIMONENE 240-280 Plant BML
LINALOL 850-7,890 Plant BML
LINALYL-ACETATE 100-150 Plant BML
LONGIFOLENE 1 Plant BML
MYRCENE 20-90 Plant BML
P-CYMENE 130-340 Plant BML
SABINENE 80-110 Plant BML
TERPINEN-4-OL 10-70 Plant BML
TERPINOLENE 1 Plant BML

THYMOL 190 Plant BML
TRANS-DIHYDRO-ALPHA-TERPINYL-ACETATE 1 Plant BML
TRANS-PINOCARVEOL 75 Plant BML
TRANS-PINOCARVYL-ACETATE Plant BML
TRANS-SABINENE-HYDRATE 80-110 Plant BML

Thymus orospedanus H. del VILLAR
"Orosped Thyme"
Unaggregated A

ALPHA-HUMULENE Plant JNP
ALPHA-PINENE 665 Plant JNP
ALPHA-THUJENE Plant JNP
BETA-OCIMENE Plant JNP
BETA-PINENE 1 Plant JNP
BORNEOL 875 Plant JNP
BORNYL-ACETATE 158 Plant JNP
CAMPHENE 578 Plant JNP
CAMPHOR 718 Plant JNP
CARVACROL 2,730 Plant JNP
CARVACRYL-ACETATE Plant JNP
CARYOPHYLLENE 718 Plant JNP
DELTA-3-CARENE Plant JNP
DELTA-CADINENE Plant FNF
EO 17,500 Plant FNF
GAMMA-TERPINENE 3,920 Plant JNP
LIMONENE 140 Plant JNP
LINALOL 945 Plant JNP
MYRCENE 858 Plant JNP
P-CYMENE 3,939 Plant JNP
PHELLANDRENE Plant JNP
TERPINEN-4-OL 140 Plant JNP
TERPINOLENE Plant JNP
THUJONE 1 Plant JNP
THYMOL 1,278 Plant JNP

Thymus riatarum HUMBERT & MAIRE
"'Moroccan' Thyme"
Unaggregated A

1-OCTEN-3-OL 85 Shoot JEO3:43
3-OCTANOL <0.1 Shoot JEO3:43
3-OCTENYL-ACETATE <0.1 Shoot JEO3:43
ALPHA-BISABOLOL <0.1 Shoot JEO3:43
ALPHA-CADINOL <0.1 Shoot JEO3:43
ALPHA-COPAENE 25 Shoot JEO3:43
ALPHA-HUMULENE 25 Shoot JEO3:43
ALPHA-P-DIMETHYLSTYRENE <0.1 Shoot JEO3:43
ALPHA-PHELLANDRENE 25 Shoot JEO3:43
ALPHA-PINENE 230 Shoot JEO3:43
ALPHA-TERPINENE 215 Shoot JEO3:43
ALPHA-TERPINEOL 25 Shoot JEO3:43
ALPHA-THUJENE 120 Shoot JEO3:43
BETA-BISABOLENE 575 Shoot JEO3:43
BETA-BOURBONENE <0.1 Shoot JEO3:43
BETA-CARYOPHYLLENE 300 Shoot JEO3:43
BETA-PHELLANDRENE 35 Shoot JEO3:43
BETA-PINENE 35 Shoot JEO3:43
BORNEOL 1,115 Shoot JEO3:43
BORNYL-ACETATE 215 Shoot JEO3:43
CAMPHENE 275 Shoot JEO3:43
CAMPHOR 25 Shoot JEO3:43
CARVACROL 2,675 Shoot JEO3:43
CARVONE <0.1 Shoot JEO3:43
CARYOPHYLLENE-OXIDE 205 Shoot JEO3:43
CUBENOL <0.1 Shoot JEO3:43
DELTA-CADINENE 50 Shoot JEO3:43
EO 12,000 Shoot JEO3:43
GAMMA-CADINENE 25 Shoot JEO3:43
GAMMA-TERPINENE 1,235 Shoot JEO3:43
GERMACRENE-D <0.1 Shoot JEO3:43
LIMONENE 85 Shoot JEO3:43
LINALOL 130 Shoot JEO3:43
MYRCENE 215 Shoot JEO3:43
O-METHYL-CARVACROL 95 Shoot JEO3:43
P-CYMEN-8-OL 50 Shoot JEO3:43
P-CYMENE 2,100 Shoot JEO3:43
SABINENE <0.1 Shoot JEO3:43
T-CADINOL 25 Shoot JEO3:43
TERPINEN-4-OL 275 Shoot JEO3:43
TERPINOLENE 35 Shoot JEO3:43

THYMOL 275 Shoot JEO3:43
TRANS-OCIMENE <0.1 Shoot JEO3:43
TRICYCLENE <0.1 Shoot JEO3:43

Thymus saturejoides
"Moroccan Savory Thyme"
Unaggregated A

3-OCTANOL 210 Shoot JEO5:45
ALPHA-PINENE 1,750 Shoot JEO5:45
ALPHA-TERPINENE 40 Shoot JEO5:45
ALPHA-TERPINEOL 40 Shoot JEO5:45
BETA-PINENE 310 Shoot JEO5:45
BORNEOL 3,120 Shoot JEO5:45
BORNYL-ACETATE Shoot JEO5:45
CAMPHENE 2,740 Shoot JEO5:45
CAMPHOR 10 Shoot JEO5:45
CARVACROL 20 Shoot JEO5:45
CIS-LINALOL-OXIDE <0.1 Shoot JEO5:45
CIS-SABINENE-HYDRATE <0.1 Shoot JEO5:45
DIHYDROCARVONE Shoot JEO5:45
EO (ASSUMED) 10,000 Shoot JEO5:45
GAMMA-TERPINENE <0.1 Shoot JEO5:45
LIMONENE 10 Shoot JEO5:45
LINALOL 630 Shoot JEO5:45
LINALYL-ACETATE 50 Shoot JEO5:45
METHYL-CARVACROL Shoot JEO5:45
MYRCENE 50 Shoot JEO5:45
P-CYMEN-8-OL <0.1 Shoot JEO5:45
P-CYMENE 460 Shoot JEO5:45
TERPINEN-4-OL 40 Shoot JEO5:45
TERPINOLENE 150 Shoot JEO5:45
THYMOL Shoot JEO5:45
TRANS-PINOCARVEOL 10 Shoot JEO5:45

Thymus serpyllum L.
"Creeping Thyme"

1,8-CINEOLE 1-38 Plant BML HHB
1-OCTEN-3-OL 20-306 Plant BML HHB
3-OCTANOL Plant tr BML
ALPHA-PINENE 1-362 Plant BML HHB
ALPHA-TERPINENE 21-344 Plant HHB
ALPHA-THUJENE 15-306 Plant BML HHB
APIGENIN Plant HHB
BETA-BISABOLENE 68-1,012 Plant BML HHB
BETA-PINENE 1-19 Plant BML HHB
BORNEOL 4-573 Plant BML HHB WOI
BORNYL-ACETATE Plant HHB
CAFFEIC-ACID 6,200 Plant HHB
CAMPHENE 1-19 Plant BML HHB
CARVACROL 318-10,505 Plant BML HHB WOI
CARYOPHYLLENE 135-2,292 Plant BML HHB
CITRAL Plant HHB
CITRONELLOL Plant WOI
DELTA-CADINENE 8-152 Plant BML HHB
EO 1,500-19,100 Plant HHB
GAMMA-TERPINENE 252-3,534 Plant HHB
GERANIOL Plant HHB
GERANYL-ACETATE Plant HHB
GERMACRENE-D 3-573 Plant BML HHB
LIMONENE 3-57 Plant BML HHB
LINALOL 8-134 Plant BML HHB
LINALYL-ACETATE Plant BML
LUTEOLIN-7-GLUCOSIDE Plant HHB
METHYL-CARVACROL 10-2,292 Plant BML HHB
METHYL-THYMOL 1-153 Plant BML HHB
MYRCENE 13-229 Plant BML HHB
NEROLIDOL Plant HHB
NERYL-ACETATE Plant HHB
OLEANOLIC-ACID 4,800 Plant HHB
P-CYMEN-8-OL 10-172 Plant BML HHB
P-CYMENE 237-3,285 Plant BML HHB
ROSMARINIC-ACID 23,000 Shoot APF48.103
SABINENE 1-19 Plant BML HHB
SCUTELLAREIN Plant HHB
SCUTELLAREIN-HETEROSIDE Plant WOI
SERPYLLIN Plant HHB
TANNIN 50,000-75,000 Plant HHB
TERPINEN-4-OL 10-191 Plant HHB

TERPINEOL Plant HHB
TERPINYL-ACETATE Plant HHB
THYMOL 3-57 Plant BML HHB
UNDECANOIC-ACID Plant HHB
URSOLIC-ACID 7,500-9,100 Plant HHB WOI
ZINGIBERENE Plant HHB

Thymus vulgaris L.
"Common Thyme, Garden Thyme"

2,6,6-TRIMETHYL-BICYCLO(3,1,1)-HEPTA-2-ENE Essential Oil NAP
3,3',4,4'-TETRAHYDROXY-5,5'-DI-ISO-PROPYL-2,2'-DIMETHYL-BIPHENYL 35 Leaf NAP
3,3',4,4'-TETRAONE-5,5'-DI-ISO-PROPYL-2,2'-DIMETHYL-BIPHENYL 3 Leaf NAP
3,4,4'-TRIHYDROXY-5,5'-DI-ISO-PROPYL-2,2'-DIMETHYL-BIPHENYL 1 Leaf NAP
4'-HYDROXY-5,5'-DI-ISO-PROPYL-2,2'-DIMETHYL-BIPHENYL-3,4-DIONE 9 Leaf NAP
4'5-DIHYDROXY-3',6,7-TRIMETHOXYFLAVONE Leaf NAP
4'5-DIHYDROXY-6,7,8-TRIMETHOXYFLAVONE 5.5 Leaf NAP
4'5-DIHYDROXY-7-METHOXYFLAVONE 43 Leaf NAP
4,4'-DIHYDROXY-5,5'-DI-ISO-PROPYL-2,2'-DIMETHYL-BIPHENYL-3,6-DIONE Leaf NAP
4-HYDROXYBENZOYL-GLUCOSE Leaf NAP
4-TERPINEOL 73-8,320 Plant ABS BML HHB WOI
5,4'-DIHYDROXY-6,7,8,3'-TETRAMETHOXYFLAVONE Plant CCO
5-HYDROXY-4',7-DIMETHOXYFLAVONE 1 Leaf NAP
6-HYDROXY-LUTEIN Plant CCO
8-DEMETHYL-THYMONIN Plant NAP
8-METHOXY-CIRSILINEOL 7.4 Leaf NAP
ALPHA-HYDROXY-LINOLENIC-ACID Seed WOI
ALPHA-LINOLENIC-ACID 6,900-7,452 Plant USA
ALPHA-PHELLANDRENE 50-425 Plant JAD NAP
ALPHA-PINENE 15-1,598 Plant ABS BML HHB
ALPHA-TERPINENE Plant NAP
ALPHA-TERPINEOL 36-6,500 Plant ABS BML HHB
ALUMINUM 920 Leaf PED
AMYL-ALCOHOL 13,900 Essential Oil NAP
ANETHOLE Essential Oil NAP
APIGENIN Plant PAS PCF
ASH 113,681-130,809 Plant USA
BETA-CADINENE Essential Oil NAP
BETA-CAROTENE 24-25 Plant CRC
BETA-CARYOPHYLLENE 15-605 Plant JAD NAP JE39:167
BETA-PHELLANDRENE Plant NAP
BETA-PINENE 15-420 Plant ABS BML HHB
BETA-TERPINEOL 79-673 Plant JAD NAP
BORNEOL 15-1,462 Leaf ABS BML HHB
BORNYL-ACETATE 16-795 Leaf BML HHB JBH JE39:167
BORON 34-48 Plant BOB
CAFFEIC-ACID 16,900 Leaf NAP WOI

CALCIUM 16,700 Leaf PED
CALCIUM 16,935-22,534 Plant USA
CAMPHENE 15-375 Plant BML HHB WOI JE39:167
CAMPHOR 5-45 Plant JE39:167
CAPRIC-ACID 1,200-1,296 Plant USA
CAPRYLIC-ACID 2,400-2,592 Plant USA
CAR-3-ENE 127-1,080 Plant JAD NAP
CAR-4-ENE Plant NAP
CARBOHYDRATES 693,000 Leaf PED
CARBOHYDRATES 639,400-690,552 Plant USA
CARVACROL 8-18,720 Plant BML HHB JAD JBH NAP
CARVONE Plant PAS
CHLOROGENIC-ACID Plant HHB
CHROMIUM 20 Leaf PED
CHRYSOERIOL Plant PAS
CINEOLE 10-4,590 Plant ABS HHB JE39:167
CINNAMIC-ACID Plant PAS
CIRSILINEOL Leaf CCO JBH
CIRSIMARITRIN 20 Leaf NAP
CITRAL Plant PAS
COBALT 113 Leaf PED
COPPER 8-9 Plant USA
COSMOSIIN Leaf NAP
CUMINYL-ALCOHOL Essential Oil NAP
CYSTINE 1,370-1,980 Plant USA 1/
DELTA-3-CARENE 510 Plant BML
DELTA-CADINENE Essential Oil NAP
DIOSMETIN Plant PAS
EO 4,000-34,000 Plant WOI
ERIODICTYOL Plant PAS
EUDESMOL Essential Oil NAP
EUGENOL Plant PAS
FAT 69,480-80,000 Plant USA WOI
FAT 370,000-389,000 Seed CRC WOI
FERULIC-ACID Plant PAS
FIBER 179,294-693,000 Plant CRC USA
GALLIC-ACID Plant PAS
GAMMA-TERPINENE 36-5,460 Plant ABS BML HHB
GAMMA-TERPINEOL Plant NAP
GERANIOL 0-10,660 Plant ABS BML HHB
GERANYL-ACETATE 0-3,380 Plant ABS BML HHB
HEX-5-EN-1-OL Essential Oil NAP
IRON 1,075-1,508 Plant USA
ISOBORNEOL Essential Oil NAP
ISOCHLOROGENIC-ACID Leaf NAP

ISOEUGENOL Plant PAS
ISOLEUCINE 4,680-5,054 Plant USA
ISOTHYMONIN Plant NAP
KAEMPFEROL Plant PAS
KILOCALORIES 2,990/kg Leaf PED
L-BORNEOL Plant WOI
LABIATIC-ACID Leaf PCF PED
LAURIC-ACID 2,300-2,484 Plant USA
LEUCINE 4,300-4,644 Plant USA
LIMONENE 15-5,200 Plant ABS BML HHB
LINALOL 20-17,420 Plant ABS BML JAD HHB NAP JE39:167
LINALYL-ACETATE 15-4,680 Plant ABS BML HHB
LINOLEIC-ACID 5,000-5,400 Plant USA
LITHIUM 4 Plant HHB
LITHOSPERMIC-ACID Shoot NAP
LUTEOLIN Plant PCF HHB
LUTEOLIN-7-DIGLUCOSIDE Plant HHB
LUTEOLIN-7-O-BETA-D-GLUCOSIDE Plant PAS
LYSINE 2,070-2,236 Plant USA
MAGNESIUM 4,360 Leaf PED
MAGNESIUM 1,630-2,992 Plant USA
MANGANESE 64-79 Leaf PED USA
MENTHONE Plant PAS
METHIONINE 1,370-1,980 Plant USA
MYRCENE 36-676 Plant ABS BML HHB
MYRCENOL-8 15-3,900 Plant ABS BML
MYRCENOL-8 ACETATE 15-2,600 Plant ABS BML HHB
MYRISTIC-ACID 1,500-1,620 Plant USA
N-TRIACONTANE Plant HHB
NARINGENIN Plant PAS PCF
NEROLIDOL 80 Plant BML
NIACIN 54 Plant CRC
OCIMENE Essential Oil NAP
OLEANOLIC-ACID 6,300 Plant HHB
OLEIC-ACID 4,700-5,076 Plant USA
P-COUMARIC-ACID 420 Leaf NAP
P-CYMEN-8-OL Plant NAP
P-CYMENE 146-20,800 Plant ABS HHB JBH WOI
P-HYDROXYBENZOIC-ACID Plant PAS
P-HYDROXYBENZOIC-ACID-4-BETA-D-GLUCOSIDE Leaf NAP
PALMITIC-ACID 17,200-18,576 Plant USA
PHENYLALANINE 2,410-2,603 Plant USA
PHOSPHORUS 950 Leaf PED
PHOSPHORUS 1,703-2,502 Plant USA
PHYTOSTEROLS 1,630-1,760 Plant USA

PLANTEOSE Seed NAP
POTASSIUM 9,680 Leaf PED
POTASSIUM 7,667-9,302 Plant USA
PROTEIN 99,000 Plant CRC
PROTOCATECHUIC-ACID-4-BETA-D-GLUCOSIDE Leaf NAP
RIBOFLAVIN 4-53 Plant CRC USA
ROSMARINIC-ACID 26,000 Plant FT62:166
ROSMARINIC-ACID 5,000-13,500 Shoot JAD NAP
SALICYLATES 180-1,830 Leaf JAD85:950
SALICYLIC-ACID-2-BETA-D-GLUCOSIDE Leaf NAP
SELENIUM 16 Leaf PED
SILICON 202 Leaf PED
SODIUM 1,490 Leaf PED
SODIUM 430-1,341 Plant CRC
SYRINGIC-ACID-4-BETA-D-GLUCOSIDE Leaf NAP
SYRINGOYL-GLUCOSE Leaf NAP
TANNIN 80,000-100,000 Plant HHB WOI
THIAMIN 5-6 Plant CRC
THREONINE 2,520-2,722 Plant USA
THYMOL 15-24,100 Plant ABS BML HHB JBH
THYMOL-METHYL-ETHER 1-10,000 Plant HHB
THYMONIN Leaf JBH NAP PCF
THYMUNIC-ACID Plant WOI
THYMUSAPONIN Plant WOI
THYMYL-ACETATE Plant JBH
TIN 17 Leaf PED
TRANS-4-THUJANOL Plant FNF
TRYPTOPHAN 1,860-2,009 Plant USA
TYROSINE 2,410-2,603 Plant USA
URSOLIC-ACID 15,000-18,800 Plant HHB WOI
VALINE 5,020-5,422 Plant USA
VANILLIC-ACID Plant PAS
VANILLIC-ACID-4-BETA-D-GLUCOSIDE Leaf NAP
WATER 832,000 Leaf PED
WATER 74,111-81,690 Plant USA
ZINC 15 Leaf PED
ZINC 55-74 Plant USA

Thymus x *citriodorus* (PERS.) SCHREB.
"Funk's thyme, Lemon Thyme"
Unaggregated A

1,8-CINEOLE 50 Plant FFJ10(3):225
6-METHYL-HEPTAN-3-ONE 270 Plant FFJ10(3):225
ALLO-AROMADENDRENE 21 Shoot FFJ10:379
ALPHA-PINENE 50 Plant FFJ10(3):225
ALPHA-TERPINEOL 220 Plant FFJ10(3):225
BETA-BISABOLENE 130 Plant FFJ10(3):225
BETA-BOURBONENE 20 Plant FFJ10(3):225
BETA-CARYOPHYLLENE 370 Plant FFJ10(3):225
BETA-CARYOPHYLLENE-EPOXIDE 30 Plant FFJ10(3):225
BETA-PINENE 10 Plant FFJ10(3):225
BORNEOL 20 Plant FFJ10(3):225
BORNYL-ACETATE Plant FFJ10(3):225
CAMPHENE 80 Plant FFJ10(3):225
CAMPHOR 20 Plant FFJ10(3):225
CIS-BETA-OCIMENE 20 Plant FFJ10(3):225
CITRONELLAL 10 Plant FFJ10(3):225
CITRONELLOL 30 Plant FFJ10(3):225
EO 10,000 Plant FFJ10(3):225 ASSUMED
GAMMA-TERPINENE Plant FFJ10(3):225
GERANIAL 820 Plant FFJ10(3):225
GERANIOL 6,130 Plant FFJ10(3):225
GERANYL-ACETATE 100 Plant FFJ10(3):225
GERANYL-BUTYRATE 80 Plant FFJ10(3):225
GERMACRADIENOL-I 20 Plant FFJ10(3):225
GERMACRENE-D 110 Plant FFJ10(3):225
LIMONENE Plant FFJ10(3):225
LINALOL 80 Plant FFJ10(3):225
LINALYL-ACETATE Plant FFJ10(3):225
NERAL 55 Plant FFJ10(3):225
NEROL 280 Plant FFJ10(3):225
OCT-1-EN-3-OL 40 Plant FFJ10(3):225
OCTAN-3-OL 70 Plant FFJ10(3):225
P-CYMENE 10 Plant FFJ10(3):225
SABINENE 10 Plant FFJ10(3):225
THYMOL 50 Plant FFJ10(3):225
THYMOL-METHYL-ETHER Plant FFJ10(3):225
TRANS-SABINENE-HYDRATE 10 Plant FFJ10(3):225

Thymus zygis* ssp. *sylvestris
"'Portuguese' Thyme"

1,8-CINEOLE <0.1-2,000 Shoot JEO3:409
ALPHA-PINENE 60-490 Shoot JEO3:409
ALPHA-TERPINENE <0.1-210 Shoot JEO3:409
ALPHA-TERPINEOL <0.1-295 Shoot JEO3:409
ALPHA-TERPINYL-ACETATE <0.1-295 Shoot JEO3:409
AROMADENDRENE 0-30 Shoot JEO3:409
BETA-CARYOPHYLLENE <0.1-460 Shoot JEO3:409
BETA-PINENE 20-200 Shoot JEO3:409
BORNEOL <0.1-420 Shoot JEO3:409
CAMPHENE 90-550 Shoot JEO3:409
CAMPHOR <0.1-390 Shoot JEO3:409
CARVACROL 350-3,550 Shoot JEO3:409
EO 10,000 Plant ASSUMED
GAMMA-TERPINENE 250-1,690 Shoot JEO3:409
GERANIOL 0-2,350 Shoot JEO3:409
GERANIOL-ACETATE 0-2,050 Shoot JEO3:409
LIMONENE <0.1-190 Shoot JEO3:409
LINALOL 100-4,920 Shoot JEO3:409
LINALYL-ACETATE 0-100 Shoot JEO3:409
MYRCENE <0.1-200 Shoot JEO3:409
MYRCENOL 0-50 Shoot JEO3:409
P-CYMENE 940-4,220 Shoot JEO3:409
SABINENE <0.1-40 Shoot JEO3:409
TERPINEN-4-OL <0.1-250 Shoot JEO3:409
THYMOL 350-3,550 Shoot JEO3:409

Thymus zygis L.
"Spanish Thyme"
Unaggregated A

3-OCTANOL 170 Shoot JEO5:45
ALPHA-PINENE 520 Shoot JEO5:45
ALPHA-TERPINENE 390 Shoot JEO5:45
ALPHA-TERPINEOL Shoot JEO5:45
BETA-PINENE 70 Shoot JEO5:45
BORNEOL 590 Shoot JEO5:45
BORNYL-ACETATE 20 Shoot JEO5:45
CAMPHENE 540 Shoot JEO5:45
CAMPHOR Shoot JEO5:45
CARVACROL 810 Shoot JEO5:45
CIS-LINALOL-OXIDE 50 Shoot JEO5:45
CIS-SABINENE-HYDRATE 20 Shoot JEO5:45
DIHYDROCARVONE <0.1 Shoot JEO5:45
EO 10,000 Shoot JEO5:45 ASSUMED
GAMMA-TERPINENE 20 Shoot JEO5:45
LIMONENE Shoot JEO5:45
LINALOL 480 Shoot JEO5:45
LINALYL-ACETATE 70 Shoot JEO5:45
METHYL-CARVACROL 10 Shoot JEO5:45
MYRCENE 170 Shoot JEO5:45
P-CYMEN-8-OL Shoot JEO5:45
P-CYMENE 5,060 Shoot JEO5:45
TERPINEN-4-OL Shoot JEO5:45
TERPINOLENE 10 Shoot JEO5:45
THYMOL 500 Shoot JEO5:45
TRANS-PINOCARVEOL Shoot JEO5:45

Thymus zygis L.
"Spanish Thyme"
Unaggregated B

(E)-BETA-OCIMENE 75 Shoot JEO5:153
1,8-CINEOLE <0.1 Shoot JEO5:153
ALLOAROMADENDRENE <0.1 Shoot JEO5:153
ALPHA-PINENE 32 Shoot JEO5:153
ALPHA-TERPINEOL 22 Shoot JEO5:153
ALPHA-TERPINYL-ACETATE 20 Shoot JEO5:153
ALPHA-THUJENE <0.1 Shoot JEO5:153
BETA-CARYOPHYLLENE <0.1 Shoot JEO5:153
BETA-PINENE 8 Shoot JEO5:153
BORNEOL <0.1 Shoot JEO5:153
CAMPHENE <0.1 Shoot JEO5:153
CAMPHOR 19 Shoot JEO5:153
CARVACROL 385 Shoot JEO5:153
DELTA-3-CARENE <0.1 Shoot JEO5:153
EO 8,000 Shoot JEO5:153
GAMMA-TERPINENE 75 Shoot JEO5:153
GERANIOL 14 Shoot JEO5:153
GERANYL-ACETATE 28 Shoot JEO5:153
LIMONENE <0.1 Shoot JEO5:153
LINALOL 13 Shoot JEO5:153
LINALYL-ACETATE 18 Shoot JEO5:153
MYRCENE 42 Shoot JEO5:153
P-CYMEN-8-OL 55 Shoot JEO5:153
P-CYMENE 825 Shoot JEO5:153
SABINENE <0.1 Shoot JEO5:153
TERPINEN-4-OL 185 Shoot JEO5:153
THYMOL 5,920 Shoot JEO5:153
TRANS-SABINENE-HYDRATE 10 Shoot JEO5:153

Trichostemma dichotomum L.
"Blue Curls"
Unaggregated A

1-OCTEN-3-OL 31 Shoot JEO2:149
ALLO-AROMADENDRENE 15 Shoot JEO2:149
ALPHA-HUMULENE 11 Shoot JEO2:149
ALPHA-PHELLANDRENE 0.6 Shoot JEO2:149
ALPHA-PINENE 35 Shoot JEO2:149
ALPHA-TERPINENE 5.5 Shoot JEO2:149
ALPHA-THUJENE 35 Shoot JEO2:149
BETA-CARYOHPYLLENE 70 Shoot JEO2:149
BETA-MYRCENE 26 Shoot JEO2:149
BETA-PHELLANDRENE 18 Shoot JEO2:149
BETA-PINENE 23 Shoot JEO2:149
CARYOPHYLLENE-OXIDE 2 Shoot JEO2:149
EO 800 Shoot JEO2:149
GAMMA-ELEMENE 7.7 Shoot JEO2:149
GAMMA-TERPINENE 11 Shoot JEO2:149
LIMONENE 5 Shoot JEO2:149
LINALOL 2.7 Shoot JEO2:149
P-CYMENE 0.3 Shoot JEO2:149
SABINENE 430 Shoot JEO2:149
TERPINEN-4-OL 13 Shoot JEO2:149
TERPINOLENE 3 Shoot JEO2:149
TRANS-BETA-OCIMENE 0.9 Shoot JEO2:149

Biological Activities of Mint Phytochemicals

(+)-CARVONE ACE-Inhibitor LGR; Antialzheimeran? LGR

(-)-MENTHOL Termiticide 382/

1,8-CINEOLE ACE-Inhibitor LGR; Allelopathic 450/; Allergenic M&R; Anesthetic; Anthelminthic JBH; Antiallergic JFH33(6):569; Antibronchitic; Anticatarrh M29; Anticholinesterase? LGR; Antihalitosic AH12(4):17; Antilaryngitic; Antipharyngitic; Antirhinitic; Antiseptic JBH; Antitussive; Bactericide JEO7:271; CNS-Stimulant; Choleretic; Counterirritant M29; Dentifrice M29; Expectorant JBH; Flavor FEMA 1-200 ARC; Fungicide; Hepatotonic; Herbicide IC50=78 uM TOX; Hypotensive; Insectifuge 382/; Nematicide NIG; Perfume ARC; Rubefacient M29; Sedative JE25:165 LD50=2,480 (orl rat) AEH

2-METHOXY-4-VINYL-PHENOL Flavor FEMA 0.2-10 ARC; Perfumery ARC

2-OCTANOL Perfumery M11

3-OCTANOL Flavor FEMA ARC; Perfumery ARC

4-(3,4-DIHYDROXY-BENZOYL-OXY-METHYL)-PHENYL-BETA-D-GLUCOPYRANOSIDE Cancer-Preventive 525/

4-TERPINEOL Flavor FEMA 20-100 ARC; Insectifuge JNP56:935; Perfumery ARC

5,4'-DIHYDROXY-6,7,8,3'-TETRAMETHOXYFLAVONE Fungistatic PC29(4):1103

ACACETIN Aldose-Reductase-Inhibitor JBH; Allergenic JBH; Antiallergic PAM; Antihistamine JBH; Antimutagenic EMP6:235; Iodothyronine-Deiodinase-Inhibitor JBH

ACETALDEHYDE Addictive 8 ug/kg SN145:294; Additive 8 ug/kg SN145:294; Fungicide; Perfumery FEMA 4-25 ARC; Respiraparalytic LD50=1,900-1,930 (orl rat) BBE CRC LD50=500 (ipr rat) BBE

ACETIC-ACID Acidulant JBH; Antiotitic MAR; Antivaginitic 1-2%; Bactericide 5,000 ppm; Expectorant MAR; Fungicide M29; Keratitigenic M&R; Mucolytic M29; Osteolytic; Perfumery FEMA 6,000-40,000 ppm ARC;

Protisticide M29; Spermicide MAR; Ulcerogenic HG20:9; Verrucolytic LD50=3,310 (orl rat)

ACETONE Allergenic M&R; CNS-Depressant M29; Narcotic M29 LD50=5,300 (orl rbt)

ACETOPHENONE Bactericide; Fungicide PC29:1093; Hypnotic JBH; Perfumery FEMA <10,000 ARC; Perfumery JBH; Soporific ARC LD50=900 (orl rat)

ACTEOSIDE Aldose-Reductase-Inhibitor IC50=0.39uM BBB56:2062; Antileukemic; Antinephritic 30 mg/kg orl rat JJP66:47; Cytotoxic ED50=2.6 ug/ml JNP53:456

AESCULETIN Analgesic PMP24:98; Anti-Capillary-Fragility WIC; Antiarrhythmic; Antiasthmatic? PMP24:98; Antidysenteric HG24:37; Antiinflammatory PMP24:98; Antimutagenic EMP6:235; Bactericide; Cancer-Preventive 525/; Cardiodepressant; Choleretic FT5:1986; Dermatitigenic WIC; Fungicide NIG; Hypertensive CCO; Lipoxygenase-Inhibitor PAM; Musculotropic WIC; Myorelaxant WOI; UV-Screen WIC

AESCULIN Anti-Capillary-Fragility BIS; Antiarthritic; Antihistaminic; Cardiotonic; Diuretic; Hypertensive CCO; Lipoxygenase-Inhibitor PAM; Sunscreen BIS

AJUGALACTONE Juvabional JBH 455

AJUGAREPTANSONE Jubabional 382/

ALANINE Antioxidant PC27:969; Cancer-Preventive 525/; Oxidant PC27:969

ALPHA-AMYRIN Antitumor; Cytotoxic 50-400

ALPHA-BISABOLOL Antiarthritic LRN-MAR91; Antiburn IWU; Antiinflammatory W&W; Antipyretic LRN-MAR91; Antiseptic WIC; Antispasmodic LRN-MAR91; Antitubercular; Antiulcer IWU; Bactericide IWU; Candidicide; Cosmetic FT61:254.1990; Fungicide IWU; Musculotropic WIC; Protisticide JNP39:420; Spasmolytic HG20:40; Vulnerary LRN-MAR91 LD50=15 ml/kg (orl rat) LRN-MAR91

ALPHA-CAROTENE Colorant JBH; Previtamin-A

ALPHA-CURCUMENE Antiinflammatory; Antitumor

ALPHA-HUMULENE *Antitumor JNP55:999; Perfumery JBH*

ALPHA-HUMULENE-EPOXIDE Antitumor JNP55:999

ALPHA-LINOLENIC-ACID Antimenorrhagic PAM; Antiprostatitic PAM; Cancer-Preventive 525/
ALPHA-PHELLANDRENE Dermal JBH; Flavor FEMA 10-130 ARC; Insectiphile PEP; Irritant JBH; Perfumery JBH

ALPHA-PINENE Allelochemic; Allergenic M&R; Antiflu EMP5:195; Antiinflammatory IJO15(4):194; Antiviral EMP5:195; Bactericide RIE12:5; Cancer-Preventive 525/; Coleoptiphile; Expectorant MIK; Flavor FEMA 15-150 ARC; Herbicide IC50=30 uM TOX; Insectifuge (50 ppm) 382/; Insectiphile JSPR22:141; Irritant JBH; Perfumery ARC; Sedative LRN-JUN90; Tranquilizer LRN-JUN90

ALPHA-SELINENE Perfumery ARC

ALPHA-TERPINENE Flavor FEMA 1-40 ARC; Insectifuge; Perfumery ARC

ALPHA-TERPINEOL Anticariogenic MIC=800-1,600 ug/ml JAF41:110G; Antiseptic JBH; Bactericide MIC=800-1,600 ug/ml JAF41:110G; Flavor FEMA 5-40 ARC; Motor-Depressant BVC:162; Nematicide MLC=1 mg/ml SZ44:183; Perfumery JBH; Sedative BVC:162; Termiticide MUO24:1

ALPHA-THUJONE Insecticide; Larvicide JNP49:941.1986

ALUMINUM Antisilicotic MPT; Antivaginitic MAR; Candidicide MAR; Encephalopathic M29

AMYGDALIN Antiinflammatory; Antitussive OMM; Cancer-Preventive 525/; Cyanogenic 52/; Expectorant OMM

AMYL-ALCOHOL Flavor FEMA 35-3,500 ppm ARC
LD50=200 (orl mus) LD50=3,030 (orl rat)

ANETHOLE Allergenic M&R; Antihepatotic JBH; Antiseptic JAF39:2290; Bactericide JNP49:156; Cancer-Preventive 525/; Carminative M11; Dermatitogenic M&R; Digesative MIK; Estrogenic AJM7:29; Expectorant

M29; Fungicide JNP49:156; Gastrostimulant JE26:65; Immunostimulant EMP1:124; Insecticide; Lactogogue; Leucocytogenic EMP1:124; Spasmolytic JBH LD50=2,090 (orl rat) LD50=3,000 orl mus AEH LD50=900 (ipr rat)

ANISALDEHYDE Antimutagenic EMP6:235; Insecticide LD50=1,510 (orl rat) 15

APIGENIN Antiaggregant 411/; Antiallergic PAM; Antiarrhythmic PT4(3):118; Antidermatitic BIS; Antiestrogenic PAM; Antiherpetic EMP5:197; Antihistaminic 411/; Antiinflammatory (= indomethacin) WIC; Antimutagenic ID50=10-40 nM PCF; Antioxidant 411/; Antispasmodic EC50=1-5 uM 411/; Antithyroid JNM1:10; Antiviral EMP5:197; Bactericide 452/; Calcium-Blocker? BBA1115:69; Cancer-Preventive 525/; Choleretic FTS1986; Cytotoxic IC88=10 ug/ml PM57:A113; Deiodinase-Inhibitor JNM1:10; Diuretic JBH; Hypotensive JBH; Musculotropic WIC; Myorelaxant JBH; Nodulation-Signal JBH; Sedative; Spasmolytic PM56:638; Vasodilator BBA1115:69

APIGENIN-7,4'-DIMETHYL-ETHER Nodulation-Signal JBH; Spasmolytic PM56:638

APIGENIN-7-GLYCOSIDE Antiinflammatory WIC; Nodulation-Signal JBH; Spasmolytic PM56:638

AR-CURCUMENE Antirhinoviral IC50=1750? JNP57:658; Antiulcer IC45=100 mg/kg YAK112:645; Antiviral IC50=1750? JNP57:658

ARBUTIN Allelochemic IC51=1.1 mM 438/; Antiseptic 60-200 mg/man; Antitussive JBH; Artemicide TOX; Bactericide JE26:65; Candidicide PAM; Diuretic 60-200 mg/man M11; Insulin-Sparing JBH; Urinary-Antiseptic JBH

ARGININE Antidiabetic? WIC; Antiencephalopathic M29; Antihepatitic M29; Antiinfertility 4g/day WER; Antioxidant? PC27:969; Diuretic CRC; Hypoammonemic JBH; Spermigenic 4g/day WER

AROMADENDRIN Antiseptic JBH; Cancer-Preventive 525/

ARSENIC Anorexic FHN; LD orl man 70-180 mg or 0.75-2 mg/kg FHN

ASCORBIC-ACID Acidulant FEMA 6,000 ARC; Analgesic 5-10 g day WER; AntiCrohn's 50-100 mg/day/orl/man M29; Antiarthritic 1 g day WER; Antiasthmatic 1,000 mg/day WER; Antiatherosclerotic PAM; Anticataract 350 mg/day WER; Anticold 1-2 g/man/day MAR; Antidecubitic 500 mg/man/2x/day/

MAR; Antidepressant 2,000 mg/day WER; Antidote (Aluminum) EMP6:189; Antidote (Cadmium) DAS; Antidote (Lead) DAS; Antidote (Paraquat) MAR; Antieczemic 3.5-5 g/day WER; Antiedemic 1 g/man/day DAS; Antiencephalitic DAS; Antigingivitic PAM; Antiglaucomic 2 g/day WER; Antihemorrhagic 1 g/man/day DAS; Antihepatitic 2-6 g/man/day DAS; Antihepatoxic EMP6:189; Antiherpetic 1-5 g/day WER; Antihistamninic WER; Antiinfertility 1 g/day WER; Antiinflammatory PAM; Antilepric 1.5 g/man/day MAR; Antimeasles DAS; Antimigraine M29; Antimutagenic EMP6:235; Antinitrosic 1 g/man/day MAR; Antiobesity 1g 3 x day WER; Antiorchitic DAS; Antiosteoarthritic 1 g 2x day WER; Antiosteoporotic 500 mg day WER PAM; Antioxidant 100 ppm; Antiparkinsonian 1g 2 x day WER; Antiparotitic DAS; Antiperiodontitic 1g 2 x day WER; Antipneumonic PAM; Antipodriac DAS; Antipoliomyelitic DAS; Antiscorbutic 10 mg/man/day MAR; Antiseptic MIC=3.3-217 mg/ml MAR; Antishingles DAS; Antiulcer WER; Antiviral 1-5 g/day WER; Asthma-preventive 1,000mg/day/orl PAM; Bactericide DAS; Cancer-Preventive 525/; Cold-preventive 1-2 g/day DAS; Detoxicant; Diuretic 700 mg/man/orl; Febrifuge EMP6:189; Fistula-Preventive PAM; Hypocholesterolemic 300-1,000 mg/day DAS; Hypotensive 1,000 mg/man/day SNE137:292; Inteferonogenic PAM; Lithogenic DAS; Mucolytic 1 g/woman/day MAR; Uricosuric 4 g/man/day DAS; Urinary-Acidulant M29; Viricide DAS; Vulnerary PAM LD50=518 (ivn mus) 15 PTD=1-5g RDA=30-95mg/day RDA=60-100 mg/man/day HSC25:1523

ASIATIC-ACID Collagenic WER317

ASPARAGINE Antisickling PM56(1):41; Diuretic HG17

ASPARTIC-ACID Antimorphinic MAR; Neuroexcitant MAR; Roborant M11

AUCUBIN Antidote (Amanitin) 200 mg/kg ipr mus ACM:276; Antioxidant; Bactericide JE26:65; Cathartic ED50=390 mg/kg W&W; Diuretic JBH; Hepatoprotective FTS1986; Laxative 411/; Paralytic WIC; RNA-Inhibitor ACM:275; Uricosuric 411/

AZULENE Antiallergic IWU; Antiinflammatory PEP; Antiseptic 6-500 VAL; Antiulcer ED50=150 (orl); Bactericide 500 ppm VAL; Febrifuge HG20:40

BAICALEIN Aldose-Reductase-Inhibitor IC50=0.98 ppm CPB37:3153; Aldose-Reductase-Inhibitor IC62=10 uM PC23:1885; Antiacne PMP25:46; Antiaggregant ED50=120-200 mg/kg AAA39:C-49; Antiallergic 411/; Antiasthmatic; Antidermatitic; Antihistaminic; Antiinflammatory 411/; Antileukemic 2 ug/ml EMP5:225; Antileukemic IC50=0.26 ug/ml LS55:1061;

Antiproliferative EJP251:91; Antipyretic KCH; Antiretroviral PMP25:46; Antitoxin(Strychnine) 10 mg mus KCH; Astringent M11; Bactericide SAY; Choleretic KCH; Diuretic; HIV-RT-Inhibitor 2 ug/ml EMP5:225; HIV-RT-Inhibitor IC50=<1 ug/ml JNP53(5):1239; Hypocholesterolemic; Triglycerolytic

BAICALIN Aldose-Reductase-Inhibitor IC42=10 uM PC23:1885; AntiHIV CMB39:119; Antiaggregant; Antiallergic 411/; Antianaphylactic CPB38:208; Antiasthmatic; Antihepatotic 500 mg man day KCH SAY; Antihistaminic; Antiinflammatory 411/; Antilaryngitic KCH; Antinephritic KCH; Antipelvitic KCH; Antipharyngitic KCH; Antipyretic SAY; Antithrombotic 0.1-1 mM JBH; Antitonsilitic KCH; Choleretic KCH; Diuretic JE26:65; Hypocholesterolemic; Hypotensive SAY; Sedative SAY; Sialidase-Inhibitor LRN-JAN93; Tranquilizer; Triglycerolytic; Vasodilator CRC OMM LD50-3000 (ipr mus) AEH

BAICALIN-A AntiHIV CMB39:119

BENZALDEHYDE Allergenic M&R; Anesthetic; Antipeptic; Antiseptic; Antispasmodic; Antitumor; Flavor FEMA 150-800 ARC; Insectifuge 50 ppm JSPR22:141; Motor-Depressant BVC:162; Narcotic M11; Nematicide MLC=1 mg/ml SZ44:183; Sedative BVC:162 LD50=1,000 (orl gpg) M11 LD50=1,300 (orl rat)

BENZOIC-ACID Allergenic MAR; Anesthetic; Antiotitic MAR; Antiseptic M29; Bactericide 33-1250 ppm; Choleretic JBH; Expectorant; Febrifuge; Flavor FEMA 250 ARC; Fungicide M11; Insectifuge 382/; Phytoalexin JBH; Uricosuric; Vulnerary JE26:65 ADI=5 mg/kg MAR LD50=2,370 (orl mus)

BENZYL-ACETATE Emetic; Flavor FEMA 35-800 ARC; Laxative LD50=2,490 (orl rat) M11

BENZYL-ALCOHOL Allergenic M29; Anesthetic MAR; Antiodontalgic MAR; Antipruritic; Antiseptic M29; Flavor Fema 200-1,200 ARC; Sedative JA6(4):40 LD50=1,230 (orl rat) LD50=3,100 (orl rat) M11

BENZYL-BENZOATE Acaricide M11; Allergenic M29; Antiasthmatic; Antidysmenorrheic; Antispasmodic; Antitumor ED50=8-15 PC29(5):1667; Artemicide LC50=2ppm PC29(5):1667; CNS-Stimulant M29; Flavor FEMA 40-300 ARC; Hypotensive; Insectifuge M29; Myorelaxant; Pediculicide M29; Scabicide M11 ADI=5 mg/kg? MAR LD50=1,000 (orl gpg) M11 LD50=1,400 (orl mus) M11 LD50=1,700 (orl rat) M11 LD50=1,800 (orl rbt) M11 LD50=270 (orl rat)

BENZYL-ISOBUTYRATE Flavor FEMA 25 ARC

BENZYL-ISOVALERATE Flavor FEMA 50-200 ARC

BETA-BISABOLENE Abortifacient JE32:167; Antirhonoviral IC50=1,800? JNP57:658; Antiulcer IC57=100 mg/kg YAK112:645; Antiviral IC50=1,800? JNP57:658; Perfumery ARC

BETA-CAROTENE Allergenic M&R; Androgenic? PAM; AntiPMS JAF34:409; Antiacne PAM; Antiaging; Antiasthmatic PAM; Anticarcinomic PAM; Anticoronary 50mg/man/2 days CMR3/4/91; Antihyperkeratotic DAS PAM; Antiichythyotic PAM; Antileukoplakic PAM; Antilupus 150 mg/man/day/2 mos MAR; Antimastitic PAM; Antimutagenic EMP6:235; Antioxidant PAM; Antiozenic; Antiphotophobic 30-300 mg man/day M29; Antipityriasic PAM; Antiporphyric 30-300 mg/man/day MAR; Antipsoriac PAM; Antiradicular PAM; Antistress DAS; Antitumor HSC25:1495; Antiulcer 12 mg/3x/day/man/orl PAM; Antixerophthalmic DAS; Cancer-Preventive 22 ppm 505/; Colorant JBH; Immunostimulant 180 mg/man/day/orl PAM; Mucogenic PAM; Phagocytotic PAM; Thymoprotective PAM; Ubiquiot JBH LD50=>1,000 mg/kg (orl rat) PAM PTD=15-30 mg/day RDA=2.25-7.8 mg/day

BETA-ECDYSONE Hypocholesterolemic; Lipolytic

BETA-ELEMENE Anticancer (Cervix) MPT

BETA-EUDESMOL Antianoxic ED60=300 orl mus CPB38:2033; Antipeptic CPB34.3854.1986; CNS-Inhibitor; Hepatoprotective FT5:1986

BETA-IONONE Anticariogenic MIC=100 ug/ml JAF41:1103; Antitumor JAF43:2144; Bactericide MIC=200 ug/ml JAF41:1103; Cancer-Preventive 525/; Fungicide NIG JAF43:2144; Hypocholesterolemic JAF43:2144; Nematicide MLC=1 mg/ml SZ44:183 LD50=4,590 (orl rat)

BETA-MYRCENE Cancer-Preventive 525/; Perfumery M11

BETA-PHELLANDRENE Expectorant JBH; Perfumery M11

BETA-PINENE Allergenic M&R; Flavor FEMA 15-600 ARC; Herbicide TOX; Insectifuge 382/; Perfumery ARC

BETA-SANTALENE Antiinflammatory; Perfumery ARC

BETA-SELINENE Perfumery ARC

BETA-SESQUIPHELLANDRENE Antirhinoviral IC50=0.44 uM JNP57:658; Antiulcer IC40=50mg/kg YAK112:645

BETA-SITOSTEROL Androgenic; Anorexic MAR; Antiadenomic M11; Antiandrogenic JE28:221; Antiestrogenic JE28:221; Antifeedant 382/; Antifertility JE28:221; Antigonadotrophic JE28:221; Antihyperlipoproteinaemic JBH; Antiinflammatory JE28:221; Antileukemic; Antimutagenic 250 ug/mL JAF37:1363; Antiophidic 2.3 mg mus EMP5:363; Antiprogestational JE28:221; Antiprostatadenomic M11; Antiprostatitic M29; Antitumor; Antiviral PS75:161; Artemicide LC50=110ppm PC29(5):1667; Bactericide QJC28:155; Cancer-Preventive 525/; Candidicide QJC28:155; Estrogenic; Gonadotrophic JE28:221; Hepatoprotective PMP23:46; Hypocholesterolemic 9-3330 g/man/day/orl MAR; Hypoglycemic JE27:243; Hypolipidemic M29; Spermicide JE28:221; Ubiquiot JBH; Viricide PS75:161 LDlo=>10,000 (inj rat) JE28:221

BETA-SITOSTEROL-D-GLUCOSIDE Antitumor; CNS-Depressant RAA; CNS-Stimulant; Convulsant; Hypoglycemic; Spasmolytic JE28:221 LD50=62 (ipr mus)

BETA-TERPINEOL Insectifuge 382/; Perfumery ARC

BETA-THUJONE Insectifuge

BETAINE Abortifacient; Antigastritic M29; Antihomocystinuric M29; Antimyoatrophic; Bruchiphobe 382/; Emmenagogue; Hepatoprotective M11; Lipotropic LD50=830 (ivn mus)

BETONICINE Hemostat 411/

BETULIN Antifeedant JAF38:493; Antiinflammatory PM61:9; Antitumor JBH; Aphidifuge JAF38:493; Cytotoxic 600 ppm; Hypolipemic PMP23:46

BETULINIC-ACID AntiHIV JNP57:243; Anticarcinomic JBH; Antiinflammatory PM61:9; Antimelanomic NYT3/28/95; Antitumor JBH; Antiviral JNP57:243; Cytotoxic 50-100 ppm BPCC

BORNEOL Analgesic; Antibronchitic JBH; Antiinflammatory; CNS-Toxic JBH; Febrifuge; Flavor FEMA<1 ARC; Hepatoprotectant; Herbicide IC50=470 mM 438/; Herbicide IC50=470 uM TOX; Inhalant JBH; Insectifuge 382/; Nematicide MLC=1 mg/ml SZ44:183; Perfumery JBH; Spasmolytic ED50=0.008 mg/ml FT59:465 LDlo=2,000 (orl rbt)

BORNYL-ACETATE Antifeedant 382/; Bactericide; Expectorant JBH; Flavor FEMA 70-80 ARC; Insectifuge; Sedative LRN-OCT91; Spasmolytic ED50=0.09 mg/ml FT59:465; Viricide

BORNYL-FORMATE Flavor FEMA <1 ARC

BORON Androgenic 3 mg/man/day; Antiosteoarthritic WER; Antiosteoporotic 3 mg/man/day; Estrogenic 3 mg/man/day

CADINENE Antifeedant 382/; Flavor FEMA 400-4,000 ARC; Perfumery JBH

CAFFEIC-ACID Allergenic M&R; Analgesic PMP23:46; Anti-Tumor-Promoter IC42=10 uM CR48:5941; AntiHIV EMP5:207 EMP6:189; Antiadenoviral EMP5:207; Antiaggregant JBH; Anticarcinogenic EMP6:189; Antiedemic EMP6:189; Antiflu EMP5:207; Antigonadotropic JNM1:10; Antihemolytic 25 uM PC36:579; Antihepatotoxic PM56(2):171; Antiherpetic 50 ug/ml V&D EMP5:206; Antihypercholesterolemic EMP6:189; Antiinflammatory JBH; Antimutagenic PCF:18; Antinitrosaminic PCF:18; Antioxidant 50 um PC27:969; Antioxidant IC57=30 ppm PCF:17; Antiperoxidant IC50=44 uM PM57:A54; Antiradicular 10 uM PC36:579; Antiseptic JE26:65; Antispasmodic EC50=3.4-15 uM PR4:73; Antistomatitic EMP5:207; Antithiamin PCF:54; Antithyroid JNM1:10; Antitumor PCF:19; Antiulcerogenic JBH EMP6:189; Antivaccinia EMP5:207; Antiviral AVR14:323; Bactericide PAM; CNS-Active WIC; Cancer-Preventive 525/; Carcinogenic 2% (diet) PCF:35; Cholagogue WIC; Choleretic 411/; Clastogenic JBH; Co-carcinogenic PCF:35; DNA-Active JBH; Diuretic WIC; Fungicide MIC=0.4 mg/ml PC54:211 JE26:76; Hepatoprotective ACM:210; Hepatotropic; Histamine-Inhibitor; Insectifuge EB48:111; Leukotriene-Inhibitor; Lipoxygenase-Inhibitor IC27=5 mM JAF38:688; Metal-Chelator PCF:25; Ornithine-Decarboxylase-Inhibitor PCF:19; Prostaglandigenic RWG; Sedative 500 mg RWG; Spasmolytic EC50=3.4-15 uM PR4:73; Sunscreen IC50=2.5 mg/l FT64:134; Sunscreen IC91=5 mg/l FT64:134; Sunscreen IC98=25 mg/l FT64:134; Tumorigenic 505/; Viricide IC50=62.5 ug/ml EMP5:253; Vulnerary JE26:65

CALCIUM AntiPMS 1 g day WER; Antiallergic 500 mg/day DAS WER; Antianxiety DAS; Antiatherosclerotic 500 mg/day WER; Antidepressant DAS; Antidote (Aluminum) DAS; Antidote (lead) DAS; Antihyperkinetic DAS; Antiinsomniac DAS; Antiosteoporotic 500 mg day WER; Antiperiodontitic 750 mg day WER; Antitic DAS, Hypocholesterolemic 500 mg day WER; Hypotensive 1 g/day WER PTD=12,000 mg/day RDA=400-1200 mg/day

CAMPHENE Antilithic? JBH; Antioxidant JA6(4):33; Flavor FEMA 15-175 ARC; Hypocholesterolemic? JBH; Insectifuge 382/; Spasmogenic FT59:465

CAMPHOR Allelopathic 450/; Analgesic MAR; Anesthetic JE26:65; Antiacne NIG; Antiemetic 100-200 mg man orl KCH; Antifeedant IC50=5,000 ppm diet 438/; Antifibrositic MAR; Antineuralgic MAR; Antipruritic M11; Antiseptic M11; CNS-Stimulant JE26:65; Cancer-Preventive 525/; Carminative JPP46:16; Convulsant RJH; Cosmetic JBH; Counterirritant MAR; Decongestant NIG; Deliriant M11; Ecbolic; Emetic; Expectorant M29; Fungicide ED50=2.7 mM TOX; Herbicide IC50=3.3-180 mM 438/; Insectifuge M11; Irritant JBH; Nematicide MLC=1 mg/ml SZ44:183; Oculoirritant M&R; Respirainhibitor M11; Respirastimulant KCH; Rubefacient MAR; Spasmolytic ED50=0.075 mg/ml FT59:465; Stimulant RJH; Verrucolytic NIG LD50=3,000 (ipr mus) M11 LDlo=1,000 mg/chd MAR LDlo=50 JNP50:1017 LDlo=900 (ipr rat) CRC

CAPRIC-ACID Fungicide; Irritant M&R
LD50=129 (ivn mus) M11

CAPRYLIC-ACID Candidicide 300 mg/man/12 x/day CPA; Fungicide M11; Irritant M&R LD50=10,080 (orl rat) M11 LD50=600 (ivn mus)

CARNOSIC-ACID Antimutagenic EMP6:235; Antioxidant

CARNOSOL 5-Lipoxygenase-Inhibitor IC50 (uM)=2 DRS LEH; Anti-Tumor-Promoter PCF:25; AntiHIV JNP56:1426; Antibiotic JE31:43; Antihepatotoxic PM56(2):171; Antiinflammatory PCF:25; Antilipoperoxidant PM56(2):171; Antimutagenic EMP6:235; Antioxidant PCF:25; Antiradicular PM56(2):171; Antitumor PCF:25; Cancer-Preventive 525/; Candidicide JE31:43; Cyclooxygenase-Inhibitor IC50 (uM) =16 DRS LEH; Fungicide JE31:43; Metal-Chelator PCF:25; Ornithine-Decarboxylase-Inhibitor PCF:25; Protease-Inhibitor JNP56:1426

CARNOSOLIC-ACID AntiHIV IC90=0.32 ug/ml JNP56:1426; Antihepatotoxic PM56(2):171; Antilipoperoxidant PM56(2):171; Antioxidant; Antiradicular PM56(2):171; Cytotoxic TC90=0.36 ug/ml JNP56:1426; Protease-Inhibitor IC90=0.08 ug/ml JNP56:1426

CARVACROL ACE-Inhibitor LGR; Allergenic M&R; Anesthetic; Anthelminthic JBH; Antialzheimeran? LGR; Anticholinesterase? LGR; Antidiuretic LRN-MAR92; Antiinflammatory IC50=4 uM; Antioxidant JEO5:153; Antiplaque MIC=39-625 ug/ml 451/; Antiradicular 600 x thymol JEO5:153; Antiseptic 1.5 x phenol M11 JBH; Bactericide MIC 170-290 PMP25:184; Bactericide MIC=39-625 450/; Carminative; Enterorelaxant

JAR4:22; Expectorant; Flavor FEMA 10-125 ARC; Fungicide JBH; Nematicide MLC=1 mg/ml M11 SZ44:183; Prostaglandin-Inhibitor; Spasmolytic 411/; Tracheorelaxant; Vermifuge LD50=810 (orl rat) PM&T20:213.1986 PM184.1986 LDlo=100 (orl rbt) PM&T20:213.1986 PM184.1986

CARVEOL CNS-Stimulant

CARVONE Allergenic M&R; Antiseptic 1.5 x phenol W&W; CNS-Stimulant; Cancer-Preventive 525/; Carminative JPP46:16; Flavor FEMA 80-3,000 ARC; Insecticide 382/; Insectifuge 382/; Motor-Depressant BVC:162; Nematicide MLC=1 mg/ml SZ44:183; Perfumery JBH; Sedative BVC:162; Vermicide LD50=1,640 (orl rat) M11

CARYOPHYLLENE Anticariogenic MIC=>1,600 ug/ml JAF41:1103; Antiedemic CPB38:2283; Antifeedant 500 ppm TOX; Antiinflammatory IC50=100 uM CPB38:2283; Antitumor JNP55:999; Bactericide JAF40:2328; Flavor FEMA 20-200 ARC; Insectifuge 382/; Perfumery M11; Spasmolytic; Termitifuge 382/

CARYOPHYLLENE-EPOXIDE Antifeedant 1mM (diet) TOX; Attifuge CIB; Fungistat CIB; Insectifuge CIB; Larvicide IC60=5 mM 438/

CARYOPHYLLENE-OXIDE Antiedemic CPB38:2283; Antifeedant ISA14:93; Antiinflammatory CPB38:2283; Antitumor JNP55:999; Insecticide ISA14:93

CATECHIN Allelochemic IC86=1 mM 438/; Antialcoholic 2,000 mg/man/day PAM; Antiarthritic PAM; Anticariogenic; Antiedemic PAM; Antiendotoxic PAM; Antifeedant 382/; Antiflu PAM; Antihepatitic 1 g/3x/day/man PAM; Antihepatotoxic M11; Antiherpetic PAM; Antihistaminic 1,000 mg 5x/day/man PAM; Antihyperlipedemic JNP54:218; Antileukemic IC50=>10 ug/ml LS55:1061; Antiosteotic 500 mg/3x/day PAM; Antioxidant PM56(6):695; Antiperiodontal PAM; Antiplaque PAM; Antiradicular IC50=8 uM PM56(6):695; Antisclerodermic PAM; Antiulcer 1 g/5x/day/man/orl PAM; Antiviral V&D; Astringent M11; Bactericide MIC=>1,000 ug/ml PM57:A42; Cancer-Preventive; Carcinogenic; Fungicide ED50=2.9-4.6 ug/ml 438/; Hemostat; Hepatoprotective ACM:253; Hypocholesterolemic JNP54:218; Immunostimulant PAM; Lipoxygenase-Inhibitor IC96=5 mM JAF38:688; Viricide PAM PM57:A42

CATECHOL Allelochemic IC50=0.65 mM 438/; Allergenic M&R; Antigonadotropic; Antioxidant 505/; Antiseptic JBH; Antistomatitic EMP5:197; Antithiamin PCF:54; Antithyreotropic PM55; Antiviral EMP5:197; Cancer-Preventive; Carcinogenic 0.8% diet 505/; Carcinogenic 2% diet PCF:35;

Co-carcinogenic PCF:35; Convulsant JBH; Dermatitigenic JBH; Tumor-promoter <0.2% diet 505/

CEDRENE Cancer-Preventive 525/

CEDROL Perfumery JBH; Termiticide MOB24:301.1989

CHAVICOL Fungicide; Nematicide HG17

CHLORINE Antiherpetic M29; Antiseptic M29; Bactericide M29; Candidicide M29; Fungicide M29; Viricide M29

CHLOROGENIC-ACID Allergenic LEL 500; Analgesic PMP23:46; Anti-Tumor-Promoter IC25=10 uM CR48:5941; AntiEBV EMP6:189; AntiHIV EMP6:189; Antifeedant 382/; Antigonadotropic JNM1:10; Antihemolytic 10 uM PC36:579; Antihepatotoxic PM56(2):171; Antiherpetic EMP5:194; Antihypercholesterolemic EMP6:189; Antiinflammatory PCF:19; Antimutagenic PCF:19; Antinitrosaminic PCF:18; Antioxidant IC53=200 ppm PCF:21; Antioxidant IC80=12 uM PC27:969; Antiperoxidant IC50=36 uM PM57:A54; Antipolio V&E; Antiradicular 10 uM PC36:579; Antiseptic PMP23:46; Antithyroid JNM1:10; Antitumor JBH; Antiulcer EMP6:189; Antiviral V&D; Bactericide JE26:65; CNS-Active WIC; Cancer-Preventive 525/; Cholagogue WIC; Choleretic KCH; Clastogenic JAF38:805; Diuretic WIC; Fungicide NIG; Hepatoprotective KCH; Histamine-Inhibitor; Immunostimulant EMP1:124; Insectifuge EB48:111; Interferon-Inducer EMP1:124; Juvabional 382/; Larvistat JBH; Leukotriene-Inhibitor; Lipoxygenase-Inhibitor IC23=5 mM JAF38:688; Metal-Chelator PCF:25; Ornithine-Decarboxylase-Inhibitor PCF:19; Oviposition-Stimulant JBH

CHOLINE Antialzheimeran 5-16 g/man/day M28 M29; Antichoreic DAS; Anticirrhotic 6,000 mg/man/day; Anticystinuric; Antidiabetic; Antidyskinetic 150-200 mg/kg/man/day MAR; Antimanic 15-30 g/man/day/orl PAM; Hepatoprotective; Hypotensive; Ileorelaxant FT1990:483; Lipotropic JE26:65 LD50=200 (scu cat) LD50=400 (ipr rat) LD50=6,700 (orl rat)

CHROMIUM Amphiglycemic? DAS; Antiatherosclerotic 20 ug/day WER DAS; Anticorneotic DAS; Antidiabetic DAS; Antidote (Lead) DAS; Antiglycosuric DAS; Antitriglyceride 20 ug/day WER; Hypocholestrolemic 20 ug/day WER; Insulinogenic

CHRYSIN Antiaggregant; Antiallergic PAM; Antimutagenic EMP6:235; Antispasmodic BP40(10):2227; Antithrombic; Anxiolytic? PBB47:1; Cancer-Preventive 525/; Fungicide 2.5 nM PC38:599; Myorelaxant; Termitifuge 382/

CHRYSOERIOL Aldose-Reductase-Inhibitor IC31=10uM PC23:1885; Antimutagenic EMP6:235; Antirhinoviral EMP5:197; Antiviral EMP5:197; Cancer-Preventive 525/

CINNAMALDEHYDE Allelochemic IC100=2.5 mM 438/; Allergenic M&R; Antiaggregant 200 uM EMP5:343; Antiherpetic AEH; Antimutagenic HG24:40; Antiulcer 500 mg/kg/orl; Antiviral AEH; CNS-Depressant 411/; CNS-Stimulant 411/; Cancer-Preventive 525/; Choleretic 500 mg/kg/orl; Chronotropic AEH; Circulostimulant; Cytotoxic ED50 5-60 ug/ml AEH; Febrifuge; Fungicide 1.7-250 ppm; Herbicide JBH; Histaminic AEH; Hypoglycemic 411/; Hypotensive 411/; Hypothermic AEH; Inotropic AEH; Insecticide; Mutagenic AEH; Nematicide 100 ug/ml SZ44:183; Perfumery; Sedative; Spasmolytic 411/; Sprout-Inhibitor PER.COM.SFV; Teratogenic AEH; Tranquilizer JNP39:420 ADI=700 ug/kg M29 LD50=132 (ivn mus) AEH LD50=2,225 (orl mus) AEH LD50=610 (ipr mus) AEH

CINNAMIC-ACID Allergenic M&R; Anesthetic; Antiinflammatory PMP25:46; Antimutagenic EMP6:235; Bactericide MAR; Cancer-Preventive 525/; Choleretic; Dermatitigenic JBH; Flavor FEMA 4-40 ARC; Fungicide; Herbicide; Laxative; Lipoxygenase-Inhibitor RWG; Spasmolytic JBH; Vermifuge JE26:65

CIRSILINEOL Antimalarial PM55(7):654; Cancer-Preventive 525/; Spasmolytic JBH

CIRSILIOL Lipoxygenase-Inhibitor JBH

CIS-3-HEXENOL Flavor FEMA 1-5 ARC

CIS-ANETHOLE Toxic 15 x trans-anethole ARC

CITRAL Allergenic M&R; Antiallergic KCH; Antihistamine KCH; Antiseptic 5.2 x phenol W&W; Antishock KCH; Bactericide JE26:65; Bronchorelaxant KCH; Cancer-Preventive 525/; Expectorant KCH; Flavor FEMA 4-170 ARC; Fungicide ABS32:49; Herbicide IC50=115 uM TOX; Nematicide IC52=100 ug/ml NIG; Nematicide MLC=100-260 ul/ml SZ44:183; Perfumery M11; Sedative ED 1-32 mg/kg W&W; Teratogenic JNP39:420 ADI 0.5 gm/kg AEH ADI=500 ug/kg MAR LD50=4,960 (orl rat)

CITRONELLAL Allergenic M&R; Antiseptic 3.8 x phenol W&W; Bactericide JE26:65; Embryotoxic; Flavor FEMA <1 ARC; Insectifuge M11; Motor-Depressant BVC:162; Mutagenic? AEH; Nematicide MIC 1 mg/ml SZ44:183; Perfumery; Sedative ED=1 mg/kg W&W; Teratogenic JNP39:420

CITRONELLOL Allergenic M&R; Bactericide JE26:65; Candidicide; Flavor FEMA 1-50 ARC; Fungicide HG17; Herbicide IC50=160 uM TOX; Nematicide MLC=100 ug/ml SZ44:183; Perfumery M11; Sedative W&W ADI=500 ug/kg MAR LD50=4,000 (ims mus)

CITRONELLYL-ACETATE Flavor FEMA <1-700 ARC

CITRONELLYL-BUTYRATE Flavor FEMA <1-2 ARC

CITRONELLYL-FORMATE Flavor FEMA 4-100 ARC

COBALT Cardiomyopathogenic DAS; Erythrocytogenic DAS

COLEONE-A Antiseptic JBH

COLEONOL Hypotensive 0.5 mg/kg ivn cat W&W

COPPER Antiarthritic DAS; Antidiabetic 2-4 mg/day WER; Antiinflammatory WER; Antinociceptive WER; Contraceptive MAR; Hypocholesterolemic DAS; Schizophrenigenic DAS ADI=2-5 mg/man DAS ADI?=<500 ug/kg MAR PTD=100 mg/day RDA=30-80 ug/kg M29

COSMOSIIN Aldose-Reductase-Inhibitor IC40=1 uM PC23:1885; Aldose-Reductase-Inhibitor IC99=10 uM PC23:1885; Bactericide MIC=500 ug/ml PM57:A43; Candidicide MIC=500 ug/ml PM57:A43

COUMARIN Allelochemic IC100=2 mM 438/; Antiaggregant; Antiandrogenic PM56(6):671; Antibrucellosic JPP42:194; Antiedemic CPB38:2283; Antiinflammatory CPB38:2283; Antimelanomic PM56(6):671; Antimitotic IJP33:7; Antimutagenic EMP6:235; Antipsoriac IJP33:7; Antitumor PM56(6):671; Bruchiphobe; Cancer-Preventive 5-25 ug/ml PM1987:526; Cardiodepressant; DME-Inhibitor IC50=57.5 uM ACM:134; Emetic; Fungicide JBH; Hemorrhagic JBH; Hepatotoxic EMP6:165; Hypnotic EB44:68.1990; Hypoglycemic 250-1,000 mg/kg orl EMP6:165; Immunostimulant PM56(6):671; Juvabional 382/; Lymphocytogenic 100 mg/day JPP42:194; Ovicide; Phagocytogenic PM56(6):671; Piscicide JBH; Rodenticide JBH; Sedative EB44:68.1990 LD50=202 (orl gpg) M11 LD50=293 (orl rat) CRC LD50=680 (orl rat) M11

CRATEGOLIC-ACID Digestive OMM

CRYPTOTANSHINONE Antiacne; Antiaggregant IC50=5-100 ug/ml; Antianginal JNP54:1583; Antiburns; Anticarbuncular; Antiischemic JNP54:1583; Antiosteomyelitic; Antitonsilitic; Bactericide

CUMIN-ALCOHOL Flavor FEMA 0.5-35 ARC
LD50=1,020 (orl rat)

CUMINALDEHYDE Bactericide; Flavor FEMA <1 ARC; Fungicide FT62:86; Larvicide AYL; Perfumery JBH LD50=1,390 (orl rat) M11

CUPARENE Antiinflammatory

CYANIDIN Allelochemic IC90=1 mM 438/; Antioxidant JJS40:138; Pigment JBH

CYASTERONE Anabolic JBH; Antigonadotrophic JBH; Juvabional 450/; Proteinogenic

CYSTINE Adjuvant M11; Antihomocystinuric M29

D-LIMONENE Anti-Tumor-Promoter HH2:12; Anticancer (Breast) HH2:12; Cancer-Preventive HH2:12; Flavor FEMA 30-2,300 ARC; Hypocholesterolemic HH2:12; Insecticide; Insectifuge 382/; Litholytic MAR LDlo=4,600 (orl rat)

DAUCOSTEROL Antileukemic; Antitumor; Hypoglycemic EMP6:158; Spasmolytic 20 mg/kg MPI LD50=3,730 (orl rat)

DECANOIC-ACID Flavor FEMA <1-1 ARC; Nematicide SZ45:270
LD50=129 (ivn mus)

DECANOL Nematicide MLC 100-700 ug/ml SZ44:183

DELPHINIDIN Allelochemic IC81=1 mM 438/; Allergenic M&R; Antioxidant JJS40:138; Cancer-Preventive 525/; Pigment JBH

DELTA-3-CARENE Allergenic? M&R; Antiinflammatory ip IJO15(4):194; Bactericide RIE12:5; Dermatitigenic M&R; Flavor FEMA ARC; Insectifuge 382/

DELTA-CADINENE Anticariogenic MIC800 ug/ml JAF41:1103; Bactericide MIC800 ug/ml JAF41:1103

DELTA-CADINOL Termiticide MUO24:1

DELTA-DODECALACTONE Flavor FEMA <1-5 ARC

DIACETYL Flavor FEMA 3-50 ARC
LD50=1,580 (orl rat)

DIETHYL-MALONATE Flavor FEMA 2-20 ARC

DIHYDROCARVEOL Flavor FEMA 5-500 ARC

DIHYDROCARVYL-ACETATE Flavor FEMA 5-500 ARC

DIHYDROCOUMARIN Cancer-Preventive 525/; Flavor FEMA 5-80 ARC

DIHYDROCUMINYL-ALCOHOL Flavor FEMA 0.5-50 ARC

DILLAPIOL Bactericide MIC=45.9->50 nM PM59:546; Insecticide JBH; Insecticide-Synergist JBH; Molluscicide PM55(7):620; Sedative ED50=1.57 CPB36:3153

DIMETHYL-SULFIDE Flavor FEMA 0.1-2 ARC

DIOSMETIN Aldose-Reductase-Inhibitor IC17=10 uM PC23:1885; Antimutagenic IC50=10-40 nM PCF; Antirhinoviral EMP5:197; Antiviral EMP5:197; Cancer-Preventive 525/

DIOSMIN Anti-Capillary-Fragility 300mg/2x/day/woman/ MAR; Antihemorrhoidal; Antiinflammatory JBH; Antimetrorrhagic FT55.1984

DIOSPHENOL Antiseptic WIC; Bactericide BIS; Diuretic JBH; Peticide LD50=170 (orl rat) M11 LD50=212 (orl rat) M11

DIPENTENE Allergenic M&R; Bactericide; Expectorant JBH; Irritant JBH; Sedative JBH; Viricide

DODECANAL Flavor FEMA 1-100 ARC

DOLICHODIAL Insectifuge JBH; Lachrymatory JBH

ELEMICIN Antiaggregant IC50=360 uM JE29:179; Antidepressant ihl JA6(4):36; Antifeedant 382/; Antihistaminic; Antiserotonic IJO15(4):199; Antistress ihl JA6(4):36; DNA-Binder JBH; Hallucinogenic RJH; Hypotensive ihl JA6(4):36; Insecticide 382/; Insectifuge 382/; Neurotoxic RJH; *Schistosomicide GMJ*

ELLAGIC-ACID ACE-Inhibitor JNP51:357; Abortifacient 1.2 mg/kg ivn mus AFR27:149; Aldose-Reductase-Inhibitor IC50=0.2 uM CPB38:2733; AntiHIV IC90=200 ug/ml JNP54:152; Antianaphylactic AFR27:149; Anticataract CPB37:2531; Antiinflammatory AFR27:149; Antimutagenic PCF:19; Antioxidant >50 x tocopherol PCF:142; Antioxidant IC98=30 ppm PCF:17; Antiperoxidant IC50=29 uM PM57:A54; Antithyroid JNM1:10; Antitumor; Antiviral EMP5:224; Astringent M29; Cancer-Preventive 3ppm 505/; Cancer-Preventive ACM:359; Cytotoxic 4.6 ug/ml ACM:359; Deiodinase-Inhibitor JNM1:10; HIV-RT-Inhibitor EMP5:224; Hemostatic M11; Hepatoprotective IC55=30 ug/ml CPB38:2201; Juvabional 382/; Sunscreen 0.5% NAP; Xanthine-Oxidase-Inhibitor IC50=3.1 uM CPB38:1225

ENMEIN Antimutagenic EMP6:235; Antitumor; Bactericide OMM

ERIODICTYOL Antifeedant IC52=<1000 ppm diet 438/; Antioxidant AEH; Antiperoxidant PM57:A110; Bactericide JBH; Cancer-Preventive 525/; Expectorant LRN-MAR91; Juvabional JBH

ESTRAGOL Anesthetic; Anticonvulsant; Myorelaxant FT61:254.1990

ESTRAGOLE Antiaggregant IC50=320 uM PR4(3):93; Cancer-Preventive 525/; Carcinogenic 4.4 uM/15 mos. scu mus AEH; DNA-Binder JBH; Flavor FEMA 4-150 ARC; Hepatocarcinogenic 411/; Hypothermic JBH; Insecticide 382/; Insectifuge 382/; Perfumery M11 LD50=1,250 (orl mus) M11 LD50=1,820 (orl rat) M11

ETHANOL Allergenic M&R; Anesthetic 4,000 orl cat BBE; Anesthetic 6,000 orl rat BBE; Anhidrotic M29; Antipruritic MAR; Antiseptic M11; Bactericide MAR; CNS-Depressant MAR; Hepatotoxic M29; Hypnotic Antidote=Naloxone M29; Hypocalcemic 4,000 orl rat BBE; Neurolytic M11; Rubefacient M29; Sclerosant M29; Teratogenic 6,000 orl dog BBE; Tremorilytic MAR; Ulcerogenic M29 LD50=6,000 orl dog BBE LD50=6,300 (orl rbt) LD50=7,060-10,600 (orl rat) M11 LDlo=2,000 (orl chd)

ETHYL-ACETATE Antispasmodic; CNS-Depressant M29; Carminative; Flavor FEMA 20-1,500 ARC; Respirotoxic >400 ppm ARC; Spasmolytic; Stimulant

ETHYL-BUTYRATE Flavor FEMA 30-1,4000 ARC

ETHYL-CINNAMATE Flavor FEMA 4-40 ARC; Nematicide MLC=1,000 uM SZ44:183

ETHYL-ISOBUTYRATE Flavor FEMA 20-200 ARC

ETHYL-LAURATE Flavor FEMA <1-40 ARC

ETHYL-MYRISTATE Flavor FEMA 6-30 ARC

ETHYL-OLEATE Flavor FEMA 0.1-40 ARC

ETHYL-PALMITATE Flavor FEMA 20-50 ARC

EUGENOL Allergenic M&R; Analgesic M11; Anesthetic 200-400 PR4:93; Antiaggregant IC50=0.3uM PR4:93; Anticonvulsant JBH; Antiedemic 100 PR4:95; Antifeedant 382/; Antiinflammatory 11 uM; Antimitotic JBH; Antimutagenic EMP6:235; Antinitrosating PCF:16; Antioxidant IC65=30 ppm PCF:17; Antiprostaglandin 11 uM PM1986:184; Antiradicular EC50=2 ul/l PMP22:231; Antiseptic 3 ml/man/day; Antithromboxane LRN-JUL87; Antitumor JNP55:999; Antiulcer; Apifuge 382/; Bactericide 500 ppm VAL; CNS-Depressant JBH; Cancer-Preventive 525/; Candidicide PR4:93; Carcinogen? NIG; Carminative JPP46:16; Choleretic; Cytotoxic 25ug/mL PR4:93; Dermatitigenic M&R; Enterorelaxant JAR4:22; Febrifuge 3 ml/man/day; Flavor FEMA 10-500 ARC; Fungicide PMP23:46; Herbicide PMP23:46; Insectifuge 382/; Irritant MAR; Juvabional 382/; Larvicide JE26:65; Motor-Depressant BVC:162; Nematicide MLC=2,000 ug/ml SZ44:183; Neurotoxic RJH; Perfumery ARC; Sedative JE26:72 BVC:162; Trichomonistat IC50=10 ug/ml NIG; Trypsin-Enhancer LRN-DEC93; Ulcerogenic 750 PR4:93; Vermifuge ADI=2.5 mg/kg MAR LD50=2,680 (orl rat) CRC(FNS) LD50=3,000 (orl mus) LDlo=500 (orl rat)

EUGENOL-METHYL-ETHER CNS-Depressant JBH; Cancer-Preventive 525/; DNA-Binding JBH; Hypothermic JBH; Myorelaxant JBH LD50=1,179 (orl rat)

EUGENYL-ACETATE Antiaggregant ED=15 uM PM56(6):501; Antiinflammatory; Antiprostaglandin ED50=3 uM PM1986:184; Antispasmodic CRC; Perfumery FEMA 0.5-200 ppm ARC; Spasmolytic CRC

EUPATILIN Lipoxygenase-Inhibitor JBH

EUPATORIN Anticarcinomic ED50=4.6 BPCC; Antimalarial PM55(7):654; Cytotoxic 411/; Emetic AEH

FARNESENE Pheromone JBH

FARNESOL Allergenic M&R; Flavor FEMA 0.1-3 ARC; Juvabional 450/; Nematicide MLC=100 ug/ml SZ44:183; Perfumery M1

FENCHONE ACE-Inhibitor LGR; Antialzheimeran? LGR; Anticholinesterase? LGR; Counterirritant JBH; Flavor FEMA 0.1-5 ARC; Perfumery M11; Secretolytic BIS LD50=4,400 (orl rat) LD50=6,160 (orl rat) M11

FENCHYL-ALCOHOL Flavor FEMA 0.2-5 ARC

FERULIC-ACID Allelopathic JCE17:865; Analgesic ACM:69; Anti-Tumor-Promoter IC46=10 uM CR48:5941; Antiaggregant CPB38:1620; Antidysmenorrheic CPB38:1620; Antiestrogenic JBH; Antihepatotoxic JBH; Antiinflammatory PCF:19; Antimitotic JBH; Antimutagenic PCF:18; Antinitrosaminic PCF:18; Antioxidant 3,000 uM PC27:969; Antioxidant IC51=200 ppm PCF:17; Antiserotonin JBH; Antispasmodic; Antitumor JBH; Antiviral V&D; Arteriodilator CPB38:1620; Bactericide JBH; Cancer-Preventive 525/; Candidicide JBH; Cardiac CPB38:1620; Cholagogue PHZ46:156; Choleretic 411/; Fungicide NIG; Hepatotropic; Herbicide AB68:195; Hydrocholeretic JBH; Immunostimulant EMP1:124; Insectifuge EB48:111; Metal-Chelator PCF:25; Ornithine-Decarboxylase-Inhibitor PCF:19; Phagocytotic JBH EMP1:124; Preservative M11; Prostaglandigenic RWG; Spasmolytic; Sunscreen CMR9/10/92; Uterosedative 30-100 mg/kg ivn rat CPB38:1620 LD50=416 (ivn mus) CPB38:1620 LDlo=1,200 (par mus)

FIBER Antidiabetic; Antiobesity; Antitumor PAM; Antiulcer PAM; Cancer-Preventive 525/; Cardioprotective; Hypocholesterolemic; Hypotensive 10g/man/day/orl PAM; Laxative RDA=20-35g/day HSC25:1488

FORMALDEHYDE Allergenic M&R; Anesthetic M29; Anticystitic M29; Antiplantar M29; Antiseptic M29; Antiviral M29; Fungicide M29; Irritant M29; Neoplastic; Scolicide M29; Viricide M29 LD50=800 (orl rat) LDlo=36 (orl wmn)

FORMIC-ACID Antiseptic; Antisyncopic; Astringent M11; Corrosive ARC; Counterirritant M11; Flavor FEMA 500-2,500 ARC; Fungitoxic JBH; Irritant M&R; Preservative ARC; Toxic JBH ADI=3 mg/kg MAR LD50=1,100 (orl mus) M11 LD50=1,210 (orl rat)

FORSKOLIN Adenylate-Cyclase-Activator M11; Amebicide 1000 ug/ml QJC28(1):1; Antiaggregant IC50=6 ug/ml EMP5:334 411; Antiglaucomic M29; Antithrombotic PMP25:46; Cardiotonic 411/; Hypotensive 10 mg/kg orl rat W&B 411; Inotropic 411/; Spasmolytic; Vasodilator 411/

FRUCTOSE Antialcoholic JBH; Antidiabetic JBH; Antihangover; Antiketotic M11; Antinauseant JBH; Neoplastic CRC; Sweetener 2 x glucose JBH

FUMARIC-ACID Acidulant MAR; Antidermatitic WIC; Antihepatocarcinogenic IC100=1% (diet) CPB38:2012; Antioxidant M11; Antipsoriac WIC; Antitumor 411/; Flavor FEMA 50-3,600 ARC ADI=6 mg/kg LD50=200 (ipr mus)

FURFURAL Antiseptic; Flavor FEMA 1-30 ARC; Fungicide; Insecticide LD50=127 (orl rat)

GALACTOSE Sweetener 0.32 x sucrose MAR

GALLIC-ACID ACE-Inhibitor JNP51:357; AntiHIV PCF:14; Antiadenovirus EMP5:194; Antiallergenic IAA94:262; Antianaphylactic JBH; Antiasthmatic W&B; Antibacterial IAA94:262; Antibronchitic IAA94:262; Anticarcinomic ED50=3; Antifibrinolytic EMP1:53; Antiflu EMP5:194; Antihepatotoxic PCF:93; Antiherpetic EC50=>10 ug/ml PCF:14; Antiiflammatory PM58(6):499; Antimutagenic EMP6:235; Antinitrosaminic AFR27:149; Antioxidant IC44=33 ppm PCF:218 MAR; Antiperoxidant IC50=69 uM PM57:A54 PCF:93; Antipolio EMP5:194; Antiseptic 411/; Antitumor JBH; Antiviral AVR14:323; Astringent 411/; Bacteristatic PR4(5):198; Bronchodilator JBH; Cancer-Preventive 525/; Carcinogenic; Choleretic JBH; Cyclooxygenase-Inhibitor RWG; Floral-Inhibitor JBH; Hemostat JE26:65; Immunosuppressant AFR27:149 RWG; Insulin-Sparer JBH; Myorelaxant JBH; Nephrotoxic M29(P.779); Styptic; Viricide JBH; Xanthine-Oxidase-Inhibitor IC50=24 uM CPB38:1225 LD50=4,500 scu rat AFR27:149 LD50=5,000 (orl rbt) M11

GAMMA-DECALACTONE Cancer-Preventive 525/

GAMMA-TERPINENE Antioxidant JA6(4):33; Flavor FEMA 1-40 ARC; Insectifuge; Perfumery ARC

GARDENIN-D Antiperoxidant IC50=140 uM PM57:A110

GENKWANIN Bactericide 452/; Purgative

GENTISIC-ACID Analgesic M11; Antifeedant; Antiinflammatory M11; Antirheumatic GMJ; Bactericide JBH; Viricide JBH LD50=800 (orl rat)

GERANIAL Bactericide JE12:279.1984

GERANIOL Allergenic M&R; Anticariogenic MIC=400 ug/ml JAF41:1103; Antiseptic 7 x phenol W&W; Antitumor JAF43:2144; Bactericide MIC=400 ug/ml JAF41:1103; Cancer-Preventive 525/; Candidicide; Embryotoxic; Emetic 3 x ipecac LRN-SEP97; Flavor FEMA 1-10 ARC; Fungicide IC93=2 mM 438/; Herbicide IC100-2,000 uM TOX; Insectifuge 50 ppm JSPR22:141; Insectiphile M11; Nematicide IC86=100 ug/ml NIG; Nematicide MLC=1,000 ug/ml SZ44:183; Perfumery M11; Sedative W&W; Spasmolytic JEO2:185 LD50=3,600 (orl rat)

GERANYL-BUTYRATE Flavor FEMA 1-10 ARC; Perfumery ARC

GERANYL-CAPROATE Flavor FEMA 1-3 ARC; Perfumery ARC

GERANYL-ISOVALERATE Flavor FEMA 4-10 ARC; Perfumery ARC

GERANYL-PROPIONATE Flavor FEMA 1-70 ARC; Perfumery ARC

GERMACRENE-D Pheromonal

GLUCOSE Acetylcholinergic; Antiedemic JBH; Antihepatotoxic M11; Antiketotic M11; Antivaricose JBH; Hyperglycemic M11; Memory-Enhancer SN138:189.1990

GLUTAMIC-ACID Antalkali? 500-1,000 mg/day/orl/man M29; Antiepileptic; Antihyperammonemic M29; Antilithic PAM; Antiprostatitic 125-250 mg 3 x day WER; Antiretardation JBH; Anxiolytic M11; Neurotoxic JBH ADI=120 mg/kg MAR ADI=5-12 g/man/day MAR

GLYCINE Antiacid M29; Antialdosteronic PAM; Antidote (Hypoglycin-A) M29; Antiencephalopathic M29; Antigastritic M29; Antiprostatitic 125-250 mg 3 x day WER; Antipruritic M11; Antisickling PM56(1):41; Antiulcer M29; Cancer-Preventive 525/; Neuroinhibitor WER; Uricosuric

GLYCOLIC-ACID Cholesterolytic; Diuretic; Hepatotonic FT5:1986; Irritant M11 LD50=1,950 (orl rat) M11

GUAIACOL Anesthetic; Antidermatitic JBH; Antieczemic JBH; Antiesophagitic M29; Antiseptic M29; Antitubercular; Bactericide 8,000 ppm VAL; Expectorant 0.3-0.6 ml/man MAR; Flavor FEMA <1-1 ARC; Insectifuge 382/ LD50=725 (orl rat) M11

GUAIAZULENE Antiallergic M29; Antiinflammatory W&W; Antileprotic M29; Antipeptic; Antipyretic M29; Antiulcer CPB38:3355

HARPAGIDE Analgesic; Antiarthritic; Antiinflammatory PM55(7):626; Calcium blocker IWU LD50=3,200 (ipr)

HARPAGOSIDE . Analgesic W&W; Antiarrhythmic IWU; Antiinflammatory BIS; Calcium blocker IWU LD50=1,000 (ipr)

HERNIARIN Allergenic BIS; Antiallergic LRN-MAR91; Bactericide JBH; Cholagogue FT5:1986; Fungicide JBH; Hepatoprotective YHJ37:182

HESPERETIN Antifeedant JBH; Antilipolytic PM59:508; Bactericide MIC=500 ug/ml PM57:A43 JBH; Cancer-Preventive 525/; Fungistatic PC29(4):1103; Hepatoprotective ACM:253; Viricide JBH

HESPERIDIN Anti-DNA; Anti-RNA; Antiallergenic YAK111:193; Antibradykinic EMP5:366; Antiflu EMP5:197; Antiinflammatory IC47=50 mg/kg scu JPP46:118; Antiinflammatory IC63=100 mg/kg scu JPP46:118; Antioxidant IC50=8 ppm PC37:1463; Antistomatitic; Antivaccinia EMP5:197; Antiviral V&D; Capillariprotective M29; Catabolic AFR27:149; Choleretic; Vasopressor LD50=1,000 (ipr mus)

HEXANAL Flavor FEMA <1-5 ARC LD50=4,890 (orl rat)

HEXANOIC-ACID Acidulant JBH; Flavor FEMA 2-450 ARC

HEXANOL Antiseptic; Flavor FEMA 0.2-25 ARC
LD50=720 (orl rat)

HEXENAL Bactericide OMM; Fungicide ABS

HEXYL-ACETATE Flavor FEMA 3-25 ARC

HEXYL-BUTYRATE Flavor FEMA 2-10 ARC

HISPIDULIN Antiaggregant IC50=13 ug/ml EMP5:333; Antihepatotoxic 32 mg/kg PM53:37; Antitumor LEL; Cancer-Preventive 525/

HISPIDULOSIDE Aldose-Reductase-Inhibitor ED50=1-10 uM; Anticataract PM1986:239

HISTIDINE Antiarteriosclerotic JE26:65; Antinephritic JBH; Antioxidant PC27:969; Antiulcer JE26:65; Antiuremic JBH; Essential JBH; Oxidant PC27:969 ADI=15 g/day/orl MAR

HYDROQUINONE Allelochemic IC100=2.7 mM 438/; Allergenic M11; Antilithic; Antimalarial; Antimelanomic AFR27:149; Antimelasmic M29; Antimenorrhagic; Antimitotic JBH; Antinephritic; Antioxidant 505/; Antipertussive; Antiseptic; Antithyreotropic; Antitumor JBH; Astringent; Bactericide JBH; Carcinogenic PCF:35; Co-carcinogenic PCF:35; Convulsant 1 g orl LRN-SEP87; Cytotoxic JBH; Depigmentor M11; Emetic 1 g orl LRN-SEP87; Flavor FEMA 5-50 ARC; Herbicide IC41=50 ug/ml 455/; Hypertensive JBH; Irritant MAR; Trypanosomicide LD50=320 (orl rat) M11 LDlo 5 g orl hmn LRN-SEP87

HYMENOXIN Cancer-Preventive 525/; Mutagenic AEH

HYPEROSIDE Anti-Capillary-Fragility TOX; Antidermatitic; Antiflu AYL; Antihepatotoxic TOX; Antiinflammatory 1/4 indomethacin PM57:A131; Antioxidant PC33:557; Antiviral AYL; Bactericide MIC=250-500 ug/ml PM57:A43; Cancer-Preventive 525/; Capillarifortificant JBH; Capillarigenic JBH; Diuretic; Hypotensive ivn dog HG22:18; Viricide AYL; cAMP-Inhibitor IC50=0.14 mg/ml PM57:A133; cAMP-Phosphodiesterase-Inhibitor PAM

HYPOLAETIN Cyclooxygenase-Inhibitor IC50 (uM) =70 DRS LEH; Lipoxygenase-Inhibitor IC50 (uM)=4.5 DRS LEH

HYPOLAETIN-8-GLUCOSIDE Antiaggregant; Antiinflammatory PT4(3):118; Antiulcer ED50=24-57 PM1985:70

INDOLE-3-ACETIC-ACID Allelochemic JAF42:2048; Cancer-Preventive 525/; Herbicide JAF42:2048; Hypoglycemic EMP6:157; Insulinase-Inhibitor EMP6:157; Insulinotonic EMP6:157

INOSITOL Antialopecic; Anticirrhotic; Antidiabetic? M29; Antineuropathic? 2 g day JBH; Cholesterolytic 2000 mg/man/day; Lipotropic JBH; Sweetener JBH RDA=1000mg/man/day

IODINE Acnegenic DAS; Antigoiter DAS; Antithyrotoxic DAS; Dermatitigenic DAS; Goitrogenic DAS; Thyrotropic DAS PTD=2mg/day RDA=40-200ug/day

IRIDOIDS Stomachic BIS

IRON Antiakathisic M29; Antianemic M29; Anticheilitic DAS; Antimenorrhagic 100 mg/day/wmn/orl PAM PTD=100mg/day RDA=6-30mg/day

ISOBORNEOL Insectifuge; Motor-Stimulant BVC:162; Nematicide MLC=1 mg/ml SZ44:183

ISOBORNYL-ACETATE Insectifuge

ISOBUTYRALDEHYDE Perfumery; LD50=2,810 (orl rat) LD50=3,700 (orl rat) M11

ISOBUTYRIC-ACID Irritant M1; LD50=28 (orl rat)

ISOCARYOPHYLLENE Perfumery JBH

ISOCHLOROGENIC-ACID Antioxidant PCF:54; Antiseptic JNP46(5); Cancer-Preventive 525/

ISOEUGENOL Allergenic M&R; Antiaggregant IC50=0.72uM PR4(3):93; Antioxidant IC79=30 ppm PCF:17; Bacteristat JAF40:2328; Cancer-Preventive 525/; Candidistat JAF40:2328; Cytotoxic 25ug/mL PR4(3):93; Fungistatic 100 ug/mL PR4(3):93; Motor-Depressant BVC:162; Motor-Stimulant BVC:162; Sedative BVC:162; Trichomonistat IC50=10 ug/ml NIG LD50=1,560 (orl rat) M11

ISOLEUCINE Antiencephalopathic? M29; Antipellagric MAR; Essential JBH ADI=5 g/man/day MAR

ISOPULEGONE Flavor FEMA 5-15 ARC; Hepatotoxic AEH; Perfumery ARC; Pulmonotoxic AEH

ISOQUERCITRIN Aldose-Reductase-Inhibitor JBH; Antifeedant 450/; Antioxidant PC33:557; Bactericide JBH; Cancer-Preventive 525/; Capillarigenic; Diuretic 10 ppm HHB; Hypotensive CPB38:1049; Insectiphile JBH

ISOROSMANOL Antioxidant

ISOSAFROLE Carcinogenic CRC; Hepatotrophic JBH; Perfumery JBH LD50=1,340 (orl rat)

ISOVALERALDEHYDE Flavor FEMA 0.5-5 ARC
LD50=3,200 (orl rat) M11

ISOVALERIC-ACID Flavor FEMA 1-16 ARC; Perfumery M11; Sedative? M11; Tranquilizer KCH LD50=1,120 (ivn mus) M11 LDlo=3,200 (orl rat)

ISOVITEXIN Antioxidant =tocopherol PCF:10; Antioxidant >BHA NIG; Cancer-Preventive 525/

JASMONE Anticariogenic MIC=800-1,600 ug/ml JAF41:1103; Bactericide MIC=800-1,600 ug/ml JAF41:1103; Cancer-Preventive 525/; Insectiphile JBH; Nematicide MLC=1 mg/ml SZ44:183; Perfumery JBH

KAEMPFEROL 5-Lipoxygenase-Inhibitor IC50 (uM)=20 DRS LEH; Anti-Tumor-Promoter CPB38:774; Antiallergic RWG; Antifertility 250mg/kg day/60days/orl rat PMP23:193; Antihistaminic 411/; Antiimplantation JE32:167; Antiinflammatory 20 mg/kg; Antileukemic IC50=3.1 ug/ml LS55:1061; Antilymphocytic JBH; Antimutagenic ID50=10-40 nM PCF; Antioxidant IC50-40uM PC27:969; Antiradicular JBH; Antispasmodic 411/; Antiulcer FT61:243.1990; Antiviral JE32:141; Bactericide JBH; Cancer-Preventive 525/; Choleretic; Cyclooxygenase-Inhibitor IC50 (uM) =20 DRS LEH; Diaphoretic? LRN-DEC90; Diuretic; HIV-RT-Inhibitor IC50=50-150 ug/ml JNP54:142; Hypotensive FT1990:483; Iodothyronine-deiodinase-Inhibitor JBH; Lipoxygenase-Inhibitor JBH; Mutagenic JBH; Natriuretic; Spasmolytic 411/; Teratologic; Viricide JNP55:1732; cAMP-Phosphodiesterase-Inhibitor AFR27:149 LD50=5,000 (ivn dog) 15/

L-MENTHOL Flavor FEMA 35-1,200 ARC; Perfumery 5,000-25,000 ARC

L-PULEGONE Antiinflammatory

LABIATIC-ACID Antioxidant

LAGOCHILIN Sedative 30 mg/man CRC

LAGOCHILIN-TETRAACETATE Sedative CRC
LD50=131 (ivn mus) M11

LEONURINE Curaroid KCH; Diuretic KCH; Myostimulant KCH; Uterotonic

LEUCINE Antiencephalopathic M29; Essential JBH

LIGNIN AntiHIV; Anticancer AFR27:149; Anticoronary AFR27:149; Antidiarrheic M29; Antinitrosaminic AFR27:149; Antioxidant AFR27:149; Antiviral CPB38:3031; Bactericide; Chelator AFR27:149; Hypocholesterolemic AFR27:149; Laxative AFR27:149; Viricide CPB38:3031

LIMONENE ACE-Inhibitor 453/; Allergenic (1/20th carene M&R; Antialzheimeran? 453/; Anticancer; Antiflu EMP5:195; Antilithic M29(P.1064); Antimutagenic EMP6:235; Antitumor JAF43:2144; Antiviral EMP5:195; Bactericide RIE12:5; Cancer-Preventive 525/; Candidistat JAF40:2328; Enterocontractant JAR4:22; Expectorant JBH; Fungiphilic JAF43:2283; Fungistat JAF40:2328; Herbicide IC50=50mM TOX; Insecticide 453/; Insectifuge 382/; Irritant M11; Nematicide IC=100 ug/ml NIG; Sedative ED=1-32 mg/kg W&W; Spasmolytic ED50=0.197 mg/ml FT59:465; Viricide LD50=4,600 (orl rat)

LIMONENE-OXIDE ACE-Inhibitor 453/; Antialzheimeran? 453/; Insecticide 453/

LINALOL Allergenic M&R; Antiallergic KCH; Anticariogenic MIC=1,600 ug/ml JAF41:1103; Antihistamine KCH; Antiseptic 5 x phenol W&W; Antishock KCH; Antiviral JE26:65; Bactericide MIC=1600 ug/ml JAF41:1103; Bactericide JE26:65; Bronchorelaxant KCH; Cancer-Preventive 525/; Candistat JAF40:2328; Expectorant KCH; Flavor FEMA 2-40 ARC; Fungicide ABS32:49; Insectifuge 382/; Motor-Depressant BVC:162; Nematicide MLC=1 mg/ml SZ44:183; Perfumery M11; Sedative ED=1-32 mg/kg W&W; Spasmolytic KEP2:185; Termitifuge 382/; Tumor-Promoter; Viricide JE26:65 LD50=2,790 (orl rat)

LINALYL-ACETATE Flavor FEMA 2-15 ARC; Motor-Depressant BVC:162; Perfumery M11; Sedative BVC:162; Spasmolytic ED50=0.21 mg/ml FT59:465 ADI=200 ug/kg MAR ADI=500 ug/kg MAR

LINALYL-CAPROATE Flavor FEMA 3-15 ARC; Perfumery ARC

LINALYL-FORMATE Flavor FEMA 1-15 ARC; Perfumery ARC

LINARIN Aldose-Reductase-Inhibitor IC20=10uM PC23:1885

LINOLEIC-ACID Anti-MS EMP6:189; Antianaphylactic EMP6:189; Antiarteriosclerotic JE26:65; Antiarthritic EMP6:189; Anticoronary JBH; Antieczemic EMP6:189; Antifibrinolytic EMP1:53; Antigranular; Antihistaminic EMP6:189; Antiinflammatory EMP6:189; Antimenorrhagic PAM; Antiprostatitic PAM; Cancer-Preventive 525/; Carcinogenic SN146:421; Hepatoprotective; Immunomodulatory EMP6:189; Insectifuge 382/; Metastatic SN146:421; Nematicide NIG

LITHIUM AntiPMS DAS; Antidepressant DAS; Antihyperthyroid DAS; Antimanic DAS; Antipsychotic DAS; Antischizophrenic DAS; Deliriant DAS; Nephrotoxic DAS

LITHOSPERMIC-ACID Antigonadotrophic JNM1:10; Antithyroid JNM1:10; Cardiotonic

LUTEOLIN Aldose-Reductase-Inhibitor EC=3ppm FT64:131; Anticataract EC=3ppm TRA64:131; Antidermatic BIS; Antifeedant IC52=<1000 ppm diet 438/; Antiherpetic EMP5:197; Antihistaminic 411/; Antiinflammatory (=indomethacin) WIC; Antimutagenic ID50=2-5 nM PCF; Antioxidant 411/; Antipolio EMP5:197; Antispasmodic 411/; Antitussive; Antiviral EMP5:197; Bactericide MIC=500 ug/ml PM57:A43; Cancer-Preventive 525/; Choleretic; Deiodinase-Inhibitor JNM1:10; Diuretic PT5:1986; Iodothyronine-Deiodinase-Inhibitor JBH; Spasmolytic 411/; Succinoxidase-Inhibitor JBH; Xanthine-Oxidase-Inhibitor IC50=0.11 ug/ml CPB38:1772

LUTEOLIN-3'-O-(3"-O-ACETYL)-BETA-D-GLUCURONIDE Antioxidant IC50=ca 25 ppm PC37:1463

LUTEOLIN-3'-O-(4"-O-ACETYL)-BETA-D-GLUCURONIDE Antioxidant IC50=14 ppm PC37:1463

LUTEOLIN-7-GLUCOSIDE Aldose-Reductase-Inhibitor IC73=10uM; IC31=1uM PC23:1885; Antidermatic BIS; Antiherpetic EMP5:204; Antiinflammatory BIS; Antiviral EMP5:204

LUTEOLIN-7-GLUCURONIDE Aldose-Reductase-Inhibitor IC88=10uM; IC68=1uM PC23:1885; Antioxidant JAF43:2162
LUTEOLINIDIN Pigment JBH

LYSINE Antialkalotic M29; Antiherpetic 0.5-3 g/day PAM WER; Essential JBH; Hypoarginanemic 250 mg/kg/ M29 LD50=181 (ivn mus)

MAGNESIUM AntiPMS 400-800 mg/day/wmn/orl PAM WER; Antiaggregant 400 mg/day PAM WER; Antianginal 400 mg/day WER; Antianorectic DAS; Antianxiety 400 mg/day WER; Antiarrhythmic 400 mg/day DAS WER; Antiarthritic DAS; Antiasthmatic PAM; Antiatherosclerotic 400 mg/day PAM WER; Anticonvulsant FAC; Antidepressant DAS; Antidiabetic 400 mg/day EMP6:162 WER; Antidysmenorrheic 100 mg 4 x day WER; Antiepileptic 450 mg/day WER DAS; Antihyperkinetic DAS; Antihypoglycemic DAS; Antiinflammatory 100 mg 4 x day WER; Antiinsomniac DAS; Antilithic PAM; Antimastalgic PAM; Antimigraine 200 mg/day WER; Antineurotic DAS; Antiosteoporotic 500-1,000 mg/day/wmn/orl PAM WER; Antispasmophilic 500 mg/day WER; CNS-depressant FAC; Ca-Antagonist RWG; Hypocholesterolemic 400 mg/day WER; Hypotensive 260-500 mg/day SN138:189 WER; Myorelaxant 100 mg 4 x day WER; Uterorelaxant 100 mg 4 x day WER; Vasodilator WER ADI=400-800 mg DAS PTD=6,000mg/day RDA=40-400mg/day

MALIC-ACID Bacteristat; Bruchiphobe 382/; Hemopoietic; Laxative? LRN-SEP91; Sialogogue M29 LDlo=1,600 (orl rat)

MANGANESE Antialcoholic FHN; Antianemic MAR; Antidiabetic 3-5 mg/day WER; Antidiabetic EMP6:162; Antidiscotic DAS; Antidyskinetic MAR; Antiepileptic 450 mg/day WER; Antiototic DAS RDA=4mg/day DAS

MANOOL Termitifuge 382/

MARRUBIIN Antiarrhythmic 411/; Choleretic WOI; Expectorant 411/; Secretogogue LRN-JUN88 LD50=370 mg/kg LRN-JUN88

MASLINIC-ACID Antihistaminic; Antiinflammatory CPB34:2164.1986

MENTHOFURAN Hepatotoxic LRN-JAN92; Nematicide MLC=1 mg/ml SZ44:183; Pulmonotoxic AEH

MENTHOL Allergenic M29; Analgesic MAR; Anesthetic 2,000 ppm JPP46:16; Antiaggregant IC50=750 TRA42:511; Antiallergic JFH33(6):569; Antibronchitic MAR; Antidandruff ALH; Antihalitosic AH12(4):17; Antihistaminic JFH33(6):569; Antiinflammatory MAR; Antineuralgic MAR; Antiodontalgic MAR; Antipruritic JPP46:16; Antirheumatic MAR; Antiseptic 4 x phenol W&W; Antisinusitic MAR; Bactericide JPP46:16; Bradycardic 65 mg/3/x day/woman/ MAR; Bronchomucolytic ACM:200; Bronchomucotropic JPP46:16; Bronchorrheic ACM:200; CNS-Stimulant KCH; CNS-depressant; Calcium-Antagonist JAR4:22; Carminative JPP46:16; Ciliotoxic JPP46:16; Congestant JPP46:16; Convulsant RJH; Counterirritant; Decongestant? 11mg/man JPP42:652; Dermatitigenic M&R; Diaphoretic KCH; Enterorelaxant

JAR4:22; Expectorant JPP46:16; Gastrosedative M11; Irritant JPP46:16; Myorelaxant M29; Nematicide MLC=1 mg/ml SZ44:183; Neurodepressant JPP46:16; Neuropathogenic 40-100 mg/day/rat JPP46:16; Nociceptive JPP46:16; Perfumery M11; Refrigerant JPP46:16; Rubefacient; Spasmolytic ED50=0.01 mg/ml FT59:465; Vibriocide LD50=700-3,180 orl rat M11 JPP46:16 LDlo=2,000 mg/man/? MAR

MENTHONE Analgesic; Antiseptic 2.25 x phenol W&W; Cancer-Preventive 525/; Sedative JE25:165; Spasmolytic ED50=0.044 mg/ml FT59:465

MENTHYL-ACETATE Perfumery M11; Spasmolytic ED50=0.105 mg/ml FT59:465

MENTHYL-VALERATE Sedative M11

MERCURY Nephrotoxic PAM

METHANOL Antidote=Sodium-bicarbonate M29 LDlo=100-200ml/man MAR LDlo=340 (orl man)

METHIONINE Anticataract PAM; Antidote (Acetaminophen) 10g/ 6hr/man/orl M29; Antidote (Paracetamol) 10g/ 16 hr/man/orl M29; Antieczemic JE26:65; Antihepatotic JBH; Antioxidant PC27:969; Antiparkinsonian 1-5 g day WER; Cancer-Preventive 525/; Emetic M29; Essential JBH; Glutathionigenic M29; Hepatoprotective ACM:207; Lipotropic M11; Urine-Acidifier 200 mg/3x day/man/orl MAR; Urine-Deodorant M29

METHYL-2-METHYLBUTYRATE Flavor FEMA 5-20 ARC; Perfumery ARC

METHYL-CHAVICOL Hepatocarcinogenic BIS; Insecticide MPI JER17:21

METHYL-CINNAMATE Flavor FEMA 2-40 ARC; Insecticide; Nematicide MLC=1 mg/ml SZ44:183; Perfumery ARC

METHYL-ISOEUGENOL Anesthetic JBH; Antiaggregant IC50=56ug/mL PR4(3):93; Antihistaminic JBH; Bactericide JBH; Cancer-Preventive 525/; Candidicide PR4(3):93; Expectorant JBH; Flavor FEMA 5-100 ARC; Herbicide JBH; Insect-attractant; Nematicide MLC=1 mg/ml SZ44:183; Perfumery ARC; Spasmolytic JBH; Toxic JBH

METHYL-NONYL-KETONE Flavor FEMA 0.5-5 ARC; Irritant? ARC; Perfumery ARC

METHYL-SALICYLATE Allergenic MAR; Analgesic JE26:65; Antiinflammatory PM56(6):660; Antipyretic; Antiradicular JPP42:205; Antirheumatalgic MAR; Cancer-Preventive 525/; Carminative JE26:65; Counterirritant M11; Dentifrice ARC; Flavor FEMA 70-8,000 ARC; Insectifuge JCE20:2847; Perfumery ARC ADI=500 ug/mg MAR LD50=1,110 (orl mus) LD50=30ml (orl man) M11 LD50=4ml (orl chd) LD50=887 (orl rat) M11 LDlo=170 (orl hmn)

METHYL-TANSHINONATE Antiaggregant IC50=25-50 ug/ml OHAI,1987

MUCILAGE Cancer-Preventive 525/; Demulcent

MYRCENE Allergenic M&R; Analgesic EMM18:28; Antimutagenic EMM18:28; Antinociceptive 10-20 mg/kg ipr mus JPP42:877; Antinociceptive 20-40 mg/kg scu mus JPP42:877; Antioxidant JA6(4):33; Bactericide; Flavor FEMA 0.5-9 ARC; Fungicide TOX; Insectifuge; Perfumery JBH; Spasmolytic PM&T20:213.1986

MYRISTIC-ACID Cancer-Preventive 525/; Cosmetic JBH; Lubricant JBH; Nematicide NIG LD50=43 (ivn mus) M11

MYRISTICIN Amphetaminagenic; Anesthetic; Antiaggregant IC50=250uM JE29:179; Antidepressant JA6(4):34; Antiinflammatory JA6(4):34; Antioxidant 25-100 mg/kg/orl JA6(4):34; Antistress ihl JA6(4):34; Cancer-Preventive 10 mg/mus/orl/day JA6(4):34; Diuretic VET; Flavor FEMA 0.1-10 ARC; Hallucinogenic 52/; Hypnotic IWU; Hypotensive ihl JA6(4):34; Insecticide; Insecticide-Synergist JBH; MAO-Inhibitor JBH; Neurotoxic RJH; Oxytocic WIC; Paralytic OMM; Psychoactive RJH; Sedative 300 mg/kg ipr JA6(4):36; Serotoninergic 1,000 ppm orl JA6(4):36; Spasmolytic JBH; Tachycardic 411/; Uterotonic VET LDlo=570 (orl cat)

MYRTENAL Herbicide TOX; Perfumery ARC

MYRTENYL-ACETATE Perfumery ARC

N-BUTYRALDEHYDE Flavor FEMA <1-5 ARC

N-HENTRIACONTANE Antiinflammatory; Cosmetic JBH; Diuretic

N-NONACOSANE Antimutagenic EMP6:235

N-VALERALDEHYDE Flavor FEMA 1-5 ARC
LD 50=3,200 (orl rat)

NAPHTHALENE Cataractagenic ARC
LD50=1,780 (orl rat) LDlo=100 (orl chd) LDlo=2g/orl/chd M29
NARINGENIN Aldose-Reductase-Inhibitor IC25=1uM; IC78=10uM PC23:1885; Antiaggregant JBH; Antihepatotoxic JBH; Antiherpetic EMP5:197; Antiinflammatory 20 ppm JNP39:420; Antileukemic IC50=>10 ug/ml LS55:1061; Antimutagenic ID50=50-100 nM PCF:13; Antioxidant PCF:54; Antiperoxidative JBH; Antispasmodic JBH; Antiulcer FT5:1990; Antiviral EMP5:197; Bactericide MIC=250-500 ug/ml PM57:A43; Cancer-Preventive 525/; Choleretic; Decarboxylase-Inhibitor JBH; Fungicide 2.5mM PC38:599 PC29(4):1103; Fungistatic PC29(4):1103; Serotonin-Inhibitor JBH; cAMP-Phosphodiesterase-Inhibitor PAM

NARINGIN Aldose-Reductase-Inhibitor WER; Antidermatitic; Antiinflammatory; Antimutagenic EMP6:235; Antineuropathic WER; Antioxidant; Antiviral V&D; Cancer-Preventive 525/; Flavor FEMA ARC; Myorelaxant TOX; Oviposition-Stimulant JBH

NARIRUTIN Oviposition-Stimulant JBH

NEOMENTHOL Perfumery ARC; Spasmolytic ED50=0.033 mg/ml FT59;465

NEPETALACTONE Feliphile JBH; Herbicide; Insectifuge 382/; Sedative

NEPETIN Aldose-Reductase-Inhibitor ED50=1 uM; Anticataract; Antitumor

NEPETRIN Aldose-Reductase-Inhibitor ED50=1-10 uM; Analgesic; Antiangiotensic; Antibradykinetic; Anticataract; Antiinflammatory FT62:372; Antipyretic FT62:372; Cancer-Preventive 525/; Radioprotective; Spasmolytic

NERAL Bactericide IWU

NEROL Bactericide HG17; Flavor FEMA 1-20 ARC; Perfumery M11

NEVADENSIN Antiinflammatory 75 mg/kg FT5:1990; Antitubercular 0.2 mg/ml JBH; Antitussive ACM:200; Expectorant

NIACIN Allergenic WER; Anti-Meniere's JAD; Antiacrodynic; Antiallergic 50 mg/2x/day WER; Antiamblyopic; Antianginal; Antichilblain JBH; Anticonvulsant 3 g/day WER; Antidermatitic DAS; Antidysphagic; Antiepileptic 3 g/day WER; Antihistaminic 50 mg/2x/day WER; Antihyperactivity 1.5-6 g/day WER; Antiinsomnic 1 g/day WER; Antineuralgic; Antiparkinsonian 100 mg day WER; Antipellagric DAS; Antiscotomic; Antispasmodic 100 mg/2x day WER; Antivertigo; Cancer-Preventive 525/; Hepatoprotective; Hypoglycemic; Hypolipidemic RWG; Sedative PAM; Serotonergic PAM; Vasodilator M29 LD50=5,000 (orl mus) LDlo=4,000 (scu mus) PTD=1,000mg/day RDA=5-20mg/day

NONANOIC-ACID Flavor FEMA 1-15 ARC; Perfumery ARC LD50=3,200 (orl rat)

NOOTKATONE Antiulcer 20 mg/kg orl rat CPB38:3053; Flavor FEMA 3-50 ARC; Perfumery ARC

O-CRESOL Allelochemic IC50=0.36 mM 438/; Antimutagenic EMP6:235; Antiseptic M11; Cancer-Preventive 525/; Rodenticide LD 50=620 cut rat AFR27:149 LD50=1,350 (orl rat) M11 LD50=650 (scu rat) AFR27:149

OCIMENE Perfumery ARC

OCTANOIC-ACID Candidicide JBH; Flavor FEMA 3-20 ARC; Fungicide JBH; Perfumery JBH

OLEANOLIC-ACID Abortifacient; Anticariogenic; Antifertility PMP24:98; Antihepatotoxic PM56(2):171; Antiinflammatory 40 mg/kg ipr JNP54:455; Antioxidant BCI28:735; Antisarcomic; Cancer-Preventive 525/; Cardiotonic; Diuretic; Hepatoprotective ACM:211; Uterotonic JE15:16 JNP:1167.1986

OLEIC-ACID Allergenic M&R; Anemiagenic; Cancer-Preventive 525/; Choleretic 5 ml/man; Dermatitigenic M&R; Flavor FEMA 1-30 ARC; Insectifuge 382/; Irritant M11; Percutaneostimulant CPB38:3086; Perfumery ARC LD50=230 (ivn mus) M11 M11 LDlo=50 (ivn cat)

ORIENTIN Antiinflammatory FT61:263.1990; Cancer-Preventive 525/; cAMP-Phosphodiesterase-Inhibitor PAM

OROXYLIN-A Antiaggregant; Antimutagenic EMP6:235; Antithrombic

OXALIC-ACID Antiseptic; CNS-paralytic JBH; Fatal M29; Hemostatic M11; Irritant M&R; Renotoxic M11 LDlo=700-30,000 orl hmn FNF NIG

P-ANISALDEHYDE Antimutagenic EMP6:235; Cosmetic JBH; Flavor FEMA 0.5-75 ARC; Fungicide JBH; Insecticide JBH; Nematicide MLC=1 mg/ml SZ44:183; Perfumery JBH LD50=1,510 (orl rat) 15/

P-COUMARIC-ACID Allelopathic JBH; Antifertility AEH; Antihepatotoxic JBH; Antioxidant IC24=30 ppm PCF:17; Antiperoxidant IC50=>100 uM PM57:A54; Antitumor JBH; Bactericide PHZ46:156; Cancer-Preventive 525/; Choleretic; Cytotoxic JBH; Diaphoretic? LRN-DEC90; Fungicide NIG; Lipoxygenase-Inhibitor IC11=5 mM JAF38:688; Prostaglandigenic RWG; Prostaglandin-Synthesis-Inhibitor

P-CRESOL Allelochemic IC50=1.13 mM 438/; Antimutagenic EMP6:235; Antiseptic JBH; Cancer-Preventive 525/; Flavor FEMA <1 ARC; Parasiticide JBH; Rodenticide LD50=1,100 cut rat AFR27:149 LD50=1,100 (cut rat) AFR27:149 LD50=1,454 (orl rat) CRC LD50=750 (orl rat) M11 LD50=861 (orl mus)

P-CYMENE Analgesic MAR; Antiflu; Antirheumatalgic MAR; Bactericide JE26:65; Flavor FEMA 12-250 ARC; Fungicide TOX; Herbicide IC50=50 uM 438/; Insectifuge; Viricide LD50=4,750 (orl rat) M11

P-HYDROXYBENZOIC-ACID Antimutagenic EMP6:235; Antioxidant PCF:54; Antisickling PM56(6):681; Bactericide; Cancer-Preventive 525/; Fungistat EC50=607 ug/ml NIG; Immunosuppressive RWG; Phytoalexin NIG; Prostaglandagenic JBH; Ubiquiot JBH

P-METHOXYCINNAMALDEHYDE Herbicide 5 mM

PACHYPODOL Viricide JBH

PALMITIC-ACID Antifibrinolytic EMP1:53; Flavor FEMA 1 ARC; Hemolytic BIS; Lubricant JBH; Nematicide NIG; Soap JBH LD50=57 (ivn mus) M11

PALMITOLEIC-ACID Soap JBH

PARAFFIN Antihemorrhoidal M29; Antiproctalgic M29; Emollient M29; Laxative 45 mL/day/orl/man/ M29

PATCHOULI-ALCOHOL Antiplaque MIC=39-2,500 ug/ml 451/; Bactericide MIC=39->2,500 450/; Fungicide 451/

PECTIN Antiatheromic 15 g/man/day HSC25:1485; Antidiabetic 10 g/man/day/orl PAM; Antidiarrheic M11; Antitumor; Antitussive; Antiulcer; Bactericide; Cancer-Preventive 525/; Demulcent; Hemostat; Hypocholesterolemic HSC25:1485; Peristaltic LDlo=1,800 (unk mus)

PELARGONIDIN Antiherpetic EMP5:197; Antiviral V&D EMP5:197; Cancer-Preventive 525/; Pigment JBH

PELARGONIN Pigment JBH

PEONIDIN Larvistat JBH; Pigment JBH

PERILLALDEHYDE Bactericide MIC=500->1000 ug/ml JAF40:2328; Candidicide MIC=500 ug/ml JAF40:2328; Fungicide MIC=250-500 ug/ml JAF40:2328; Nematicide MLC=1 mg/ml SZ44:183; Sedative

PERILLYL-ALCOHOL Anticancer BOIK1995; Bactericide JAF40:2328; Candidicide JAF40:2328; Fungicide JAF40:2328; Nematicide MLC=1 mg/ml SZ44:183

PETUNIDIN Larvistat JBH; Pigment JBH

PHELLANDRENE Hyperthermic HG22:45; Spasmogenic FT59:465; Tumor-promoter AEH

PHENOL Anesthetic JE26:65; Anodyne M29; AntiMS M29; Antihemorrhoidal MAR; Antihydrocoele M29; Antiincontinence M29; Antionychogryphotic MAR; Antiotitic MAR; Antioxidant 505/; Antiprostatitic M29; Antipyruvetic JBH; Antiseptic; Antisinusitic MAR; Antispastic MAR; Antiviral M29; Antiwrinkle MAR; Bactericide JE26:65; CNS-Depressant M29; Cancer-Preventive 525/; Carcinogenic; Emetic M29; Fungicide M29; Hemolytic 5,400 ppm MAR; Rodenticide LD50=1,040 cut rat CRC(FNS); Vasodilator; Viricide M29 LD50=400 scu rat AFR27:149 LD50=530 (orl rat) CRC(FNS) LD=10,000-30,000 orl hmn AFR27:149 LDlo=140 (orl hmn)

PHENYLACETIC-ACID Flavor FEMA 0.1-30 ARC; Perfumery ARC

PHENYLALANINE Anti-attention-deficit-disorder 587 mg/day/orl M29; Antidepressant 50-4,000 mg/day/man M28 M29 WER; Antiparkinsonian 200-500 mg/day/man MAR; Antisickling PM56(1):41; Antivitiligic 100

mg/kg/day/orl/man M29; Monoamine-Precursor PAM; Tremorigenic 1,600-12,600 mg/man/day MAR RDA=2,200 man M11

PHENYLETHYL-2-METHYLBUTYRATE Flavor FEMA 1-9 ARC; Perfumery ARC

PHOSPHORUS Antiosteoporotic DAS; Immunostimulant DAS; Osteogenic DAS PTD=12,000 mg/day RDA=300-1,200mg/day

PHYTOL Cancer-Preventive 525/

PINENE Antiseptic; Bactericide; Expectorant JE26:65; Fungicide; Herbicide IC50=30mM 438/; Perfumery M11; Spasmogenic FT59:465; Spasmolytic FT59:465

PIPERITENONE Perfumery ARC; Toxic 10,000/man ARC

PIPERITENONE-OXIDE Insecticide JEO4:57

PIPERITONE Antiasthmatic; Flavor FEMA 1-2 ARC; Herbicide IC50=30mM 438/; Herbicide IC50=75 uM TOX; Insectifuge (> DEET) JNP56:935; Perfumery ARC

PODOPHYLLOTOXIN Abortifacient; Anticondyolomata M29; Antifertility JE32:167; Antiherpetic IC88=414 ppb JNP45:725; Antileukemic 5.4 ppm; Antimeasles V&D; Antimelanomic ED50=>20 ug/ml JE32:117; Antimitotic W&W; Antisarcomic ED50=0.012 ug/ml JE32:117; Antitumor 2 ppm iv; Antitumor (Breast) ED50=0.041 ug/ml JE32:117; Antitumor (Colon) ED50=0.011 ug/ml JE32:117; Antitumor (Lung) ED50=0.0181 ug/ml JE32:117; Antiviral IC88=1 uM 458/; Cathartic JBH; Embryocide 411/; Fungicide; Herbicide 50 ppm; Insecticide 1-20 ppm; Irritant W&R; Juvabional 1,200 ppm; Larvicide 5 ppm 454/; Piscicide 1-20 ppm; Purgative; Viricide IC88=1 uM 458/ LD50=1.7 ivn cat AEH LD50=15 (ipr rat) M11 LD50=3 ims rat AEH LD50=33 ipr mus AEH LD50=4 ims cat AEH LD50=8.7 (ivn rat) M11 LD50=90 (orl mus) 15/

POGOSTONE Antiplaque MIC=39-625 ug/ml 451/; Bactericide MIC=39-625 450/; Fungicide 451/

POTASSIUM Antiarrhythmic DAS; Antidepressant DAS; Antifatigue DAS; Antihypertensive DAS; Cardiotoxic 18,000 mg/man/day; Spasmolytic DAS

PREHISPANOLONE Antiplatelet BJP103:1719

PROPIONALDEHYDE Flavor FEMA 5-12 ARC; Perfumery ARC LD50=1,400 (orl rat) M11 LDlo=800 (orl rat)

PROPIONIC-ACID Flavor FEMA 1-36,000 ARC; Fungicide 2,500 -36,000 ARC M29; Perfumery LD50=1,510 (orl rat) LD50=4,290 (orl rat) M11

PROTOCATECHUIC-ACID Antiarrhythmic ACM:407; Antiasthmatic; Antihepatotoxic JBH; Antiherpetic; Antiinflammatory JBH; Antiischemic JBH; Antiophidic EMP5:363; Antioxidant ABC53:519; Antiperoxidant IC50=>100 uM PM57:A54; Antispasmodic EC50=4.6-17 uM PR4(2):73; Antitussive ACM:200; Antiviral; Bactericide PR4(5):198; Fungicide 500ug/ml 438/; Immunostimulant EMP1:124; Phagocytotic EMP1:124; Prostaglandigenic RWG; Ubiquiot JBH; Viricide

PRUNASIN Cyanogenic 52/

PRUNIN Fungicide 2.5 mM PC38:599; Hypocholesterolemic JNP54:218; Hypoglycemic PM57:208; Hypolipemic PM57:208

PULEGONE ACE-Inhibitor 453/; Antialzheimeran? 453/; Antihistaminic JPP42:295; Antipyretic JE25:165; Avifuge; Bactericide PM57:347; Cancer-Preventive 525/; Candidicide PB57:347; Cerebrotoxic orl man JAR4:24; Encephalopathic 80 mg/kg orl rat LRN-JAN92; Flavor FEMA 5-30 ARC; Fungicide PM57:347; Glutathionalytic LRN-JAN92; Hallucinogenic RJH; Hepatotoxic PM56(2):226; Herbicide IC50=1.5 uM TOX; Insecticide 453/; Insectifuge; Nephrotoxic RJH; Neurotoxic RJH; Perfumery ARC; Pulifuge; Pulmonotoxic AEH; Sedative JE25:165 LD50=150 (ipr mus)

PYRIDINE CNS-Depressant M11; Flavor FEMA 0.02; Irritant M11 LD50=1,580 (orl rat) M11 LD50=891 (orl rat)

QUERCETIN 5-Lipoxygenase-Inhibitor IC50 (uM)=4 DRS LEH; ATPase-Inhibitor NIG; Aldose-Reductase-Inhibitor CPB38:297; Allelochemic IC82=1 mM 438/; Allergenic JBH; Anti-Tumor-Promoter PAM; Anti-lipo-peroxidant IC67=50 PM56(2):171 PC27:972; AntiCrohn's 400 mg/man/3x/day PAM; AntiPMS 500 mg/2x/day/wmn PAM; Antiaggregant IC50=55 uM EMP5:333; Antiallergic IC50=14 uM JIM127:546; Antianaphylactic PR4(5):201; Antiasthmatic IC50=14 uM JIM127:546; Anticarcinomic (Breast) IC50=1.5 uM MED; Anticataract PM56(3):254; Anticolitic 400 mg/man/3x/day PAM; Antidermatitic PAM; Antidiabetic PAM; Antiencephalitic EMP5:199; Antiestrogenic PAM; Antifeedant IC52=<1000 ppm diet 438/; Antiflu V&D; Antigastric RR21:85; Antigonadotropic JBH; Antihepatotoxic PM56(2):171; Antiherpetic PAM; Antihistaminic 411 WER; Antihydrophobic V&D; Antihypertensive KCH; Antiinflammatory (20 mg/kg) 150 mg/kg FT5:1990; Antileukemic IC50=10 uM EMP5:225; Antileukemic IC50=>10 ug/ml LS55:1061; Antileukotrienic PAM; Antimalarial IC50=1-6.4 ug/ml MPT; Antimutagenic ID50=2-5 nM PCF; Antimyocarditic EMP5:199; Antioxidant IC96=300 ppm PCF; Antiperiodontal PAM; Antipermeability; Antiperoxidant PM57:A110; Antipharyngitic PAM; Antiplaque PAM; Antipodriac PAM; Antipolio PAM; Antiprostanoid PCF:48; Antipsoriac PAM; Antiradicular IC50=4.6uM PM56(6):695; Antispasmodic 411/; Antithiamin PCF:54; Antitumor HG22:9; Antiviral IC50=10 uM EMP5:225; Bactericide JBH; Bradycardiac KCH; Calmodulin-Antagonist PAM; Cancer-Preventive 525/; Capillariprotective M11; Carcinogenic 40,000 ppm (diet) mus NIG; Catabolic AFR27:149; Cyclooxygenase-Inhibitor IC50 (uM) =16 DRS LEH PCF:49; Cytotoxic ED50=70ug/ml PM56(6):677; Cytotoxic IC82=100 ug/ml PM57:A113; Deiodinase-Inhibitor JNM1:10; Diaphoretic? LRN-DEC90; HIV-RT-Inhibitor IC50=<1 ug/ml JNP53(5):1239; Hemostat KCH; Hepatomagenic 5,000 ppm (diet) rat PCF; Hypoglycemic 100 mg/kg orl rats JE27:243; Insulinogenic PAM; Juvabional 438/; Larvistat 8,000 ppm diet 438/; Lipoxygenase-Inhibitor IC11=1.25 mM JAF38:688; Lipoxygenase-Inhibitor IC50 (uM)=3.5 DRS LEH; Mast-Cell-Stabilizer; Mutagenic HG22:18; Ornithine-Decarboxylase-Inhibitor PCF:14; Protein-Kinase-C-Inhibitor PCF:14; Spasmolytic PR4(5):201; Teratologic; Tumorigenic 0.1% diet orl rat/yr 505/; Tyrosine-Kinase-Inhibitor EMP6:170; Vasodilator KCH; Viricide JNP55:1732; Xanthine-Oxidase-Inhibitor IC50=>0.4 ug/ml CPB38:1772; cAMP-Phosphodiesterase-Inhibitor PAM LD50=160 (orl mus) M11 LD50=>2,000 orl rat PAM

QUERCITRIN Aldose-Reductase-Inhibitor IC50=2 uM CPB38:2733; Aldose-Reductase-Inhibitor IC55=0.1uM PM56(3):254; Antiarrhythmnic; Anticataract PM1986:239; Antiedemic KCH; Antifeedant IC52=<1000 ppm diet 438/; Antiflu AYL; Antihemorrhagic JBH; Antihepatotoxic JBH EMP6:189; Antiherpetic EMP5:198; Antiinflammatory KCH; Antimutagenic ID50=2-5 nM PCF; Antioxidant IC50=120 uM PC27:969; Antipurpuric; Antispasmodic JE26:65; Antithrombogenic; Antitumor; Antiulcer JBH; Antiviral EMP5:198; Bactericide JBH; CNS-Depressant; Cancer-Preventive 525/; Cardiotonic; Choleretic; Detoxicant; Diuretic KCH; Dye M11; Hemostat JBH; Hepatotonic; Hypoglycemic EMP6:189; Hypotensive; Insectiphile JBH; Paralytic; Spasmolytic JE26:65; Vasopressor JE26:65; Viricide JE26:65 LDl0=200 (ipr mus)

RAFFINOSE Flatugenic PFH46:77

RETUSIN Antitumor; Cytotoxic ED50=>25 ug/ml JNP54:900

RHAMNETIN Aldose-Reductase-Inhibitor IC23=1uM; IC57=10uM PC23:1885; Cancer-Preventive 525/; Dye M11

RIBOFLAVIN Antiarabiflavinotic 2-10 mg/orlday MAR; Anticarpal-Tunnel 50 mg/day WER; Anticataract 15 mg/day WER; Anticheilotic; Antidecubitic; Antiglossitic DAS; Antikeratitic; Antimigraine; Antipellagric; Antiphotophobic DAS; Cancer-Preventive 525/ LD50=340 (ipr mus) M11 LD50=5,000 (scu rat) LD50=560 (ipr rat) PTD=1,000 mg/day RDA=0.4-1.8mg/day

ROSMANOL Antihepatotoxic PM56(2):171; Antilipoperoxidant PM56(2):171; Antioxidant; Antiradicular PM56(2):171; Cancer-Preventive 525/

ROSMARIDIPHENOL Antioxidant PC27:969; Cancer-Preventive 525/

ROSMARINIC-ACID Antianaphylactic 1-100 mg/kg orl IJI10:729; Anticomplementary 1/2 aspirin IJI10:729 PM57:A48; Antiedemic 0.316-3.16 mg/kg ims IJI10:729; Antigonadotropic JNM1:10 JBH; Antihemolytic IC70=5-10 um/1 IJI10:729; Antihepatotoxic PM56(2):171; Antiherpetic 50 ug/ml V&D; Antiinflammatory IJI10:729; Antileukotrienic; Antilipoperoxidant PM56(2):171; Antioxidant EC50=2.7 ug/ml FT62:170; Antipulmonotic VET94; Antiradicular PM56(2):171; Antishock IJI10:735; Antithyreotropic; Antithyroid JNM1:10; Antiviral V&D; Bactericide PM55:663; Cancer-Preventive 525/; Deiodinase-Inhibitor JNM1:10; Viricide 411/

ROSMARIQUINONE Antioxidant; Cancer-Preventive 525/

RUTIN Aldose-Reductase-Inhibitor CPB38:297; Anti-Capillary-Fragility 20-100 mg orl man M7; Anti-Tumor-Promoter CPB38:774; Anti-tumor-promoter CPB38:774; Antiapoplectic; Antiatherogenic; Anticataract; Anticonvulsant KCH; Antidermatitic BIS; Antidiabetic CPB38:297; Antiedemic 411/; Antierythemic; Antifeedant 450/; Antiglaucomic 60 mg/day WER; Antihematuric; Antihepatotoxic TOX; Antiherpetic EMP5:198; Antihistaminic; Antihypertensive KCH; Antiinflammatory 20 mg/kg; Antimalarial IC50=>100 ug/ml JE15:204; Antimutagenic ID50=2-5 nM PCF; Antinephritic; Antioxidant IC28=30 ppm PCF; Antioxidant IC50=120uM PC27:969; Antipurpuric; Antiradicular JBH; Antithrombogenic; Antitrypanosomic 100 mg/kg PM57:A44; Antitumor PCF:267; Antivaricosity JBH; Antiviral V&D; Bactericide JBH; Cancer-Preventive 525/; Capillariprotective M11; Catabolic AFR27:149; Estrogenic? EMP6:189; Hemostat KCH; Hepatomagenic 20,000 ppm (diet) rat PCF; Hypocholesterolemic KCH; Hypotensive; Insecticide TOX; Insectiphile JBH; Juvabional 382/; Larvistat IC95=4,000-8,000 ppm diet 438/; Lipoxygenase-Inhibitor IC75=2.5 mM JAF38:688; Mutagenic EMP6:189;

Myorelaxant EMP6:189; Oviposition-stimulant JBH; Spasmolytic JE26:65; Vasopressor; Viricide JBH; cAMP-Phosphodiesterase-Inhibitor PAM LD50=950 (ivn mus) M11

SABINENE Perfumery ARC

SABINOL Anthelminthic JBH; Emmenagogue JBH; Perfumery ARC; Toxic JBH

SAFROLE Anesthetic JE26:65; Antiaggregant IC50=110uM JE29:179; Anticonvulsant JBH; Antiseptic M11; Bactericide JE26:65; CNS-Depressant JBH; Cancer-Preventive 525/; Carcinogenic M11; Carminative JE26:65; Controlled HG20:15; DNA-Binder JBH; Hepatoregenerative CRC; Hepatotoxic M29; Hypothermic JBH; Nematicide MLC=1 mg/ml SZ44:183; Neurotoxic RJH; Pediculicide M11; Perfumery M11; Psychoactive RJH; Tremorigenic AFR27:149 LD50=1950 (orl rat) LD50=2,350 (orl mus) M11

SALICYLALDEHYDE Allelochemic IC50=0.51 mM 438/; Flavor FEMA 0.5-20 ARC; Perfumery ARC; Sprout-Inhibitor PER.COM.SFV LDlo=1,000 (scu rat) M11

SALICYLIC-ACID Analgesic JE26:65; Antidandruff ALH; Antidermatotic; Antieczemic; Antiichthyosic M29; Antiinflammatory PM56(6):660; Antineuralgic; Antioncychomycotic M29; Antioxidant KCH; Antiperiodic; Antipodagric; Antipsoriac; Antipyretic; Antirheumatic; Antiseborrheic; Antiseptic ALH; Antitumor PCF:48; Antitympanitic M11; Bactericide MAR; Cancer-Preventive 525/; Comedolytic M29; Cyclo-Oxygenase-Inhibitor PCF:48; Dermatitigenic JBH; Febrifuge; Fungicide M29; Hypoglycemic EMP6:158; Insectifuge EB48:111; Keratolytic JBH; Tineacide M29; Ulcerogenic LD50=500 (ivn mus) M11 LD50=891 (orl rat) LDlo=450 (orl dog)

SALVIANOLIC-ACID-A ATPase-Inhibitor JBH; Anticoagulant JBH; Antisecretory; Antiulcer 25 mg/kg ip PM56(4):360

SALVIN Antiseptic LRN-AUG92; Bactericide LRN-AUG92

SALVIN-MONOMETHYL-ETHER Antiseptic LRN-AUG92; Bactericide LRN-AUG92

SALVIOL Bactericide; Cytostatic ID50=10ug/ml PM56(1):70 LDlo=3,000 (orl mus)

SCLAREOL Allelopathic JNP54:483; Bactericide JNP54:483; Fungicide IC100=80 uM 438/; Perfumery ARC

SCUTELLAREIN Aldose-Reductase-Inhibitor IC56=10uM; IC10=1uM PC23:1885; Antihemolytic; Antileukemic IC50=10 uM EMP5:225; Antiviral IC50=10 uM EMP5:225; Cancer-Preventive 525/; Phospholipase-A2--Inhibitor FT2:103.1986

SELENIUM Analgesic 200 ug/day WER; Anorexic 52/; Antiacne 200 ug/day WER; Antiaggregant PAM; Anticirrhotic DAS; Anticoronary 200 ug/day WER DAS; Antidandruff DAS; Antidote (Mercury) DAS; Antikeshan DAS; Antileukotrienic PAM; Antimyalgic 200 ug/day WER; Antiosteoarthritic HSC25:1486; Antioxidant DAS; Antiulcerogenic JE29:25; Cancer-Preventive DAS; Depressant 52/; Fungicide DAS; Prostaglandin-Sparer JPP42:655 PTD=1mg/day RDA=10-75ug/day

SELINENE Expectorant

SERINE Cancer-Preventive 525/

SIDERITOFLAVONE Aldose-Reductase-Inhibitor; Antiarthritic; Anticataract 1 uM; Antiinflammatory JNP50:313.1987

SILICON Antiarteriosclerotic DAS

SINAPIC-ACID Antihepatotoxic JBH; Antioxidant IC27=30 ppm PCF:17; Bactericide JBH; Cancer-Preventive 525/; Fungicide JBH

SINENSETIN Fungicide JBH

SKULLCAPFLAVONE-II Antiaggregant; Cytotoxic JBH; Thrombolytic

SODIUM Hypertensive DAS

SQUALENE Antitumor JBH; Bactericide JBH; Cancer-Preventive 525/; Immunostimulant JBH; Lipoxygenase-Inhibitor CE7:297; Perfumery M11

STACHYDRINE Cardiotonic; Emmenagogue LRN-MAR91; Insectifuge EB48:111; Lactagogue LRN-MAR91; Oxytocic?; Systolic-Depressant JBH

STACHYOSE Flatugenic PFH46:77

STARCH Absorbent M29; Antidote (Iodine) JBH; Antinesidioblastosic M29; Emollient M29; Poultice M29

STEARIC-ACID Cosmetic JBH; Flavor FEMA 2-4,000 ARC; Lubricant JBH; Perfumery ARC; Suppository JBH LD50=22 (ivn rat)

STIGMASTEROL Antihepatotoxic; Antiinflammatory LRN1995; Antiophidic 2.3 mg/mus EMP5:363; Antiviral PS75:161; Artemicide LC50=110ppm PC29(5):1667; Cancer-Preventive 525/; Estrogenic MPI; Hypocholesterolemic NIG; Ovulant; Sedative PM52:499; Viricide PS75:161

STYRENE Irritant; Narcotic M11; Perfumery ARC
LD50=316 (orl mus) LD50=5,000 (orl rat) LD50=660 (ipr mus) M11 LD50=90 (ivn mus) M11

SUCROSE Aggregant PAM; Antihiccup 1 tsp M29; Antiophthalmic M29; Antioxidant M11; Atherogenic PAM; Flatugenic PFH46:77; Hypercholesterolemic PAM; Preservative M11; Sweetener 1 x sucrose M11; Triglycerigenic PAM; Uricogenic PAM; Vulnerary M29

SULFUR Acarifuge; Antiacne M29; Antidandruff M29; Antigrey M29; Antiseborrheic M29; Antiseptic M29; Comedogenic M29; Keratolytic M29; Laxative M29; Parasiticide M29; Scabicide M29

TANNIC-ACID Allergenic M&R; AntiHIV IC90=200 ug/ml JNP54:152; Anticolitic; Antidecubitic; Antidermatotic; Antidotal (Heavy Metal) M11; Antidysenteric KCH; Antiencephalitic V&D; Antienteritic V&D; Antifeedant 2-4% diet AFR27:149; Antigingivitic; Antihemorrhoidal M29; Antiherpetic CPB38:3031; Antimutagenic EMP6:235; Antinitrosaminic AFR27:149; Antiobesity (Antinutrient) KCH; Antioxidant IC56=30 ppm PCF:17; Antipharyngitic; Antipolio V&D; Antirhinitic; Antiseptic 411/; Antistomatitic; Antitonsilitic; Antiulcer KCH; Antiviral AVR14:323; Astringent 411/; Bactericide; Cytotoxic 15 ug CPB38:3031; Detoxicant KCH; Emetic M29; Flavor FEMA 1-1,000 ARC; Hemostat M11; Hepatotoxic KCH LD100=6,000 (orl mus) M11 LD50=2,250-6,000 orl mus rat rbt AFR27:149 LD50=20-80 mg/kg ivn rat AFR27:149 LD50=200-700 mg/kg ipr rat AFR27:149 LDlo=2,000 (orl mus)

TANNIN Antidiarrheic; Antidysenteric; Antimutagenic; Antinephritic CPB38:1049; Antiophidic EMP5:363; Antioxidant IC50=1.44 ug/ml CPB38:1049; Antiradicular 500 mg/kg/day orl mus CPB38:1049; Antirenitic CPB38:1049; Antiviral JE26:65; Bactericide JE26:65; Cancer-Preventive

HG22:14; Hepatoprotective; Immunosuppressant RWG; Psychotropic CPB38:1049; Viricide JE26:65

TANSHINONE-I Antiischemic JNP54:1583; Bacteristat KCH

TERPINEN-4-OL Antiallergic; Antiasthmatic CJT4:203.1982; Antiseptic FFJ9:129; Antitussive CJT; Bactericide JEO7:271; Bacteriostatic CJT; Diuretic 0.1 ml/rat AEH; Fungicide; Herbicide IC50=200 mM 438/; Herbicide IC50=22 uM TOX; Insectifuge; Nematicide MLC=1 mg/ml SZ44:183; Spermicide ED100=0.015 AHN188:83; Vulnerary LRN-JAN91 LD50=0.25 ml/kg ipr mus AEH LD50=0.75 ml/kg scu mus AEH LD50=0.78 ml/kg ims mus AEH LD50=1.85 ml/kg orl mus AEH

TERPINEOL Antiallergenic; Antiasthmatic; Antiseptic M11; Antitussive; Bactericide; Cholagogue; Expectorant; Insectifuge 382/; Perfumery M11 LD50=4,300(orl rat)

TERPINOLENE Deodorant JBH; Flavor FEMA 15-60 ARC; Fungicide TOX; Perfumery ARC

TERPINYL-ACETATE Bactericide RIE12:5; Flavor FEMA 1-260 ARC; Insectifuge 382/; Perfumery ARC

THIAMIN Analgesic 1-4 g/day WER; Antialcoholic; Antialzheimeran 100-3,000 mg day WER; Antianorectic JBH; Antiback-ache 1-4 g day WER; Antiberiberi; Anticardiospasmic; Anticolitic; Antidecubitic; Antideliriant; Antiencephalopathic M29; Antifatigue JBH; Antigastritic JBH; Antiheartburn; Antiherpetic; Antimigraine; Antimyocarditic; Antineuralgic 1-4 g day WER; Antineurasthenic; Antineuritic; Antineuropathic 50 mg WER; Antipoliomyelitic; Insectifuge 75-150 mg/man/day MAR LD50=301 (scu mus) 15 PTD=300mg/day RDA=0.3-1.6mg/day

THREONINE Antioxidant? PC27:969; Antiulcer M29; Essential JBH; Flavor FEMA 1,000 ARC

THUJONE Anthelminthic JBH; Bactericide JEO7:271; Cerebrodepressant; Convulsant 40 mg/kg RJH; Counterirritant JBH; Epileptigenic; Hallucinogenic; Herbicide IC50=22 mM 438/; Perfumery ARC; Respirainhibitor; Spasmolytic ED50=0.127 mg/ml FT59:465; Toxic ARC LD50=134 (scu mus) M11 LDlo=120 (ipr rat) 15/

THUJOPSENE Fumigant 382/

THYMOHYDROQUINONE Antiseptic ABS; Perfumery ARC

THYMOL ACE-Inhibitor LGR; Allergenic M&R; Anesthetic; Anthelminthic 411/; Antiacne NIG; Antiaggregant IC50=0.75 TRA42:511; Antialzheimeran? LGR; Antibronchitic; Anticholinesterase? LGR; Antihalitosic AH12(4):17; Antiherpetic MAR; Antiinflammatory; Antineuritic; Antioxidant 100 ppm MAR; Antiplaque MIC=39-625 ug/ml 451/; Antiradicular EC=60 JEO5:15; Antirheumatic 411/; Antiseptic 20 X PHENOL W&W; Antispasmodic IWU; Antitussive IWU; Bactericide; Carminative JPP46:16; Counterirritant NIG; Dentifrice ARC; Deodorant MAR; Dermatitigenic M&R; Enterorelaxant JAR4:22; Enterotoxic WIC; Expectorant; Flavor FEMA 2-100 ARC; Fungicide M11; Gastroirritant M29; Larvicide 10 ppm 454/; Myorelaxant JAR4:22; Nematicide MLC=0.1 mg/ml SZ44:183; Perfumery ARC; Spasmolytic 411/; Sprout-Inhibitor PER.COM.SFV; Tracheorelaxant; Urinary-Antiseptic 411/; Vermicide LD50=1,800 (orl mus) 15 LD50=980 (orl rat) M11

THYMONIN Myocontractant JBH; Spasmolytic FT2:1986

THYMOQUINONE Antiarthritic PM61:33; Antiasthmatic; Antieicosanoid PM61:33; Antihistaminic; Antioxidant PM61:33; Antirheumatic PM61:33; Bactericide JNP39:420; Carcinogenic 15/; Choleretic HHB; Cyclooxygenase-Inhibitor IC50 3.5 ug/ml PM61:33; Lipoxygenase-Inhibitor IC50 1 ug/ml PM61:33

THYMYL-ACETATE Antiseptic JBH; Carminative JBH; Counterirritant JBH; Irritant JBH; Rubefacient JBH

TIN Antiacne; Bactericide; Taenicide MAR

TOLUENE Encephalopathic M29

TORREYOL Termiticide MUO24:1

TRANS-2-HEXENAL Flavor FEMA 0.7-20 ARC; Paramecicide WOI; Pesticide WOI

TRANS-ANETHOLE Flavor FEMA 1,250-3,150ppm ARC; Sweetener JE28:103 LD50=900 (ipr rat) M10

TRANS-BETA-FARNESENE Insectifuge 382/

TRYPTOPHAN Analgesic 750mg/4 x day/orl/man/ M29; Antianxiety 500-1,000 mg/meal WER; Antidementic 3 g/day WER; Antidepressant 1-3 g/3x day/orl/man/ M29; Antidyskinetic 2-8 g/orl/wmn/day/ M29; Antihypertensive

JAF36:1079; Antiinsomniac 1-3 g/day WER; Antimanic 12g/man/day/orl PAM; Antimenopausal 6 g/day WER; Antimigraine 500 mg/man/4 x day/ M29; Antioxidant? PC27:969; Antiparkinsonian 2 g 3 x day WER; Antiphenylketonutic M29; Antipsychotic 12 g/man/day M29; Antirheumatic M29; Carcinogenic CRC; Essential JBH; Hypnotic M29; Hypoglycemic EMP6:157; Hypotensive 3 g/day WER; Insulinase-Inhibitor EMP6:157; Insulinotonic JAF36:1079; Monoamine-Precursor PAM; Prolactinogenic PAM; Sedative 3-10 g/man/day M29 PAM; Serotonigenic 6-12 g/day/orl/man/ MAR; Tumor-Promoter JAF36:1079 RDA=300-1,200 mg/day?

TYROSINE Antidepressant? M29 WER; Antiencephalopathic M29; Antiparkinsonian 100 mg/kg day WER; Antiphenylketonuric M29; Antiulcer 256 mg scu, 400 mg/kg ipr rat BBE; Cancer-Preventive 525/; Monoamine-Precursor PAM

UMBELLIFERONE Allelochemic IC94=2 mM 438/; Antihistaminic; Antiinflammatory RWG; Antimitotic 5-25 ug/ml PM1987:526; Antimutagenic EMP6:235; Antiprostaglandin RWG; Antiseptic FT1990:483; Antispasmodic LRN-MAR91; Bactericide JBH; Cancer-Preventive 5-25 ug/ml PM1987:526; Choleretic; Fungicide 411/; Lipoxygenase-Inhibitor PAM RWG141; Perfumery ARC; Sunscreen M11

UMBELLULONE Antiseptic; Circulodepressant; Hemolytic; Respiradepressant; Toxic ARC LD=0.18 ml/kg (dog)

UNDECANAL Flavor FEMA 1-60; Perfumery ARC
LD50=140 (ivn mus)

URSOLIC-ACID Antiarthritic PM57:A56; Anticholestatic 28-100 mg/kg orl PR6:74; Antidiabetic CCO; Antiedemic PM57:A56; Antihepatotoxic 5-20 mg/kg ipr PR6:74; Antihistaminic CPB39:3276; Antiinflammatory 1/3x indomethacin PM57:A56 PM61:9; Antileukemic JNP53(2):513; Antimutagenic PPB17:990; Antiobesity? IJO16:4; Antioxidant BCI28:735; Antitumor CR54:701; Antitumor-promoter PM57:A56; Antiulcer PR6:74; CNS-Depressant; Cancer-Preventive 525/; Choleretic 5-20 mg/kg orl PR6:74; Cytotoxic 50 ppm; ED50=3.75 ug/ml JNP53(2):513; Diuretic; Hepatoprotective JNP53(2):513; Hypoglycemic JE27:243; Piscicide; Potassium-Sparing 3 mg/rat IWU; Protisticide CE7:285; Sodium-Sparing 3 mg/rat IWU
UVAOL Antitumor; Cytotoxic 100-200 ppm
LD50=3,200 (orl rat)

VALERIC-ACID Antihysteric MAR; Antiinsomniac 1.2-2.4; Flavor FEMA 1-8 ARC; Hypotensive; Myorelaxant JE26:65; Perfumery M11; Sedative; Spasmolytic JE26:65; Tranquilizer LD50 1290 ivn mus JBH LD50=500 (orl rat)

VALINE Antiencephalopathic M29; Essential; Flavor FEMA 1,000-2,000 ARC

VANILLIC-ACID Anthelminthic JBH; Antifatigue JAD; Antiinflammatory JBH; Antioxidant IC21=30 ppm PCF ACM:489; Antisickling PM56(6):681; Ascaricide; Bactericide 1.5-15 mg/ml 452/; Cancer-Preventive 525/; Choleretic; Immunosuppressive RWG; Laxative MPI; Ubiquiot JBH

VANILLIN Allelochemic IC50=4.26 mM 438/; Allergenic M&R; Antimutagenic EMP6:235; Antipolio EMP5:194; Antiviral EMP5:194; Cancer-Preventive 525/; Flavor FEMA 50-20,000 ARC; Fungicide JBH; Immunosuppressive RWG; Insectifuge? MAR; Irritant M&R; Perfumery JBH MAR LD50=1,580 (orl rat) M11

VERBASCOSE Flatulent LDlo=125 (ipr mus)

VERBENONE Perfumery ARC LDlo=250 (ipr mus)

VICENIN-2 Antiinflammatory ED50=148 (ipr mus); Oviposition-Stimulant JBH LD50=>2,000 (ipr mus) FT61:277.1990

VITEXIN Aldose-Reductase-Inhibitor IC15=10uM PC23:1885; Antiarrhythmic PT4(3):118; Antibradiquinic FT61:277.1990; Antidermatitic; Antihistaminic; Antiinflammatory FT61:263.1990; Antiserotoninic; Antithyroid JBH; Cancer-Preventive 525/; Goitrogenic JBH; Hypotensive; Thyroid-Peroxidase-Inhibitor JBH; cAMP-Phosphodiesterase-Inhibitor IC50=1.6 mg/ml PM57:A133

WOGONIN Aldose-Reductase-Inhibitor IC13=10uM PC23:1885; Antiaggregant; Antifertility 10 mg/kg/day PR4(3):87; Antiimplantation JBH; Antiinflammatory IC50=75 FT5:1990; Antileukemic IC50=0.160 ug/ml LS55:1061; Antithrombic; Bactericide; Diuretic; Estrogenic 10 mg/kg/day PR4(3):87; Lipolytic; Sialidase-Inhibitor LRN-JAN93; Triglycerolytic

XANTHOPHYLL Bruchifuge; Insectifuge 382/

YLANGENE Perfumery

ZINC AntiCrohn's DAS; Antiacne 135 mg/day DAS WER; Antiacrodermatitic 8-34 mg/day/orl/chd M29; Antialopecic DAS; Antialzheimeran 50 mg day WER; Antianorexic M29; Antiarthritic? 50 mg/3x/day/orl/man M29; Anticanker 100 mg/day WER; Anticataract 30 mg/day WER; Anticoeliac MAR; Anticold 50 mg DAS; Anticolitic DAS; Antidandruff DAS; Antidote (Cadmium) LL-MAY89; Antieczemic 150 mg/day WER; Antiencephalopathic M29; Antiepileptic 100 mg/day WER; Antifuruncular 45 mg/3x/day/man MAR; Antiherpetic? 25 mg/day/ PAM; Antiimpotence M29; Antiinfective 50 mg/day WER; Antiinfertility 60 mg/day WER; Antiinsomniac DAS; Antilepric M29; Antiplaque M29; Antiprolactin PAM; Antiprostatitic 50 mg/man/day/orl PAM; Antirheumatic DAS; Antistomatitic 50 mg/man/3x/day MAR; Antitinnitic 60-120 mg/day WER; Antiulcer 50 mg/3x/day/man MAR; Antiulcer 88 mg/rat/day/15 days; Antiviral? M29; Astringent MAR; Deodorant M29; Immunostimulant DAS; Immunosuppressant 300 mg/day/6 wks/orl/man M29; Mucogenic JPP42:657; Spermigenic 60 mg/day WER; Testosteronigenic 60 mg/day WER; Trichomonicide M29; Vulnerary MAR PTD=500mg/day RDA=5-19mg/day RDA=5.5-22 M29

ZINGIBERENE Antiulcer IC54=100 mg/kg JDD6:30 HG20:23; Insecticide TOX; Perfumery ARC

ZINGIBERONE Antimutagenic JDD6:25; Antiulcer JDD6:25

References

15 -- Jim Duke's personal files.

382 -- Jacobson, M., Glossary of Plant-Derived Insect Deterrents, CRC Press, Inc., Boca Raton, FL, 213 p, 1990.

411 -- Williamson, E. M. and Evans, F. J., Potter's New Cyclopaedia of Botanical Drugs and Preparations, Revised Ed., Saffron Walden, the C. W. Daniel Co., Ltd., Essex UK, 362 pp, 1988, reprint 1989.

438 -- Lydon, J. & Duke, S., The potential of pesticides from plants, pp. 1-41 in Craker, L. & Simon, J., eds, Herbs, Spices & Medicinal Plants: Recent Advances in Botany, Horticulture, & Pharmacology, v. 4, Oryx Press, Phoenix, 1989, 267pp.

450 -- Russell, G. B., Phytochemical resources for crop protection, N. Z. J. Technol., 2, 1986, 127-134.

451 -- Osawa, K., Matsumoto, T., Maruyama, T., Takiguchi, T., Okuda, K., and Takazoe, I., Studies of the antibacterial activity of plant extracts and their constituents against periodontopathic bacteria, Bull Tokyo Dent. Coll., 31 (1), 1990, 17-21

452 -- Recio, M. C., Rios, J. L., and Villar, A., A review of some antimicrobial compounds isolated from medicinal plants reported in the literature 1978-1988, Phytotherapy Research, 3(4), 1989, 117-125,

453 -- Grundy, D. L. and Still, C. C., Inhibition of acetylcholinesterases by pulegone-1,2-epoxide, Oest. Biochem. & Physiol., 23, 1985, 383-8.

454 -- Berenbaum, M. R., The natives knew, Chemtech, (May), 1990, 275-279.

455 -- Hanke, F. J., Natural Products as a Resource for Biologically Active Compounds, Thesis, U. Cal (Santa Cruz), 1986, 252 pp

458 -- Bedows, E. and Hatfield, G. M. An investigation of the antiviral activity of Podophyllum peltatum, J. Nat. Prod., 45(6), 1982, 725-9.

505 -- Jim Duke's personal files.

52 -- Cooper-Driver, G. A., Chemical substances in plants toxic to animals, pp. 213-47 in Rechcigl, M., Jr., ed. CRC Handbook of Naturally Occurring Food Toxicants, CRC Press, Boca Raton, 1983, 339 pp.

525 -- Stitt, P. A., Why George Should Eat Broccoli, Dougherty Co, Milwaukee, WI, 1990, 399 pp.

AAA39:C-49 -- Butenko, I.G., Gladtchenko, S.V., and Galushko, S.V. Anti-inflammatory properties and inhibition of leukotriene C4 biosynthesis in vitro by flavonoid baicalein from Scutellaria baicalensis georgy roots. Agents Actions 39: C49, 1993.

AB68:195 -- Angewandte Botanik 68(5/6)=1994. Page 195.

ABC53:519 -- Kikuzaki, H. and Nakatani, N. 1988. Structure of a New Antioxidative Phenolic Acid from Oregano (Origanum vulgare L.). Agric. Biol. Chem. 53: 519-524.

ABS -- (followed by vol.num.) Acta Botanica Sinica

ABS32:49 -- Acta Botanica Sinica, 32: 49.

ACM:134 -- Advance in Chinese Medicinal Materials Research. 1985. Eds. H. M. Chang, H. W. Yeung, W. -W. Tso and A. Koo. World Scientific Publishing Co., Philadelphia Pa., page 134.

ACM:200 -- Advance in Chinese Medicinal Materials Research. 1985. Eds. H. M. Chang, H. W. Yeung, W. -W. Tso and A. Koo. World Scientific Publishing Co., Philadelphia Pa., page 200.

ACM:207 -- Advance in Chinese Medicinal Materials Research. 1985. Eds. H. M. Chang, H. W. Yeung, W. -W. Tso and A. Koo. World Scientific Publishing Co., Philadelphia Pa., page 207.

ACM:210 -- Advance in Chinese Medicinal Materials Research. 1985. Eds. H. M. Chang, H. W. Yeung, W. -W. Tso and A. Koo. World Scientific Publishing Co., Philadelphia Pa., page 210.

ACM:211 -- Advance in Chinese Medicinal Materials Research. 1985. Eds. H. M. Chang, H. W. Yeung, W. -W. Tso and A. Koo. World Scientific Publishing Co., Philadelphia Pa., page 211.

ACM:253 -- Advance in Chinese Medicinal Materials Research. 1985. Eds. H. M. Chang, H. W. Yeung, W. -W. Tso and A. Koo. World Scientific Publishing Co., Philadelphia Pa., page 253.

ACM:275 -- Advance in Chinese Medicinal Materials Research. 1985. Eds. H. M. Chang, H. W. Yeung, W. -W. Tso and A. Koo. World Scientific Publishing Co., Philadelphia Pa., page 275.

ACM:276 -- Advance in Chinese Medicinal Materials Research. 1985. Eds. H. M. Chang, H. W. Yeung, W. -W. Tso and A. Koo. World Scientific Publishing Co., Philadelphia Pa., page 276.

ACM:359 -- Advance in Chinese Medicinal Materials Research. 1985. Eds. H. M. Chang, H. W. Yeung, W. -W. Tso and A. Koo. World Scientific Publishing Co., Philadelphia Pa., page 359.

ACM:407 -- Advance in Chinese Medicinal Materials Research. 1985. Eds. H. M. Chang, H. W. Yeung, W. -W. Tso and A. Koo. World Scientific Publishing Co., Philadelphia Pa., page 407.

ACM:69 -- Advance in Chinese Medicinal Materials Research. 1985. Eds. H. M. Chang, H. W. Yeung, W. -W. Tso and A. Koo. World Scientific Publishing Co., Philadelphia Pa., page 69.

AEH -- De Smet, P.A.G.M., Keller, K., Hansel, R. and Chandler, R.F., eds. 1993. Adverse Effects of Herbal Drugs 2. Springr-Verlag, Berlin. 348pp.

AFR27:149 -- Singleton, V.L. 1981. Naturally Occurring Food Toxicants: Phenolic Substances of Plant Origin Common in Foods. Advances in Food Research 27: 149-217.

AH12(4):17 -- American Health, 12(4): 17, 1993.

AHN188:83 -- Acta Horticulturae, No.188: 83.

AJM7.29 -- Australian J. Med Herbalism, 7: 29.

ALH -- A.L.L. Hunting. Encyclopedia of Shampoo Ingredients. Micelle Press, Cranford NJ. 467 pp.

APF48:103 -- Jim Duke's personal files.

ARC -- Aloe Research Council - Duke writeup of non-peer reviewd book by Coats and draft by Henry NOTE: Most ARC numerical data were in mg/dl. I know of no other plant where zinc is more prevalent than other minerals and protein. These data should be viewed as suspect.

AVR14:323 -- Antiviral Research, 14: 323.

AYL -- Leung, A.Y., Encyclopedia of Common Natural Ingredients Used in Food, Drugs, and Cosmetics, John Wiley & Sons, New York, 1980.

BBA1115:69 -- Biochimica & Biophysica Acta, 1115: 69, 1991.

BBB56:2062 -- Murata, M., Yamakoshi, Y., Homma, S., Arai, K., and Nakamura, Y. 1992. Macrocarpals, Antibacterial Compounds from Eucalyptus, Inhibit Aldose Reductase. Biosci. Biotech. Biochem., 56(12): 2062-2063, 1992.

BBE -- Borchard, R. E., Barnes, C. D., and Eltherton, L. G. 1991. Drug Dosage in Laboratory Animals: A Handbook. (3rd Ed.) The Telford Press, Inc., P. O. Box 287, Caldwell NJ 07006.

BCI28:735 -- Balanehru, S. and Nagarajan, B. 1992. Intervention of Adriamycin Induced Free Radical Damage. Biochemistry International 28(4): 735-744, 1992.

BIS -- Bisset, N.G., ed. 1994. Herbal Drugs and Phytopharmaceuticals. CRC Press. Boca Raton, FL. 566 pp. NOTE: English translation of Wichtl, 1984,1989).

BJP103:1719 -- Jim Duke's personal files.

BML -- Lawrence, B.M., Essential Oils 1976-1977, Essential Oils 1978, Essential Oils 1979-1980. NOTE: Three annual compilations on essential oils. Percentages multiplied by 250-10,000 ppm and rounded to get fresh and dry weights.

BOB -- Betting on Boron, Unpublished draft by J. A. Duke on file at USDA, draft and papers relating to boron percentages. Includes Internat. Z. Vit. Ern. Forschung 43:1973 (boron).
BOIK1995 -- Boik, John. 1995. Cancer & natural medicine. Oregon Medical Press, Princeton, MN. 315 pp.

BP40(10):2227 -- Biochem Pharmacol, 40(10): 2227.

BPCC -- Biosyn. Prod. Cancer Chemotherapy (Petit et al)

BVC:162 -- Jime Duke's personal files.

CAL -- Chen, H.C. and Lin, S.M. 1988. Determination of Mineral Elements in Certain Crude Drugs (Part 1), Kaohsiung J. Med. Sci., 4: 259-272.

CCO -- Chemical Constituents of Oriental Herbs (3 diff. books)

CE7:285 -- Chemical Express, 7: 285, 1992.

CE7:297 -- Chemical Express, 7: 297, 1992.

CIB -- CIBA (Chadwick, D.J. and Marsh, J. 1990. Bioactive Compounds from Plants. CIBA Foundation Symposium 154. John Wiley & Sons, Ltd. Chichester UK. 242 pp.

CJT -- Chin. J. Tuber. Respir. Dis.

CJT4:203.1982 -- Chin. J. Tuber. Respir. Dis., 4:203, 1982.

CMB39:119 -- Cell Mol. Biol. Res. (Vol 39=1993)

CMR3/4/91 -- Chemical Marketing Reporter; a weekly tabloid, 3/4/91, p.16.

CMR9/10/92 -- Chemical Marketing Reporter; a weekly tabloid, 9/10/92.

CPA -- Can. Pharm. Assoc. Medication. 1988

CPB34:2164.1986 -- Chem & Pharm. Bull., 34: 2164, 1986.

CPB34:3854.1986 -- Chem & Pharm. Bull., 34: 3854, 1986.

CPB36:3153 -- Chem & Pharm. Bull., 36: 3153.

CPB37:2531 -- Shimizu, M., Horie, S., Terashima, S., Ueno, H., et al. 1989. Studies on Aldose Reductase Inhibitors from Natural Products.II. Active Components of a Paraguayan Crude Drug "Para-parai mi," Phyllanthus niruri. Chem. Pharm. Bull. 37(9): 2531-2532, 1989.

CPB37:3153 -- Kohda, H., Tanaka, S., Yamaoka, Y., Yahara, S., Nohara, T., Tanimoto, T., Tanaka, A. 1989. Studies on Lens-Aldose-Reductase Inhibitor in Medicinal Plants.II Active Constituents of Monochasma savatierii FRANCH, et MAXIM. Chem. Pharm. Bull. 37(11):3153-3154

CPB38:1049 -- Uchida, U., Ohta, H., Niwa, M., Mori, A., Nonaka, G-i., Nishioka, I., and Zaki, M. 1989. Prolongation of Life Span of Stroke-Prone Spontaneously Hypertensive Rats (SHRSP) Ingesting Persimmon Tannin. Chem. Pharm. Bull. 38(4): 1049-1052, 1990.

CPB38:1225 -- Hatano, T., Yasuhara, T., Yoshihara, R., Agata, I., Noro, T., and Okuda, T. 1989. Effects of Interaction of Tannins with Co-existing Substances.VII. Inhibitory Effects of Tannins Related Polyphenols on Xanthine Oxidase. Chem. Pharm. Bull. 38(5): 1224-1229

CPB38:1620 -- Ozaki, Y. and Ma, J-P. 1989. Inhibitory Effects of Tetramethylpyrazine and Ferulic Acid on Spontaneous Movement of Rat Uterus in Situ. Chem. Pharm. Bull. 38(6): 1620-1623, 1990.

CPB38:1772 -- Chem. & Pharm. Bull. 38: 1772.

CPB38:2012 -- Akao, M. and Kuroda, K. 1990. Inhibitory Effect of Fumaric Acid on Hepatocarcinogenesis by Thioacetamide in Mice. Chem. Pharm. Bull. 38(7): 2012-2014, 1990.

CPB38:2033 -- Chem. & Pharm. Bull. 38: 2033.

CPB38:208 -- Abe, K-i., Inoue, O., and Yumioka, E. 1989. Biliary Excretion of Metabolites of Baicalin and Baicalein in Rats. Chem. Pharm. Bull. 38(1): 208-211, 1990.

CPB38:2201 -- Mayumi Ito (nee Someya), et al. 1990. Hepatoprotective Compounds from Canarium album and Euphorbia nematocypha. Chem. Pharm. Bull. 38(8): 2201-2203, 1990.

CPB38:2205 -- Chem. & Pharm. Bull. 38: 2205.

CPB38:2283 -- Shimizu,*M., et al. 1990. Anti-inflammatory Constituents of Topically Applied Crude Drugs. IV.1) Constituents and Anti-inflammatory Effect of Paraguayan Crude Drug "Alhucema" (Lavandula latifolia Vill.)2). Chem. Pharm. Bull. 38(8): 2283-2284, 1990.

CPB38:2733 -- Terashima, S., Shimizu, M., Nakayama, H., Ishikura, M., et al. 1990. Studies on Aldose Reductase Inhibitors from Medicinal Plant of "Sinfito," Potentilla candicans, and Further Synthesis of Their Related Compounds. Chem. Pharm. Bull 38(10):2733-2736, 1990

CPB38:297 -- Jin, G-Z., Yamagata, Y., and Tomita, K.-i. 1989. Structure of Rutin Pentamethanol. Chem. Pharm. Bull. 38(2): 297-300, 1990.
CPB38:3031 -- Oh-Hara, T., Kawazoe, Y., and Sakagami, H. 1990. Lignified Materials as Potential Medicinal Resources.III. Diversity of Biological Activity and Possible Molecular Species Involved. Chem. Pharm. Bull. 38(11): 3031-3034, 1990.

CPB38:3053 -- Yamahara, J., Li, Y.H., and Tamai, Y. 1990. Anti-ulcer Effect in Rats of Bitter Cardamon Constituents. Chem. Pharm. Bull. 38(11): 3053-3054, 1990.

CPB38:3086 -- Seki, T., Toeda, C., Kawaguchi, T., Juni, K., Sugibayashi, K., and Morimoto, Y. 1990. Enhanced Transdermal Delivery of Zidovudine in Rats and Human Skin. Chem. Pharm. Bull. 38(11): 3086-3089, 1990.

CPB38:3355 -- Yanagisawa, T., Kosakai, K., Tomiyama, T., Yasunami, M., and Takase, K. 1990. Studies on Anti-ulcer Agents. II. Synthesis and Anti-ulcer Activities of 6-Isopropylazulene-1-sodium Sulfonate Derivatives. Chem. Pharm. Bull. 38(12): 3355-3358, 1990.

CPB38:774 -- Yasukawa, K., Takido, M., Takeuchi, M., Sato, Y., Nitta, K., and Nakagawa, S. 1989. Inhibitory Effects of Flavonol Glycosides on 12-O-Tetradecanoylphorbol-13-acetate-Induced Tumor Promotion. Chem. Pharm. Bull. 38(3): 774-776, 1990.

CPB39:3276 -- Tsuruga, T., Chun, Y.-T., Ebizuka, Y., and Sankawa, U. 1991. Biologically Active Constituents of Melaleuca leucadendron: Inhibitors of Induced Histamine Release from Rat Mast Cells. Chem. Pharm. Bull. 39(12): 3276-3278, 1991.

CR48:5941 -- Cancer Research, 48: 5941.

CR54:701 -- Cancer Research, 54: 701, 1994.

CRC -- CRC Handbook of Medicinal Herbs and/or CRC Handbook of Proximate Analyses

CRC(FNS) -- Jim Duke's personal files.

DAS -- Davies, S., and Stewart, A. 1990. Nutritional Medicine. Avon Books, New York. 509pp.

DAZ129:155 -- Deutsche Apot. Zeit. 129(4):155-163. W. Schulze et al. Die Melisse. NOTE: <0.1=Trace.

DIE -- Die Nahrung. Pino, J., Rosado, A., Goire, I., Roncal, E., and Garcia, I. 1993. Analysis of the Essential Oil from Cuban Basil. Die Nahrung 37:(5): 501-504. NOTE: Also published in JEO6:89, 1994.

EB41:474 -- Economic Botany, a publication of the Society for Economic Botany, The New York Botanical Garden. 41: 474.
EB44:68.1990 -- Schultes, R.E. De Plantis Toxicariis e Mundo Novo Tropicale Commentationes XXXVI. Justicia (Acanthaceae) as a Source of an Hallucinogenic Snuff. Economic Botany 44(1): 61-70, 1990.

EB48:111 -- Tunon, H., Thorsell, W., and Bohlin, L. 1993. Mosquito Repelling Activity of Compounds Occurring in Achillea millefolium L. (Asteraceae). Economic Botany 48(2): 111-120, 1994.

EB48:60 -- Alkire, B.H., Tucker, A.O., and Maciarello, M.J. 1994. Tipo (Minthostachys mollis (Lamiaceae): An Ecuadorian Mint. Econ. Bot. 48(1): 60-64. NOTE: EO=pers. comm. AOT

EJP251:91 -- Huang, H-C., Wang, H-R., and Hsieh, L-M. 1993. Antiproliferative effect of baicalein, a flavonoid from a Chinese herb, on vascular smooth muscle cell. European J. of Pharmacology 251. 91-93, 1994.

EMM18:28 -- Kauderer, B., Zamith, H., Paumgartten, F.J.R., and Speit, G. Evaluation of the Mutagenicity of B-Myrcene in Mammalian Cells In Vitro. Environmental and Molecular Mutagenesis 18: 28-34, 1991.

EMP1:124 -- Economic & Medicinal Plant Research, 1: 124.

EMP1:53 -- Economic & Medicinal Plant Research, 1: 53.

EMP5:194 -- Economic & Medicinal Plant Research, 5: 194.

EMP5:195 -- Economic & Medicinal Plant Research, 5: 195.

EMP5:197 -- Economic & Medicinal Plant Research, 5: 197.

EMP5:198 -- Economic & Medicinal Plant Research, 5: 198.

EMP5:199 -- Economic & Medicinal Plant Research, 5: 199.

EMP5:204 -- Economic & Medicinal Plant Research, 5: 204.

EMP5:207 -- Economic & Medicinal Plant Research, 5: 207.

EMP5:224 -- Economic & Medicinal Plant Research, 5: 224.

EMP5:225 -- Economic & Medicinal Plant Research, 5: 225.

EMP5:253 -- Economic & Medicinal Plant Research, 5: 253.

EMP5:333 -- Economic & Medicinal Plant Research, 5: 333.

EMP5:343 -- Economic & Medicinal Plant Research, 5: 343.
EMP5:363 -- Economic & Medicinal Plant Research, 5: 363.

EMP5:366 -- Economic & Medicinal Plant Research, 5: 366.

EMP6:157 -- Economic & Medicinal Plant Research, 6: 157.

EMP6:158 -- Economic & Medicinal Plant Research, 6: 158.

EMP6:162 -- Economic & Medicinal Plant Research, 6: 162.

EMP6:165 -- Economic & Medicinal Plant Research, 6: 165

EMP6:170 -- Economic & Medicinal Plant Research, 6: 170.

EMP6:189 -- Economic & Medicinal Plant Research, 6: 189.

EMP6:235 -- Economic & Medicinal Plant Research, 6: 235.

FAC -- Factos on File.

FFJ10(3):199 -- Pedro, L.G., et al. 1995. Composition of the Essential oil of Micromeria varia Benth. ssp. thymoides (Sol. ex Lowe) Perez var. thymoides, and endemic species of the Madeira Archipelago. flav. & Fragr. J. 10(3): 199-202.

FFJ10(3):225 -- Stahl-Biskup, E. and Holthuijzen, J. 1995. Essential oil and glycosidally bound volatiles of lemon-scented thyme, Thymus x citriodorus (Pers.) Schreb. Flav. & Fragr. J. 10: 225-229.

FFJ10:379 -- Vila, R., et al. 1995. Composition and study of the variability of the essential oil of Thymus funkii Cousson. Flav. & Fragr. J. 10(6): 379-383.

FFJ6:153 -- Flavour and Fragrance Journal, 6: 153

FFJ6:154 -- Flavour and Fragrance Journal, 6: 154.

FFJ6:69 -- Flavour and Fragrance Journal, 6: 69.

FFJ6:72 -- Flavour and Fragrance Journal, 6: 72.

FFJ6:80 -- Flavour and Fragrance Journal, 6: 80.

FFJ8:127 -- Velasco-Negueruela,A., Perez-Alonso,M.J., Jiminez,S.M. and Garcia,F.M. 1993. The Volatile Constituents of Acinus alpinus (L.) Moench ssp. meridionalis (Nyman). P.W. Ball Growing in Spain. Flav. & Frag. J. 8:127-130.)

FFJ8:331 -- Biondi, D., Cianci, P., Geraci, C. and Ruberto, G. 1993. Antimicrobial Activity and Chemical Composition of Essential Oils from Sicilian Aromatic Plants. Flav. & Frag. J. 8: 331-7.

FFJ9:129 -- Flav. & Frag. J. 9: 129.

FFJ9:201 -- Casnigueral,S., Iglesias,J., Vila,R., Virgili,A. and Ibanez,C.1994. The Essential Oil from Leaves of Salvia canariensis L. Flav. & Frag. J. 9:201-204. S. Canigueral, Facultat de Farmacia, Universitat de Barcelona, Ave. Diagonal 643,E-08028, Barcelone Spain

FFJ9:29 -- Soliman, F.M., El-Kashoury, M.M., Fathy, M.M. and Gonaid, M.H. 1994. Analysis and Biological Activity of the Essential Oil of Rosmarinus officinalis L. from Egypt. Flavour and Fragrance J. 9: 29-33.

FHN -- Forest H. Nielsen, USDA, Grand Forks, various publications.

FNF -- Father Nature's Farmacy: The aggregate of all these three-letter citations.

FT1990:483 -- Singla*, A.K. Pathak, K. 1989. Phytoconstituents of Euphorbia species. Fitoterapia 61(6): 483-508, 1990.

FT2:103.1986 -- Somasundaram, S., Sadique, J. 1985. Anti-hemolytic Effect of Flavonoidal Glycosides of Clerodendron inerme: An in vitro Study. Fitoterapia 57(2): 103, 1986.

FT2:1986 -- Fitoterapia No.2-1986.

FT55.1984 -- Fitoterapia No.55-1984.

FT59:465 -- Fitoterapia No.59-1984.

FT5:1986 -- Fitoterapia No.5-1986.

FT5:1990 -- Fitoterapia No.5-1990.

FT60:277 -- Fitoterapia No.60: 277.

FT60:349 -- Fitoterapia No.60: 349.

FT61:243.1990 -- Fitoterapia No.61: 243, 1990.

FT61:254.1990 -- Fitoterapia No.61: 254, 1990.

FT61:263.1990 -- Fitoterapia No.61: 263, 1990.

FT61:277.1990 -- Fitoterapia No.61: 277, 1990.

FT62:166 -- Fitoterapia No.62: 166.

FT62:170 -- Parveen, N. 1991. Antifilarial Activity of Vitex negundo Against Setaria cervi. Fitoterapia, 62(2): 163, 1991.

FT62:372 -- Pathak, D., Pathak, K., Singla*, A.K. 1991. Flavonoids as medicinal agents - Recent advances. Fitoterapia 62(5): 62, 1991.

FT62:86 -- Singh*, G., Upadhyay, R.K. 1990. Fungitoxic Activity of Cumaldehyde, Main Constituent of the Cuminum cyminum Oil. Fitoterapia 62(1): 86, 1991.

FT64:131 -- Shin*, K.H., Chung, M.S., Chae, Y.J., Yoon, K.Y., Cho, T.S. 1992. A survey for aldose reductase inhibition of herbal medicines. Fitoterapia 64(2): 130, 1993.

FT64:134 -- Lazarova*, G., Kostova, I., Neychev, H. 1992. Photodynamic damage prevention by some hydroxycoumarins. Fitoterapia 64(2): 134-136, 1993.

FTS1986 -- Jim Duke's personal files.

GEO -- Guenther, E., The Essential Oils, 6 volumes, D. van Nostrand, New York, 1948-1952.

GMJ -- Grenand, P., Moretti, C., and Jacquemin, H. 1987. Pharmacopees Traditionnelles en Guyane. l'ORSTROM, Paris. 569 pp.

HEG -- Hegnauer's Chemotaxonomie der Pflanzen, 8 vols., Birkhauser Verlag, Berlin.

HG17 -- HerbalGram No. 17.

HG20:15 -- HerbalGram No. 20, page 15.

HG20:40 -- HerbalGram No. 20, page 40.

HG20:9 -- HerbalGram No. 20, page 9.

HG22:14 -- HerbalGram No. 22 - Spring 1990, page 14.

HG22:18 -- Blumenthal, M. "Traditional Herbal Medicines" in Current Drug Category. HerbalGram No. 22 - Spring 1990, page 18.

HG22:45 -- HerbalGram No. 22 - Spring 1990, page 45.

HG22:9 -- HerbalGram No. 22 - Spring 1990, page 9.

HG24:37 -- HerbalGram No. 24, page 37.

HG24:40 -- HerbalGram No. 24, page 40.

HH2:12 -- Herbal Healthline, 2: 12.

HHB -- List, P.H. and Horhammer, L., Hager's Handbuch der Pharmazeutischen Praxis, Vols. 2-6, Springer-Verlag, Berlin, 1969-1979.

HSC25:1485 -- Cerda, J.J. Pectin in Health and Disease. HortScience, 25(12): 1485, 1990.

HSC25:1486 -- Levander, O.A. Fruit and Vegetable Contributions to Dietary Mineral Intake in Human Health and Disease. HortScience, 25(12): 1486, 1990.

HSC25:1488 -- Anderson, J.W. Dietary Fiber and Human Health. HortScience, 25(12): 1488, 1990.

HSC25:1495 -- Simon, P.W. Carrots and Other Horticultural Crops as a Source of Provitamin A Carotenes. HortScience, 25(12): 1495, 1990.

HSC25:1523 -- Kyle, D., Bingham, S., and Radmer, R. Long-chain Omega-3 Polyunsaturated Fatty Acids: Prospects for Introduction into Horticultural Food Plants. HortScience, 25: 1523, 1990.

IAA94:262 -- Dorsch, W., Wagner, H. New Antiasthmatic Drugs from Traditional Medicine? Int Arch Allergy Appl Immunol 94: 262-265, 1991.

ID25:487 -- Jim Duke's personal files.

ID25:488 -- Jim Duke's personal files.

IJI10:729 -- Int. J. Immunopharmacology. Vol 10: 729, 1988.

IJI10:735 -- Int. J. Immunopharmacology. Vol 10: 735, 1988.

IJO15(4):194 -- Internat. J. Oriental Med. 15(4): 194, 1990+

IJO15(4):199 -- Internat. J. Oriental Med. 15(4): 199, 1990+

IJO16:4 -- Internat. J. Oriental Med. 16: 4.

IJP33:7 -- Pandeya, S.N. 1970. Biological Activity of Adamantyl Derivatives. The Indian Journal of Pharmacy, 33: 7.

IP35:51 -- Indian Perfumer, 35: 51.

IP35:52 -- Indian Perfumer, 35: 52.

ISA14:93 -- Bettarini, F., Borgonovi, G.E., et al. 1991. Antiparasitic Compounds from East African Plants: Isolation and Biological Activity of Anonaine, Matricarianol, Canthin-6-One and Caryophyllene Oxide. Insect Sci. Applic. 14(1): 93-99, 1993.

IWU -- Iwu, M.M. 1993. Handbook of African Medicinal Plants. CRC Press, Boca Raton, FL 435 pp.

JA6(4):33 -- Jim Duke's personal files.

JA6(4):34 -- Jim Duke's personal files.

JA6(4):36 -- Jim Duke's personal files.

JA6(4):40 -- Jim Duke's personal files.

JAD -- Duke, J. A. Writeups or information summaries on approximately 2,000 economic plants, USDA, ARS, Beltsville, MD 20705.

JAD85:950 -- J. Amer. Diet. Ass. 85{8}:950. NOTE: Leaves, fruits, and other normally wet items back-calculated to fresh weight by dividing by 10.

JAF34:409 -- Bushway, R.J. 1986. Determination of alpha- and beta-carotene in some raw fruits and vegetables by high-performance liquid chromatography. J. Agric. Food Chem. 34:409.

JAF36:1079 -- J. Agric. Food Chem. 36: 1079.

JAF37:1363 -- Lawson, T., J. Nunnaly, B. Walker, E. Bresnick, D. Wheeler, and M. Wheeler. 1989. Isolation of compounds with antimutagenic activity from savoy chieftain cabbage. J. Agric. Food Chem. 37:1363.

JAF38:493 -- Lugemwa, F.N., Huang, F-Y., Bentley, M.D., Mendel, M.J., and Alford, A.R. 1990. A Heliothis zea Antifeedant from the Abundant Birchbark Triterpene Betulin. J. Agric. Food Chem. 38(2): 493-494.

JAF38:688 -- Oszmianski, J. and Lee, C.Y. 1990. Inhibitory Effect of Phenolics on Carotene Bleaching in Vegetables. J. Agric. Food Chem. 38: 688-690.

JAF38:805 -- Friedman, M. and Dao, L. 1990. Effect of Autoclaving and Conventional and Microwave Baking on the Ergot Alkaloid and Chlorogenic Acid Contents of Morning Glory (Ipomoea tricolor Cav. cv.) Heavenly Blue Seeds. J. Agric. Food Chem. 38: 805-808.

JAF39:2290 -- Kubo, I. and Himejima, M. 1991. Anethole, a Synergist of Polygodial against Filamentous Microorganisms. J. Agric. Food Chem. 39: 2290-2292.

JAF40:2328 -- Kang, R., Helms, R., Stout, M.J., Jaber, H., Chen, Z., and Nakatsu, T. 1992. Antimicrobial Activity of the Volatile Constituents of Perilla frutescens and Its Synergistic Effects with Polygodial. J. Agric. Food Chem. 40: 2328-2330.

JAF41:1103 -- Muroi, H. and Kubo, I. 1993. Combination Effects of Antibacterial Compounds in Green Tea Flavor against Streptococcus mutans. J. Agric. Food Chem. 41: 1102-1105.

JAF41:110G -- J. Agric. Food Chem. 41: 110G.

JAF42:2048 -- Korhammer, S.A. and Haslinger, E. 1994. Isolation of a Biologically Active Substance from Rhizomes of Quackgrass [Elymus repens (L.) Gould]. J. Agric. Food Chem. 42: 2048-2050.

JAF42:776 -- Kerrola, K., Galambosi, B. and Kallio, H. 1994. Volatile Components and Odor Intensity of Four Phenotypes of Hyssop (Hyssopus officinalis L.) J. Agric. Food Chem. 42: 776-781.

JAF43:2144 -- Yu, S.G., Anderson, P.J., and Elson, C.E. 1995. Efficacy of B-Ionone in the Chemoprevention of Rat Mammary Carcinogenesis. J. Agric. Food Chem. 43: 2144-2147.

JAF43:2162 -- Igile, G.O., Oleszek, W., Burda, S., and Jurzysta, M. 1995. Nutritional Assessment of Vernonia amygdalina Leaves in Growing Mice. J. Agric. Food Chem. 43: 2162-2166.

JAF43:2283 -- J. Agric. Food Chem. 43: 2283.

JAR1:11 -- J. Aromatherapy 1: 11.

JAR4:22 -- J. Aromatherapy 4: 22.

JAR4:24 -- J. Aromatherapy 4: 24.

JBH -- Jeffery B. Harborne and H. Baxter, eds. 1983. Phytochemical Dictionary. A Handbook of Bioactive Compounds from Plants. Taylor & Frost, London. 791 pp.

JCE17:865 -- Holappa, L.D., and Blum*, U. 1991. Effects of Exogenously Applied Ferulic Acid, a Potential Allelopathic Compound, on Leaf Growth, Water Utilization, and Endogenous Abscisic Acid Levels of Tomato, Cucumber, and Bean. J. of Chemical Ecology, 17(5): 865.

JCE20:2847 -- Hardie, J., Isaacs, R., Pickett,* J.A., Wadhams, L.J., and Woodcock, C.M. 1994. Methyl Salicylate and (-)-(1R,5S)-Myrtenal are Plant-Derived Repellents for Black Bean Aphid, Aphis Fabae Scop. (Homoptera: Aphididae). J. of Chemical Ecology, 20(11):2847.

JDD6:25 -- J. Drug. Dev. (Vol. 6=1993) e.g. Mustafa, T., Srivastava, K.C., and Jensen, K.B. 1993. Drug Development Report (9): Pharmacology of Ginger, Zingiber officinale. J. Drug. Dev. 6(1): 25-39.

JE12:279.1984 -- Onawunmi, G.O., Yisak, W-A., Ogunlana,* E.O. 1984. Antibacterial Constituents In The Essential Oil Of Cymbopogon citratus (DC.) Stapf. Journal of Ethnopharmacology, 12: 279-286, 1984.
JE15:204 -- Khalid, S.A., Farouk, A., Geary**, T.G., and Jensen, J.B. 1985. Potential Antimalarial Candidates From African Plants: An In Vitro Approach Using Plasmodium falciparum*. Journal of Ethnopharmacology, 15: 201-209, 1986.

JE25:165 -- Ortiz de Urbina, A.V., Martin*, M.L., Montero, M.J., Moran, A., and San Roman, *L. 1988. Sedating and Antipyretic Activity of the Essential Oil* of Calamintha sylvatica subsp. Ascendens. Journal of Ethnopharmacology, 25: 165-171, 1989.

JE26:65 -- Spring, M.A. 1988. Ethnopharmacologic Analysis of Medicinal Plants Used by Laotian Hmong Refugees in Minnesota. Journal of Ethnopharmacology, 26: 65-91, 1989.

JE27:243 -- Ivorra, M.D., Paya, M., and Villar, A. 1989. A Review of Natural Products and Plants as Potential Antidiabetic Drugs. Journal of Ethnopharmacology, 27: 243-275, 1989.

JE28:103 -- J. Ethnopharmacology, 28: 103.

JE28:221 -- Malini, T. and Vanithakumari, G. 1989. Rat Toxicity Studies With B-Sitosterol. Journal of Ethnopharmacology, 28: 221-234, 1990.

JE29:179 -- J. Ethnopharmacology, 29: 179.

JE29:25 -- Rafatullah, S., Tariq, M., Al-Yahya, M.A., Mossa, J.S., and Ageel, A.M. 1989. Evaluation of Turmeric (Curcuma longa) for Gastric and Duodenal Antiulcer Activity in Rats. J. of Ethnopharmacology, 29: 25-34, 1990.

JE31:43 -- Encarnacion Dimayuga, R., Keer Garcia, S., Halfdan Nielsen, P., and Christophersen, C. 1989. Traditional Medicine of Baja California Sur (Mexico) III. Carnosol: A Diterpene Antibiotic from Lepechinia hastata. J. Ethnopharmacology, 31: 43-48, 1991.

JE32:117 -- J. Ethnopharmacology, 32: 117.

JE32:141 -- J. Ethnopharmacology, 32: 141.

JE32:167 -- Pei-Gen, X., and Nai-Gong, W. Can ethnopharmacology contribute to the development of anti-fertility drugs? Journal of Ethnopharmacology, 32: 167-177, 1991.

JE35:105 -- Arnold, N., Bellomaria, B., Velentini G. and Rafaiani, S.M. 1991. Comparative Study on Essential Oil of Some Teucrium Species from Cyprus. J. Ethnopharm. 35: 105-113.

JE39:167 -- J. Ethnopharmacology, 39: 167.

JEO2:149 -- Tucker, A.O. and Maciarello, M.J. 1990. The Essential Oil of Trichostemma dichotomum. J. Ess. Oil Res. 2: 149-150.

JEO2:185 -- Buchbauer, G., Jirovetz*, L., Nikiforov, A., Remberg, G., Raverdino, V. 1989. Headspace-Analysis and Aroma Compounds of Austrian Hay-Blossoms (Flores Graminis, Graminis Flos) used in Aromatherapy. J. Ess. Oil Res., 2: 185-191, 1990.

JEO3:11 -- Bandoni, A.L., Mizrahi, I., and Juarez, M.A. 1991. The Essential Oil of Rosmarinus officinalis Growing in Argentina. J. Ess. Oil Res., 3: 11.

JEO3:111 -- Gbolade, A.A. and Lockwood, G.B. 1991. The Constituents of Melissa officinalis L. Cultures. J. Ess. Oil Res., 3: 111-114.

JEO3:121 -- Fleisher, A. & Fleisher, Z. 1991. Chemical Composition of Origanum syriacum L. Essential Oil. J. Ess. Oil Res. 3: 121-123.

JEO3:153 -- Stanic, G., Petricic, J., and Blazevic, N. 1991. Gas Chromatographic Investigations of Essential Oils of Satureja montana and Satureja subspicata from Yugoslavia. J. Ess. Oil Res., 3: 153-158.

JEO3:191 -- Tumen, G. 1991. The Volatile Constituents of Acinos suaveolens (Sibt. et Smith). G. Don fil. Growing in Turkey. J. Ess. Oil Res. 3:191-2.

JEO3:361 -- Carnat, A., Chossegros, A., and Lamaison, J. 1991. The Essential Oil of Satureja grandiflora (L.) Scheele from France. J. Ess. Oil Res., 3: 361-362

JEO3:365 -- Tumen, G. 1991. The Volatile Constituents of Satureja cuneifolia. J. Ess. Oil Res., 3: 365-366.

JEO3:387 -- Kirimer, N., Ozek, T., and Baser, K.H.C. 1991. Composition of the Essential Oil of Micromeria congesta. J. Ess. Oil Res., 3: 387-393.

JEO3:395 -- Manez, S., Jimenez, A., and Villar, A. 1991. Volatiles of Sideritis mugronensis Flower and Leaf. J. Ess. Oil Res., 3: 395-397.

JEO3:409 -- De Cunha, A.P. and Salguiero, L.R. 1991. The Chemical Polymorphism of Thymus zygis ssp. sylvestris from Central Portugal. J. Ess. Oil Res. 3: 409-12.

JEO3:43 -- Iglesias, J., Vila, R., Canigueral, S., Bellakdhar, and Il Idrissi, A. 1991. Analysis of the Essential Oil of Thymus riatarum. J. Ess. Oil Res. 3: 43-4.

JEO3:441 -- Perez-Alonso, M., Velasco-Negueruela, A., and Lopez-Saez, A. 1991. The Essential Oil of Cleonia lusitanica. J. Ess. Oil Res., 3: 441-442.

JEO3:445 -- Baser, K.H.C., Tumen, G., Sezik, E. 1991. The Essential Oil of Origanum minutiflorum O. Schwarz and P.H. Davis. J. Ess. Oil Res. 3: 445-446.

JEO3:477 -- Fleisher, Z. and Fleisher, A. 1991. The Essential Oil of Micromeria fruticosa *(L.) Druce subsp.* barbata (Boiss et. Ky.), P.H. Davis. Aromatic Plants of the Holy Land and the Sinai. Part VII. J. Ess. Oil Res 3: 477-479.

JEO3:57 -- Fleisher, Z. and Fleisher, A. 1991. The Essential Oils from Mentha longifolia Growing in Sinai and Israel. J. Ess. Oil Res., 3: 57.

JEO3:7 -- Akgul, A., De Pooter, H.L., and De Buyck, L.F. 1991. The Essential Oils of Calamintha nepeta subsp. glandulosa and Ziziphora clinopodioides from Turkey. J. Ess. Oil Res., 3: 7-10.

JEO4:121 -- Bestman, H.J., Rauscher, J., Vostrowsky O., Pant, A.K., Dev. V., Perihar, R. and Mathela, C.S. 1992. Constituents of the Essential Oil of Elsholtsia blanda Benth. (Labiatae). J. Ess. Oils Res. 4: 121-124 NOTE: Parenthetical values based on other studies.

JEO4:125 -- Charles, D.J., and Simon, J.E. 1992. Essential Oil Constituents of Ocimum killimandscharicum Guerke. J. Ess. Oil Res., 4: 125-128.

JEO4:139 -- Baser, K.H.C., Ozek, T., Kurkcuoglu, M. and Tumen, G. 1992. Composition of the Essential Oil of Origanum sipyleum of Turkish Origin. J. Ess. Oil Res. 4: 139-142.

JEO4:189 -- Kirimer, N., Baser, K.H.C., Ozek, T. and Kurkcuoglo, M. 1992. Composition of the Essential Oil of Calamintha nepeta subsp. glandulosa. J. Ess. Oil Res. 4:189-190

JEO4:223 -- Blazevic, N., Kalodera, Z., Petricic, J., and Plazibat, M. 1992. Essential Oil Content and Composition of Teucrium arduini L. J. Ess. Oil Res. 4: 223-225.

JEO4:231 -- Charles, D.J. and Simon, J.E. 1992. A New Geraniol Chemotype of Ocimum gratissimum L. J. Ess. Oil Res. 4: 231-234.

JEO4:295 -- Mazza, G. and Kiehn, F.A. 1992. Essential Oil of Agastache foeniculum, A Potential Source of Methyl Chavicol. J. Ess. Oil Res., 4: 295-299.

JEO4:533 -- J. Essential Oil, 4: 533.

JEO4:547 -- Pant, A.K., Dev, V., Parihar, R., Mathela,C.S., Rauscher, J., Vostrowsky, O. and Bestmann, H.J. 1992. The Essential Oil from Elsholtzia eriostachya var. pusilla. J. Ess. Oil Res. 4: 547-549.

JEO4:57 -- Shah, G.C., Bhandari, R. & Mathela, C.S. 1992. 1,2-Epoxy-p-Menthane Derivatives from some Labiatae Species. J. Ess. Oil Res. 4: 57-59.

JEO4:79 -- Ozek, T., Kirimer, N., and Baser, K.H.C. 1992. Composition of the Essential Oil of Micromeria myrtifolia Boiss. et Hohen. J. Ess. Oil Res., 4: 79-80.

JEO5:15 -- Stengele, Michael and Stahl-Biskup, Elisabeth. 1991. Glycosidically Bound Volatiles in Peppermint (Mentha piperita L.) J. Essent. Oil Res., 5: 13-19.

JEO5:153 -- Jimenez, J., Navarro, M.C., Montilla, M.P., Martin, A. and Martinez, A. 1993. Thymus zygis Oil: Its Effects on CCl4-Induced Hepatotoxicity and Free Radical Scavenger Activity. JEO5: 153-8.

JEO5:159 -- Bourrel, C., Perineau, F., Michel, G., Bessiere, J.M. 1993. Catnip (Nepeta cataria L.) Essential Oil: Analysis of Chemical Constituents, Bacteriostatic and Fungistatic Properties. J. Ess. Oil Res., 5: 159-167.

JEO5:215 -- Baser, K.H.C., Ozek, T., Akgul, A. and Tumen, G. 1993. Composition of the Essential Oil of Nepeta racemosa Lam. J. Ess. Oil Res. 5: 215-7.

JEO5:243 -- Soriano Cano, M.C., Sotomayor Sanchez, J.A., Sanchez Gomez, P. and Garcia Vallejo, M.C. 1993. Essential Oils of the Rosmarinus eriocalyx-tomentosus Complex in Southeast Spain. J. Ess. Oil Res. 5: 243-246.

JEO5:291 -- Baser, K.H.C., Ozek, T., Kirimer, N. and Tumen, G. 1993. The Occurrence of Three Chemotypes of Thymus longicaulis C. Presl subsp. longicaulis in the same Population. J. Ess. Oil Res. 5: 291-5.

JEO5:311 -- Buchbauer, G., Jorovetz, L., Wasicky, M. and Nikiforov, A. 1993. Volatile Constituents of the Headspace and Essential Oil of Plectranthus coleoides Marginatus (Labiatae). J. Ess. Oil Res. 5: 311-313.

JEO5:319 -- Velasco-Negueruela, A. et al. 1993. The Essential Oil of Salvia gilliesii Benth. J. Ess. Oil Res. 5: 319-320.

JEO5:321 -- Mallvarapu, G.R., Ramesh, S., Kaul, P.N., Bhattacharya, A.K., and Rao, B.R.R. 1993. The Essential Oil of Hyptis suaveolens (L.) Poit. J. Ess. Oil Res. 5: 321.

JEO5:397 -- Perez-Alonso, M.J. Velasco-Negueruela, A. and Lopez-Saez, J.A. 1993. The Essential Oils of Two Iberian Teucrium Species. J. Ess. Oil Res. 5: 397-402.

JEO5:425 -- Sezik, E., Tumen, G., Kirimer, N., Ozek, T., and Baser, K.H.C. 1993. Essential Oil Composition of Four Origanum vulgare Subspecies of Anatolian Origin. J. Ess. Oil Res., 5: 425-431.

JEO5:443 -- Tucker, A.O., Maciarello, M.J., Angell, S., Espaillat, J.R., and French, E.C. 1993. The Essential Oil of Lavandula x hybrida Balb. ex Ging., a Distinct Hybrid from L. x heterophylla Poir. (Labiatae). J. Ess. Oil Res. 5: 443-445. NOTE: Figures rounded somewhat but including the variance from s.d.'s

JEO5:45 -- Tantaoui-Elaraki, A., Lattaoui, N., Errifi, A. and Benjilali, B. 1993. Composition and Antimicrobial Activity of the Essential Oils of Thymus broussonettii, T. zygis and T. saturejoides. J. Ess. Oil Res. 5: 45-53.

JEO5:547 -- Tumen, G. Baser, K.H.C. and Kirimer, N. 1993. The Essential Oil of Satureja cilicica P.H. Davis. J. Ess. Oil Res. 5: 547-548.

JEO5:549 -- Zygadlo, J.A., Merino. E.F., Maestri, D.M., Guzman, C.A. and Espinar, L.A. 1993. The Essential Oils of Satureja odora and Satureja parvifolia from Argentina. J. Ess. Oil Res. 5: 549-51.

JEO5:669 -- Ozek, T., Baser, K.H.C. and Tumen, G. 1993. The Essential Oil of Sideritis athoa Papanikolaou Et Kokkini. J. Ess. Oil Res. 5. 669-670.

JEO6:27 -- Collins, J.E., Bishop, C.D., Deans, S.G. and Svoboda, K.P. 1994. Composition of the Essential Oil from the Leaves and Flowers of Monarda citriodora var. citriodora grown in the United Kingdom. J. Ess. Oil Res. 6: 27-9.

JEO6:79 -- Tucker, A.O. & Maciarello, M.J. 1994. The Essential Oil of Salvia dorisiana Standley. J. Ess. Oil Res. 6: 97-8.

JEO6:97 -- Tumen, G., Koyuncu, M., Kirimer, N., and Baser, K.H.C. 1994. Composition of the Essential Oil of Thymus cilicicus Boiss. & Bal. J. Ess. Oil Res. 6: 97-8. NOTE: Trace recorded as <0.1

JEO7:271 -- J. Ess. Oil Res. 7: 271.

JEO:459 -- Brophy, J.J., Goldsack, R.J., and Clarkson, J.R. 1993. The Essential Oil of Ocimum tenuiflorum L. (Lamiaceae) Growing in Northern Australia, J. Ess. Oil Res. 5: 459-461.

JFH33(6):569 -- J. Food Hyg. Soc. Jap. 33(6): 569.

JFM -- Morton, J.F., Major Medicinal Plants. 1977. Atlas of Medicinal Plants of Middle America. Bahamas to Yucatan. 1981.

JIM127:546 -- Jim Duke's personal files.

JJF7:47 -- Burzaco, A., Velasco-Negueruela, A. and Perez-Alonso, M.J. 1992. Essential Oil Analysis of Sideritis pauli Pau. FFJ7: 47-8. 1992.

JJP66:47 -- Jap. J. Pharmacology, 66: 47.

JJS40:138 -- Jap. J. Pharmacology, 40: 138.

JLS58:156 -- Jim Duke's personal files.

JNM1:10 -- Jim Duke's personal files.

JNP -- J. Nat. Prod.

JNP39:420 -- Farnsworth, N.R. and Cordell, G.A. A Review of Some Biologically Active Compounds Isolated From Plants as Reported in the 1974-1975 Literature. J. Nat. Prod. 39: 420, 1976.

JNP44:308 -- J. Nat. Prod. 44: 308.

JNP45:725 -- Bedows, E. and Hatfield, G.M. An Investigation of the Antiviral Activity of Podophyllum peltatum. J. Nat. Prod. 45: 725.

JNP46(5) -- J. Nat. Prod. 46(5).

JNP49:156 -- Wolf, R.B. Effects Of p-Methoxycinnamaldehyde From Star Anise And Related Cinnamic Acid Derivatives On Velvetleaf Germination. Journal of Natural Products, 49(1): 156-158, 1986.

JNP49:941.1986 -- Misra, L.N. and Singh, S.P. Alpha-Thujone, The Major Component Of The Essential Oil From Artemisia vulgaris Growing Wild In Nilgiri Hills. Journal of Natural Products, 49: 941, 1986.

JNP50:1017 -- J. Nat. Prod. 50: 1017.

JNP50:313.1987 -- Barberan, F.A.T., Manez, S., and Villar*, A. Identification Of Antiinflammatory Agents From Sideritis Species Growing In Spain. Journal of Natural Products, 50: 313, 1987.

JNP51:357 -- Ueno, H., Horie, S., et al. Chemical And Pharmaceutical Studies On Medicinal Plants In Paraguay. Geraniin, An Angiotensin-Converting Enzyme Inhibitor From "Paraparai Mi," Phyllanthus niruri. J. of Natural Products, 51(2): 357-359, 1988.

JNP53(2):513 -- Lin,* C-N., Lu, C-M., Cheng, M-K., Gan, K-H. The Cytotoxic Principles Of Solanum incanum. Journal of Natural Products, 53(2): 513-516, 1990.

JNP53(5):1239 -- Nakane, H., Fukushima, M., and Ono*, K. Differential Inhibition Of Reverse Transcriptase And Various DNA Polymerases By Digallic Acid And Its Derivatives. Journal of Natural Products, 53(5): 1234-1240, 1990.

JNP53:456 -- Pettit,* G.R., Numata, A., Takemura, T., Ode, R.H., Narula, A.S., Schmidt, J.M., Cragg, G.M., and Pase, C.P. Antineoplastic Agents, 107. Isolation Of Actetoside And Isoacteoside From Castilleja linariaefolia. J. of Natural Products, 53(2): 456-458, 1990.

JNP54:142 -- Garcia, J., Morin,* C., Nyasse, B., Sondengam, B-L., Tchouankeu, J-C., and Tsamo, E. Complete Structural Assignments Of An Ergosterol Derivative From Entandrophragma utile. Journal of Natural Products, 54(1): 136-142, 1991.

JNP54:152 -- Tan, G.T., Pezzuto, J.M., Kinghorn,* A.D., Hughes, S.H. Evaluation Of Natural Products As Inhibitors Of Human Immunodeficiency Virus Type 1 (HIV-1) Reverse Transcriptase. Journal of Natural Products, 54(1): 143-154, 1991.

JNP54:1583 -- Shimomura,* K., Kitazawa, T., Okamura, N., and Yagi, A. Tanshinone Production In Adventitious Roots And Regenerates Of Salvia miltiorrhiza. Journal of Natural Products, 54(6): 1583-1587, 1991.

JNP54:218 -- Choi, J.S., Yokozawa,* T., and Oura, H. Antihyperlipidemic Effect Of Flavonoids From Prunus davidiana. Journal of Natural Products, 54(1): 218-224, 1991.

JNP54:455 -- Aquino, R., De Feo, V., De Simone, F., Pizza, C.*, and Cirino, G. Plant Metabolites. New Compounds And Anti-Inflammatory Activity Of Uncaria tomentosa. Journal of Natural Products, 54(2): 453-459, 1991.

JNP54:483 -- Kouzi, S.A., and McChesney*, J.D. Microbial Models Of Mammalian Metabolism: Fungal Metabolism Of The Diterpene Sclareol By Cunninghamella Species. Journal of Natural Products, 54(2): 483-490, 1991.

JNP54:900 -- Arisawa, M., Hayashi, T., Shimizu, M., Morita, N., Bai, H., Kuze, S., and Ito, Y. Isolation And Cytotoxicity Of Two New Flavonoids From Chrysosplenium grayanum And Related Flavonols. J. Natural Products, 54(3): 898-901, 1991.

JNP55:1732 -- Amoros, M., Simoes, C.M.O., Girre, L., et al. Synergistic Effect Of Flavones And Flavonols Against Herpes Simplex Virus Type 1 In Cell Culture. Comparison With The Antiviral Activity Of Propolis. J. of Natural Products 55(12):1732-1740, 1992.

JNP55:999 -- Zheng, G-Q., Kenney, P.M., and Lam, L.K.T. Sesquiterpenes From Clove (Eugenia caryophyllata) As Potential Anticarcinogenic Agents. Journal of Natural Products 55(7): 999-1003, 1992.

JNP56:1426 -- J. Nat. Prod. 56: 1426.

JNP56:935 -- Bowers, W.S., Ortego, F., You, X., and Evans, P.H. Insect Repellents From The Chinese Prickly Ash Zanthoxylum bungeanum. Journal of Natural Products, 56(6): 935-938, 1993.

JNP57:243 -- Fujioka, T., Kashiwada, Y., et al. Anti-AIDS Agents 11. Betulinic Acid and Platanic Acid as Anti-HIV Principles From Syzigium claviflorum, and the Anti-HIV Activity of Structurally Related Triterpenoids. Journal of Natural Products, 57(2): 243-247, 1994.

JNP57:658 -- Denyer, C.V., Jackson, P., Loakes, D.M., Ellis, M.R., and Yound, D.A.B. Isolation of Antirhinoviral Sesquiterpenes From Ginger (Zingiber officinale). Journal of Natural Products, 57(5): 658-662, 1994.

JPP42:194 -- Gawron, A., Gorski*, G., Glowniak**, K. 1989. Increased proliferation of phytohaemagglutinin (PHA)-stimulated human leucocytes after 8-methoxypsoralen treatment. J. Pharm. Pharmacol. 42: 194-195, 1990.

JPP42:205 -- O'Connell, M.J., Webster, N.R. 1989. Hyperoxia and salicylate metabolism in rats. J. Pharm. Pharmacol. 42: 205-206, 1990.

JPP42:295 -- Ortiz De Urbina, A.V., Martin, M.L., Montero, M.J., Carron, R., Sevilla, M.A., San Roman, L. 1989. Antihistaminic activity of pulegone on the guinea-pig ileum. J. Pharm. Pharmacol. 42: 295-296, 1990.

JPP42:652 -- Eccles, R., et al. 1989. The effects of oral administration of (-)-menthol on nasal resistance to airflow and nasal sensation of airflow in subjects suffering from nasal congestion associated with the common cold. J. Pharm. Pharmacol. 42: 652-654, 1990.

JPP42:655 -- Fujita, T., Nakatani, E., Funaishi, N., Sakuma, S., Fujimoto, Y. Potent inhibition of prostaglandin inactivation in rabbit gastric antral mucosal slices by selenium ions in-vitro. J. Pharm. Pharmacol. 42: 655-657, 1990.

JPP42:657 -- Li, K.M. Effect of zinc sulphate on acetic acid-induced gastric ulceration in rats. J. Pharm. Pharmacol. 42: 657-659, 1990.

JPP42:877 -- Rao, V.S.N., Menezes, A.M.S., Viana, G.S.B. 1990. Effect of myrcene on nociception in mice. J. Pharm. Pharmacol. 42: 877-878, 1990.

JPP46:118 -- Da Silva Emim, J.A., Oliveira, A.B. and Lapa, A.J. Pharmacological Evaluation of the Anti-inflammatory Activity of a Citrus Bioflavonoid, Hesperidin, and the Isoflavonoids, Duartin and Claussequinone, in Rats and Mice. J. Pharm. Pharmacol. 46:118-122,1994

JPP46:16 -- Yamamoto, A., Umemori, S., and Muranishi, S. 1993. Absorption Enhancement of Intrapulmonary Administered Insulin by Various Absorption Enhancers and Protease Inhibitors in Rats. J. Pharm. Pharmacol. 46: 14-18, 1994.

JSG -- J.S. Glasby Dict.Pls Containing 2ndary Metabolite. 1991.

JSPR22:141 -- J. Stored. Prod. Res., 22:141, 1986.

KCH -- Huang, K. C. 1993. The Pharmacology of Chinese Herbs. CRC Press, Boca Raton, FL 388 pp.

KEP2:185 -- Jim Duke's personal files.

LEL -- Lewis and Elvin-Lewis

LGR -- L. Gracza (1985) as cited by Buchbauer re antiacetylcholinesterase.

LL-MAY89 -- Let's Live, May 1989.

LRN-AUG92 -- Lawrence Review of Natural Products, Aug-92.

LRN-DEC90 -- Lawrence Review of Natural Products, Dec-90.

LRN-DEC93 -- Lawrence Review of Natural Products, Dec-93.

LRN-JAN87 -- Lawrence Review of Natural Products, Jan-87.

LRN-JAN91 -- Lawrence Review of Natural Products, Jan-91.

LRN-JAN92 -- Lawrence Review of Natural Products, Jan-92.

LRN-JAN93 -- Lawrence Review of Natural Products, Jan-93.

LRN-JUL87 -- Lawrence Review of Natural Products, Jul-87.

LRN-JUN88 -- Lawrence Review of Natural Products, Jun-88.

LRN-JUN90 -- Lawrence Review of Natural Products, Jun-90.

LRN-MAR91 -- Lawrence Review of Natural Products, Mar-91.

LRN-MAR92 -- Lawrence Review of Natural Products, Mar-92.

LRN-OCT91 -- Lawrence Review of Natural Products, Oct-91.

LRN-SEP87 -- Lawrence Review of Natural Products, Sep-87.

LRN-SEP91 -- Lawrence Review of Natural Products, Sep-91.

LRN-SEP97 -- Lawrence Review of Natural Products, Sep-97.

LRN1995 -- Lawrence Review of Natural Products, 1995

LS55:1061 -- Life Sciences 55: 1061.

M&R -- Mitchell, J. and Rook, A.J. 1923. Botanical Dermatology: plants and plant products injurious to the skin. Vancouver: Greengrass, 1979. xiii, 787 p.

M1 -- Merck 1st Edition

M10 -- Merck 10th Edition.

M11 -- Merck 11th Edition

M29 -- Martindale's 29th

M29(P.1064) -- Martindale's 29th, p. 1064.

M29(P.779) -- Martindale's 29th, p. 779.

M7 -- Merck 7th Ed.

MAR -- Martindale's 28th

MED -- Medline

MIK -- Michael Castleman 1991. The Healing Herbs. Rodale Press, Emmaus, PA 436 pp.

MOB24:301.1989 -- Mater Org (Berl), 24:301, 1989.

MPC -- Jim Duke's personal files.

MPI -- Medicinal Plants of India NOTE: (2 vols so far).

MPT -- Medicinal and Poisonous Plants of the Tropics. Leeuwenberg, A.J.M., ed. Pudoc, Wageningen. 1987.

MUO24:1 -- Material und Organismen , 24: 1, 1989.

NAP NAPRALERT database (National Product Alert).

NIG -- Nigg, H.N. and Seigler, D.S., eds. 1992. Phytochemical Resources for Medicine and Agriculture. Plenum Press, New York. 445 pp.

NNK67:1417 -- Umemoto, K., Arai, T., Nii, H. and Furukawa, K. 1993. A New Chemotype of Mentha aquatica Containing Sesquiterpene Alcohols as Major Components Nippon Nogeikagaku Kaishi 67(10): 1417-1419. NOTE: <0.1=Trace.

NYT3/28/95 -- New York Times, 3/28/95.

OHAI,1987 -- Oriental Healing Arts Institute, 1987. (Changed to Int. J. Oriental Medicine).

OMM -- Oriental Materia Medica (three volumes by Dr. Hsu and collaborators)

PAM -- Pizzorno, J.E. and Murray, M.T. 1985. A Textbook of Natural Medicine. John Bastyr College Publications, Seattle, Washington (Looseleaf).

PAS -- Stitt, Paul. Why George should eat broccoli. NOTE: Based on NAPRALERT.

PB57:347 -- Jim Duke's personal files.

PBB47:1 -- Jim Duke's personal files.

PC23:1677 -- Phytochemistry, 23: 1677.

PC23:1885 -- Shimizu, M., Ito, T., Terashima, S., Hayashi, T., Arisawa, M., Morita, N., Kurokawa, S., Ito, K., and Hashimoto, Y. 1984. Inhibition of Lens Aldose Reductase by Flavonoids. Phytochemistry. 23: 1885-1888.

PC25:2085 -- Phytochemistry, 25: 2085.

PC25:859 -- Phytochemistry, 25: 859.

PC27:969 -- Larson, R.A. 1987. Review Article Number 30. The Antioxidants of Higher Plants. Phytochemistry. 27: 969-978.

PC29(4):1103 -- Weidenborner, M., Hindorf, H., Jha, H.C., and Tsotsonos, P. 1989. Antifungal Activity of Flavonoids Against Storage Fungi of the Genus Aspergillus. Phytochemistry. 29(4): 1103-1105.

PC29(5):1667 -- Jung, J.H., Pummangura, S., Chaichantipyuth, c., Patarapanich, C., and McLaughlin, J.L. 1989. Bioactive Constituents of Melodorum fruticosum. Phytochemistry. 29(5): 1667-1670. 1990.

PC29:1093 -- Tomas-Barberan, F., Iniesta-Sanmartin, E., Tomas-Lorente, F., and Rumbero, A. 1989. Antimicrobial Phenolic Compounds From Three Spanish Helichrysum Species. Phytochemistry. 29(4): 1093-1095. 1990.

PC29:1165 -- Velasco-Negueruela, A. and Perez-Alonso, M.J. 1990. The Volatiles of Six Teucrium Species from the Iberian Peninsula and the Balearic Islands. Phytochemistry 29(4): 1165-9.

PC29:821 -- Phytochemistry, 29: 821.

PC33:557 -- Okamura, H., Mimura, A., Yakou, Y., Niwano, M., and Takahara, Y. 1992. Antoxidant Activity of Tannins and Flavonoids in Eucalyptus rostrata. Phytochemistry. 33(3): 557-561. 1993.

PC35:83 -- Arrebola, M.L., Navaro, M.C., Jimenez, J. and Ocana, F.A. 1994. Variations in Yield and Composition of the Essential Oil of Satureja obovata. Phytochemistry 35(1): 83.

PC36:579 -- Ohnishi, M., Morishita, H., Iwahashi, H., Toda, S., Shirataki, Y., Kimura, M., and Kido, R. 1993. Inhibitory Effects of Chlorogenic Acids on Linoleic Acid Peroxidation and Haemolysis. Phytochemistry. 36(3): 579-583. 1994.

PC37:1463 -- Phytochemistry, 37: 1463.

PC38:599 -- Geibel, M. 1994. Sensitivity of the Fungus Cytospora persoonii to the Flavonoids of Prunus cerasus. Phytochemistry. 38(3): 599-601. 1995.

PCF -- Phenolic Compounds in Food and Their Effects on Health. Antioxidants & Cancer Prevention. Huang, M.T., Ho, C.T. and Lee, C.Y. eds. 1992. ACS Symposium Series 507.ACS, Washington 402 pp.

PCF:10 -- Shahidi, F. Phenolic Compounds of Brassica Oilseeds. Phenolic Compounds in Food and Their Effects on Health, Ch. 10, 132.

PCF:13 -- Maga, J.A. Contribution of Phenolic Compounds to Smoke Flavor. Phenolic Compounds in Food and Their Effects on Health, Ch.13.

PCF:14 -- Naim, M., Zehavi, U., Nagy, S., and Rouseff, R.L. Hydroxycinnamic Acids as Off-Flavor Precursors in Citrus Fruits and Their Products. Phenolic Compounds in Food and Their Effects on Health, Ch.14.

PCF:142 -- Osawa, T. Phenolic Antioxidants in Dietary Plants as Antimutagens.Phenolic Compounds in Food and Their Effects on Health, Ch.11 p.142.

PCF:16 -- Hwang, L.S., Wang, C-K, Sheu, M-J., Kao, L-S. Phenolic Compounds of Piper betle Flower as Flavoring and Neuronal Activity Modulating Agents. Phenolic Compounds in Food and Their Effects on Health, I:200.

PCF:17 -- Shahidi, F., Wanasundara, P.K.J.P.D., Hong, C. Antioxidant Activity of Phenolic Compounds in Meat Model Systems. Phenolic Compounds in Food and Their Effects on Health, Ch.17.

PCF:18 -- Ki Soon Rhee. Oilseed Food Ingredients Used To Minimize Oxidative Flavor Deterioration in Meat Products. Phenolic Compounds in Food and Their Effects on Health, Ch.18.
PCF:19 -- Hagerman, A.E. Tannin-Protein Interactions. Phenolic Compounds in Food and Their Effects on Health, Ch.19.

PCF:21 -- Huang, M-T., and Ferraro, T. Phenolic Compounds in Food and Cancer Prevention. Phenolic Compounds in Food and Their Effects on Health, Ch.2 p.21.

PCF:25 -- McEvily, A.J., Iyengar, R., and Gross, A.T. Inhibition of Polyphenol Oxidase by Phenolic Compounds. Phenolic Compounds in Food and Their Effects on Health, Ch.25.

PCF:267 -- Deschner E.E. Dietary Quercetin and Rutin. Inhibitors of Experimental Colonic Neoplasia. Phenolic Compounds in Food and Their Effects on Health, Ch.19 p.267.

PCF:35 -- Weisburger, J.H. Mutagenic, Carcinogenic, and Chemoproventive Effects of Phenols and Catechols.Phenolic Compounds in Food and Their Effects on Health, Ch.3.

PCF:48 -- Newmark, H.L. Plant Phenolic Compounds as Inhibitors of Mutagenesis and Carcinogenesis. Phenolic Compounds in Food and Their Effects on Health, Ch.4.

PCF:54 -- Pratt, D.E. Natural Antioxidants from Plant Material. Phenolic Compounds in Food and Their Effects on Health, Ch.5.

PCF:67 -- Phenolic Compounds in Food and Their Effects on Health, 67.

PCF:68 -- Phenolic Compounds in Food and Their Effects on Health, 68.

PCF:93 -- Okuda, T., Yoshida, T., and Hatano, T. Antioxidant Effects of Tannins and Related Polyphenols. Phenolic Compounds in Food and Their Effects on Health, Ch.7 p.93.

PED -- Pedersen, M. 1987. Nutritional Herbology. Pederson Publishing. Bountiful, Utah. 377 pp. NOTE: Some quantitative data, e.g., selenium, seems inaccurate. All PED data is ZMB.

PEP -- Pakistan Encyclopedia Planta Medica. 1986.

PER.COM.SFV -- Duke, J.A.. Personal Communication.

PFH46:77 -- Singh, J., Gupta, K., and Arora, S.K. 1993. Changes in the anti-nutritional factors of developing seeds and pod walls of fenugreek (Trigonella foenum graecum L.). Plant Foods for Human Nutrition. 46: 77-84, 1994.

PHYT13:2002 -- Pavannasasivam, G., & M.U.S. Sultanbawa. 1974. Betulinic acid in the Dilleniaceae and a review of its natural distribution. Phytochem. 13:2002-2006.

PHZ46:156 -- Pharmazie, 46: 156, 1991.

PM&T20:213.1986 -- Jim Duke's personal files.

PM1985:70 -- Planta Medica 1985, page 70.

PM1986:184 -- Wagner, H., Wierer, M., and Bauer, R. 1985. Planta Medica 1986: 184.

PM1986:239 -- Tomas-Barberan, F.A., Lopez-Gomez, C., Villar, A., and Tomas-Lorente, F. 1986. Inhibition of Lens Aldose Reductase by Labiatae Flavonoids. Planta Medica 1986: 239.

PM1987:268 -- Kobold, U., Vostrowsky, O., Bestmann, H.J., Bisht, J.C., Pant, A.K., Melkani, A.B. and Mathela, C.S. 1987. Terpenoids from Elsholtzia Species; II. Constituents of Essential Oil from a New Chemotype of Elsholtzia cristata. Planta Medica 1987: 268-271.

PM1987:526 -- Gawron, A. and Glowniak, K. 1987. Cytostatic Activity of Coumarins in vitro. Planta Medica 1987: 526.

PM52:499 -- Yang, L-L., Yen, K-Y., Konno, C., Oshima, Y., Kiso, Y., and Hikino, H. 1986. Antihepatotoxic principles of Wedelia chinensis herbs. Planta Medica, 52:499.

PM53:37 -- Soicke, H. and Leng-Peschlow, E. 1986. Characterisation of Flavonoids from Baccharis trimera and their Antihepatotoxic Properties. Planta Medica 53: 37, 1987.

PM55 -- Planta Medica, 55. 1986.

PM55(7):620 -- Planta Medica, 55(7): 620.

PM55(7):626 -- Planta Medica, 55(7): 626.

PM55(7):654 -- Planta Medica, 55(7): 654.

PM55:226 -- Planta Medica, 55: 226.

PM55:663 -- Planta Medica, 55: 663.

PM56(1):41 -- Ekeke, G.I. and Shode, F.O. 1988. Phenylalanine is the Predominant Antisickling Agent in Cajanus cajan Seed Extract. Planta Medica 56(1): 41, 1990.

PM56(1):70 -- Darias, V., Bravo, L., Rabanal, R., Sanchez-Mateo, C.C., and Martin-Herrera, D.A. 1988. Cytostatic and Antibacterial Activity of Some Compounds Isolated from Several Lamiaceae Species from the Canary Islands. Planta Medica 56(1): 70, 1990.

PM56(2):171 -- Joyeux, M., Rolland, A., Fleurentin, J., Mortier, F., and Dorfman, P. 1989. Tert-Butyl Hydroperoxide-Induced Injury in Isolated Rat Hepatocytes: A Model for Studying Anti-Hepatotoxic Crude Drugs. Planta Medica 56(2): 171-173, 1990.

PM56(2):226 -- Lewis, E.J. and Friedrich, E.C. 1989. The Essential Oil of Poliomintha incana. Planta Medica 56(2): 224-226, 1990.

PM56(3):254 -- Aida, K., Tawata, M., Shindo, H., Onaya, T., Sasaki, H., Yamaguchi, T., Chin, M.(Chen, Z.), and Mitsuhashi, H. 1989. Isoliquiritigenin: A New Aldose Reductase Inhibitor from Glycyrrhizae Radix. Planta Medica 56(3): 254-258, 1990.

PM56(4):360 -- Murakami, S., et al. 1989. Effect of Salvianolic Acid A, a Depside from Roots of Salvia miltiorrhiza, on Gastric H+,K+-ATPase. Planta Medica, 56(4): 360, 1990.

PM56(6):501 -- Planta Medica, 56(6): 501, 1990.

PM56(6):660 -- Planta Medica, 56(6): 660, 1990.

PM56(6):671 -- Planta Medica, 56(6): 671, 1990.

PM56(6):677 -- Matsukawa, Y., et al. The Effect of Quercetin and Other Flavonoids on Cell Cycle Progresssion and Growth of Human Gastric Cancer Cells. Planta Medica, 56(6): 677, 1990.

PM56(6):681 -- Planta Medica, 56(6): 681, 1990.

PM56(6):695 -- Planta Medica, 56(6): 695, 1990.

PM56:638 -- Planta Medica, 56: 638, 1990.

PM57:208 -- Choi, J.S., Yokozawa, T., and Oura, H. 1990. Improvement of Hyperglycemia and Hyperlipemia in Streptozotocin-Diabetic Rats by a Methanolic Extract of Prunus davidiana Stems and Its Main Component, Prunin. Planta Medica, 57: 208.

PM57:347 -- Economou, D., and Nahrstedt, A. 1991. Chemical, Physiological, and Toxicological Aspects of the Essential Oil of Some Species of the Genus Bystropogon*. Planta Medica, 57: 347, 1991.

PM57:A110 -- Planta Medica, 57: A110, 1991.

PM57:A113 -- Planta Medica, 57: A113, 1991.

PM57:A131 -- Planta Medica, 57: A131, 1991.

PM57:A133 -- Planta Medica, 57: A133, 1991.

PM57:A137 -- Menkovic, N., et al. 1991. The Essential Oil of Sideritis scardica. Pl. Med. 57. Suppl. 2. pp. A137-A132.

PM57:A42 -- Planta Medica, 57: A42, 1991.

PM57:A43 -- Planta Medica, 57: A43, 1991.

PM57:A44 -- Planta Medica, 57: A44, 1991.

PM57:A54 -- Planta Medica, 57: A54, 1991.

PM57:A56 -- Planta Medica, 57: A56, 1991.

PM57:A83 -- Planta Medica, 57: A83, 1991.

PM58(6):499 -- Kroes, B.H., et al. 1991. Anti-Inflammatory Activity of Gallic Acid. Planta Medica, 58(6): 499.

PM58:376 -- Mathela,C.S., Melkani,A.B., Bisht,J.C., Pant,A.K., Bestmann,H.J., Erler,J., Kobold,U., Rauscher,J. and Vostrowsky,O. 1992. Chemical Varieties of Essential Oils from Elsholtzia polystachya from Two Different Locations in India. Planta Medica 58: 376-379.

PM59:276 -- Planta Medica, 59: 276.

PM59:508 -- Kuppusamy, U.R., and Das, N.P. 1993. Antilipolytic Action of Hesperetin in Rat Adipocytes. Planta Medica, 59: 508.

PM59:546 -- Orjala, J., et al. 1993. Five New Prenylated p-Hydroxybenzoic Acid Derivatives with Antimicrobial and Molluscicidal Activity from Piper aduncum Leaves. Planta Medica, 59: 546.

PM61:33 -- Houghton, P.J., Zarka, R., Heras, B., and Hoult, J.R.S. 1994. Fixed Oil of Nigella sativa and Derived Thymoquinone Inhibit Eicosanoid Generation in Leukocytes and Membrane Lipid Peroxidation. Planta Medica, 61: 33.

PM61:9 -- Recio, M., et al. 1994. Investigations on the Steroidal Anti-Inflammatory Activity of Triterpenoids from Diospyros leucomelas*. Planta Medica, 61: 9.

PMP22:231 -- Lamaison, J.L., Petitjean-Freytet, C., Carnat, A.P. et Carnat, A. Activites Antiradicalaires In Vitro D'Huiles Essentielles. Plantes medicinales et phytotherapie, 22: 231-234, 1988.

PMP23:193 -- Kumar, P., Dixit*, V.P., and Khanna, P. Antifertility Studies of Kaempferol: Isolation and Identification from Tissue Culture of some Medicinally Important Plant Species. Plantes medicinales et phytotherapie, 23: 193-201, 1989.

PMP23:46 -- Bezanger-Beauquesne*, L. Noveautes En Phytotherapie. Sixieme Revue. Plantes medicinales et phytotherapie, 23: 46-76, 1989.

PMP24:193 -- Plantes Medicinales et Phytotherapie, 24: 193.

PMP24:98 -- Bezanger-Beauquesne, L. Noveautes En Phytotherapie. Septieme Revue. Plantes medicinales et phytotherapie, 24: 98, 1990.

PMP25:184 -- Chalchat, J.C. et Garry, R.P.*, Bastide, P., Fabre, F. et Malhuret, R. Correlation Composition Chimique/Activite Antimicrobienne: V - Contribution a la Comparaison de 2 Methodes de Determination Des Cmi. Plantes medicinales et phytotherapie, 25: 184-193.

PMP25:46 -- Bezanger-Beauquesne, L. Noveautes en Phytotherapie. Huitieme Revue. Plantes medicinales et phytotherapie, 25: 46-76, 1991.

POT -- Jim Duke's personal files.

PPB17:990 -- Jim Duke's personal files.

PR4(2):73 -- Ortiz de Urbina, J.J., Martin, M.L., Sevilla, M.A., Montero, M.J., Carron, R., and San Roman, L. Antispasmodic Activity on Rat Smooth Muscle of Polyphenol Compounds Caffeic and Protocathechic Acids. Phytotherapy Research 4(2): 73, 1990.

PR4(3):87 -- Singh, P., Jain, S., Bhala, M., Goyal, R.B., Jayaprakash, D., and Lohiya, N.K. Wogonin, 5,7-Dihydroxy-8-methoxyflavone as Oestrogenic and Anti-implantational Agent in the Rat. Phytotherapy Research 4(3): 87, 1990.

PR4(3):93 -- Laekeman, G.M., Van Hoof, L., Haemers, A., Vanden Berghe, D.A., Herman, A.G., and Vlietinck, A.J. Eugenol a Valuable Compound for In Vitro Experimental Research and Worthwhile for Further In Vivo Investigation. Phytotherapy Research 4(3): 93, 1990.

PR4(5):198 -- Lamikanra, A., Ogundaini, A.O., and Ogungbamila*, F.O. Antibacterial Constituents of Alchornea cordifolia Leaves. Phytotherapy Research 4(5): 198.

PR4(5):201 -- Meli, R., Autore, G., Di Carlo, G., Capasso, F. Inhibitory Action of Quercetin on Intestinal Transit in Mice. Phytotherapy Research 4(5): 201, 1990.

PR4:73 -- Phytotherapy Research, 4: 73.

PR4:93 -- Phytotherapy Research, 4: 93.

PR4:95 -- Phytotherapy Research, 4: 95.

PR6:74 -- Shukla, B., Visen, P.K.S., Patnaik, G.K., Tripathi, S.C., Srimal, R.C., Dayal, R., and Dobhal, P.C. Hepatoprotective Activity in the Rat of Ursolic Acid Isolated from Eucalyptus Hybrid. Phytotherapy Research 6: 74-79, 1992.

PS75:161 -- Abid Ali Khan, M.M., Jain, D.C., Bhakuni, R.S., Zaim, M., and Thakur, R.S. 1991. Occurrence of some antiviral sterols in Artemisia annua. Plant Science, 75: 161-165, 1991.

PT4(3):118 -- Jim Duke's personal files.

PT5:1986 -- Jim Duke's personal files.

QJC28(1):1 -- Recently became Internat. J. Crude Drug Res. 28(1,2,3,4):1990, page 1.

QJC28:155 -- Recently became Internat. J. Crude Drug Res. 28(1,2,3,4):1990, page 155.

QJC28:74 -- Recently became Internat. J. Crude Drug Res. 28(1,2,3,4):1990, page 74.

RAA -- Rizk, A.F.M. and Al-Nowaihi, A.S., The Phytochemistry of the Horticultural Plants of Qatar, Scientific and Applied Research Centre, University of Qatar.

RIE12:5 -- Revista Itiliana Eppos, 12: 5, 1994.

RIZ -- Rizk, A.F.M., The Phytochemistry of the Flora of Qatar, Scientific and Applied Research Centre, University of Qatar, Kingprint, Richmond, UK, 1986.

RJH -- R.J. Hixtable (Huxtable, R.J. (as T. Max). 1992. This and That:The Essential Pharmacology of Herbs and Spices. Tips 13:15-20 Huxtable, R.J. 1992b. The Myth of Benficent Nature. The Risk of Herbal Preparations. Anns. Intern. Med. 117(2):165-6; Huxtable, R.

RR21:85 -- Rastitel'nye Resursy, 21: 85.

RWG -- Gardner, R.W. 1994. Chemical Intolerance-Physiological Causes and Effects and Treatment Modalities. CRC Press, Boca Raton FL 163 pp.

SAY -- Foster, S. and Yue, C.-X. 1992. Herbal Emissaries: Bringing Chinese Herbs to the West. Healing Arts Press, Rochester, Vermont 05767. 356 pp.

SMO -- Suziki, A., Morimoto, I., and Okitzu, T., Elution of Metals from Crude Drugs, Shoykugaku Zasshi 36(3):190-195.

SN138:189.1990 -- Science News, 138: 189, 1990.

SN145:294 -- Science News, 145: 294.

SN146:421 -- Science News, 146: 421.

SNE137:292 -- Jim Duke's person files.

SZ44:183 -- Shoyakugaku Zasshi, 44: 183.

SZ45:270 -- Shoyakugaku Zasshi, 45: 270.

TOX -- Keeler, R.F. and Tu, A.T. eds. 1991. Toxicology of Plant and Fungal Compounds. (Handbook of Natural Toxins Vol. 6) Marcel Dekker, Inc. NY. 665 pp.

TRA -- Tramil

TRA42:511 -- Tramil, 42: 511.

TRA64:131 -- Tramil, 64: 131.

USA -- USDA's Ag Handbook 8 and sequelae)

V&D -- Vlietinck, A.J. and Dommisse, R.A. eds. 1985. Advances in Medicinal Plant Research. Wiss. Verlag. Stuttgart.

V&E -- Jim Duke's personal files.

VAL -- Valnet, J. 1982. The Practice of Aromatherapy. Translated from the 1980 French Edition by Robin Campbell and Libby Houston. Destiny Books, New York. 279 pp.

VET -- Tyler, V.E.. 1982. The Honest Herbal. George F. Stickley Co., Philadelphia PA. 263 pp.

VET94 -- Tyler, V.E.. 1994.

W&B -- Watt, J.M., and Breyer-Brandwijk, M.G. The Medicinal and Poisonous Plants of Southern and Eastern Africa. 1962.

W&R -- Jim Duke's personal files.

W&W -- Wagner & Wolff, eds. 1977. New Natural Products (RS164. I56. 176)

WBB -- Watt, J.M. and Breyer-Brandwijk, M.G. 1962. The Medicinal and Poisonous Plants of Southern and Eastern Africa, 2nd ed.

WER -- Werbach, M. 1993. Healing with Food. Harper Collins, New York, 443 pp.

WER317 -- Werbach, M. 1993. Healing with Food. p. 317. Harper Collins, New York, 443 pp.

WIC -- Wichtl, M. 1984. Teedrogen. Ein Handbuch fur Apotheker und Arzte. Wissenschaftliche Verlagsgesellschaft. mbH Stuttgart. 393 pp.

WO2 -- Wealth of India.

WOI -- ANON. 1948-1976. The Wealth of India raw materials. Publications and Information Directorate, CSIR, New Delhi. 11 volumes.

YAK106:1108 -- Yakugaku Zasshi, 106: 1108

YAK111:193 -- Yakugaku Zasshi, 111: 193, 1991.

YAK112:645 -- Yamahara, J., et al. Stomachic principles in ginger. II. Pungent and anti-ulcer effects of low polar constituents isolated from ginger, the dried rhizoma of... Yakugaku Zasshi 112: 645-655 (1992).

YHJ37:182 -- Yakhak Hoeji, 37: 182, 1993.

YYB -- Jim Duke's personal files.

ZLU197:20 -- Lagouri, V., Blekas, G., Tsimidou, M., Kokkini, S., and Boskou, D. 1993. Composition and Antioxidant Activity of Essential Oils from Oregano Plants Grown Wild in Greece. Z. Lebensm Unters Forsch 197: 20-23. NOTE: Trace scored as <0.1

ZN43:370 -- Bestmann, H.J., Kobold, U., Classen, B., Vostrowsky, O., Sah, G.C.S., Pant, A.K., Melkani, A.B. and Mathela, C.S. 1988. New Constituents of Essential Oil from Elsholtzia pilosa. Zeit. Naturforsch. 43c: 370-372.

ZNA45:1073 -- Vostrowsky, O., Garbe, W., Bestmann, H.J. and Maia, J.G.S. 1990. Essential Oil of Alfavaca, Ocimum gratissimum, from Brazilian Amazon. Zeitschr. Naturforschung 45(C): 1073-6.